① identify object

② replace object with

③ draw vectors) forces on
   object

F_1

F_2

④ then $\vec{a} = \vec{F}/m$

PHYSICS FOR TECHNICAL EDUCATION

Dale Ewen
Parkland Community College

LeRoy Heaton
Parkland Community College

PRENTICE-HALL, INC., ENGLEWOOD CLIFFS, N.J. 07632

Library of Congress Cataloging in Publication Data

EWEN, DALE, (date)
   Physics for technical education

   Includes index.
   1. Physics. 2. Engineering. I. Heaton,
LeRoy, (date) joint author. II. Title.
QC21.2.E96   530   80-17182
ISBN 0-13-674127-4

©1981 by Prentice-Hall, Inc., Englewood Cliffs, N.J. 07632

Printed in the United States of America

10 9 8 7 6 5 4 3 2 1

Editorial production/supervision by Ellen W. Caughey
Interior design by Judy Winthrop and Ellen W. Caughey
Cover design by Bill Agee
Drawings by George Morris
Manufacturing buyer: Joyce Levatino

Prentice-Hall International, Inc., *London*
Prentice-Hall of Australia Pty. Limited, *Sydney*
Prentice-Hall of Canada, Ltd., *Toronto*
Prentice-Hall of India Private Limited, *New Delhi*
Prentice-Hall of Japan, Inc., *Tokyo*
Prentice-Hall of Southeast Asia Ptd. Ltd., *Singapore*
Whitehall Books Limited, *Wellington, New Zealand*

CONTENTS

8 MOMENTUM

140

9 ROTATIONAL AND CIRCULAR MOTION

156

10 NONCONCURRENT FORCES

192

11 SIMPLE MACHINES

214

12 GEARS AND PULLEYS

239

PREFACE

Physics for Technical Education has been written for students in technology programs as well as for those in engineering technology programs. It is designed as a basic one-year course in technical or applied physics for colleges, community and junior colleges, technical institutes, and technical schools. The text can be adapted, however, to meet the needs of a shorter course or courses in programs that have lesser demand for physics.

We have developed this text with the belief that a technical physics sequence should primarily give a background of the basic scientific principles which complement the skills developed by a specific educational program. Such a sequence should also provide a sound platform for growth after completion of the formal technical program.

This text contains three features which deserve special mention:

1. The text is all metric and nearly all SI. Because many of the engineering physics courses are going metric, and as the technician and the engineering technician provide such close support for the engineer, many teachers have asked for an all metric text for the technical student.

2. A proven successful problem-solving method is presented in detail in Chapter 3 and then is used consistently throughout the text. In each set of exercises thereafter, the student is reminded by the box at the top of page xvi to use the method. The method is easily remembered and can be used and applied daily in industry, research, and development.

3. Measurements are presented as approximate numbers in Chapter 1 and used consistently throughout the text.

Sketch	Data	Basic Equation	Working Equation	Substitution
12 cm² $h = ?$ $b = 6.0$ cm	$A = 12$ cm² $b = 6.0$ cm $h = ?$	$A = bh$	$h = \dfrac{A}{b}$	$h = \dfrac{12 \text{ cm}^2}{6.0 \text{ cm}}$ $= 2.0$ cm

The text has the following rather general divisions: introduction and background, mechanics, matter, waves and sound, electricity and magnetism, light and geometric optics; and modern physics.

The first three chapters present units, systems of units, standards of measurement, computations using measurements, and give a review of mathematical skills, including vectors.

Mechanics is broken into translational kinematics, dynamics and the conservation principles of energy, and linear momentum. These principles are then extended to rotational systems. Three chapters are devoted to applied mechanics.

In Chapter 13 the atomic nature of matter is discussed and serves as a background for introducing macroscopic variables such as pressure, density, heat, and temperature to describe and predict the behavior of materials.

Six chapters have been devoted to electricity and magnetism. Static electricity has been used to develop concepts of electric field, potential difference, and capacitance. Direct current electricity permits a definition of resistance and familiarity with circuits in general. Magnetism provides the concept of inductance and the operation of motors and generators. The chapter on ac circuits provides a vehicle to bring all the concepts together in applications and in situations which are not foreign to students.

Four chapters discuss light, its reflection, its refraction, and applications of these phenomena to optical instruments. Diffraction is presented in some detail because of its importance as a research and control tool.

The last three chapters deal with modern physics. An attempt has been made to present phenomena, such as the emission of photoelectrons and the continuous spectrum from a hot body, that are applied tools in industry and illustrative of quantum effects. Bohr's model of the atom and nuclear reactors are also presented.

The authors wish to thank the many teachers and reviewers who have offered suggestions and who gave freely of their time. We especially thank Ronald J. Nelson, Neill Schurter, and LaVerne M. McFadden for their permission to use some of the art and ideas from the first edition of *Physics for Career Education*. If anyone wishes to correspond with us regarding suggestions, criticisms, or questions, please feel free to contact us directly. We are most grateful to Normandy Hathaway and Patricia Schuldt for typing the manuscript. Finally, we are grateful to our families for their patience, understanding, and encouragement, and especially grateful to Joyce Ewen and Mary Heaton for their excellent proofing assistance.

DALE EWEN

LEROY HEATON

MEASUREMENT

1.1   A SHORT HISTORY OF MEASUREMENT

People in ancient times saw the need for measurement as their civilizations and trade developed. At first, trade between people, villages, and nations was conducted by the physical exchange of items, such as animals for grain. They soon realized that standards for the necessary measurements, such as length, weight, and volume, had to be agreed upon by those involved in any trade. Later, money was also used for trade, especially when different people and items were involved in rather long series of exchanges.

Length measurements were commonly expressed in units that related to parts of the human body. Common units included the

1. cubit—the distance between the point of the elbow and the tip of the middle finger.

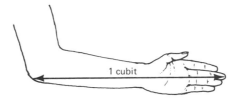
1 cubit

2. digit—the width of the first or middle finger.
3. hand—the width of a hand.

4. yard—the distance from the tip of the king's nose to the fingertips of his outstretched arm.

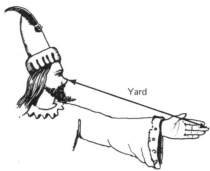

Yard

5. rod—this length was determined by having 16 men put one foot behind the foot of another man in a line.

←——————— 1 rod ———————→

Distance divided by 16 equals one foot.

6. foot—divide the distance of the rod by 16 and you had a legal foot. More commonly, the length of one's foot was used as a foot. No doubt you can think of other length measurements.

Units based on parts of the human body were, of course, not satisfactory. In an effort to remedy the confusion and disagreement during the fourteenth century, King Edward II of England proclaimed that the English inch was the length of three barley corns, round and dry, taken from the center of the ear, and laid end to end.

Balance scales for measuring weight date back more than 5000 years.

Grains, seeds, precious metals, and other agreed-upon standard weights were used to measure the weights of items that were sold or traded by weight. The "grain," a small unit of weight used for measuring and prescribing certain drugs, is still used.

During the Middle Ages new systems of weight measures developed which complicated matters even more. The troy pound, used for weighing precious metals, contained 12 ounces. The French avoirdupois pound, used for weighing more common, larger items, contained 16 ounces. The apothecary system was developed somewhat later and used for measuring and dispensing drugs. Its units are different for dry and liquid measures.

Volume measure has been expressed in the most haphazard sets of units that one can imagine. In ancient times, the volume was the size of the container that held the item to be traded or bought. Down through the ages, scores of standards, such as shells, animal horns, barrels, and jars, have been used as units for volume. Even today, the U.S. gallon and the British imperial gallon are not the same.

The U.S. units of measure constitute a complex, makeshift system of Anglo-Saxon, Roman, and French–Norman weights and measures. For example, the U.S. gallon equals the British wine gallon, which is about 20 percent smaller than the British imperial gallon. The pennyweight equaled $\frac{1}{20}$ tower ounce, or 32 wheatcorns from "the midst of the ear."

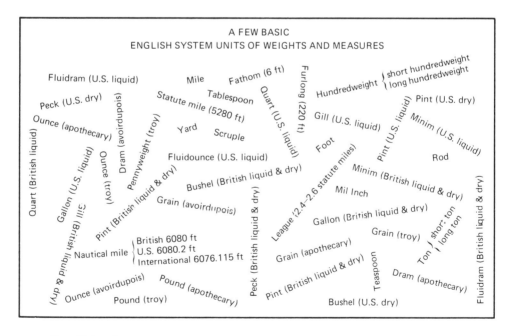

In 1790, Thomas Jefferson, as George Washington's secretary of state, proposed a more precise definition of the standard foot, to get away from the grain and anatomical references. He proposed that the foot be the length of a cylindrical iron rod of a simple pendulum of such length that a swing from one end of its arc to the other and back again would take 1 second.

Jefferson also proposed the establishment of a decimal system for measurement, similar to the recently adopted decimal coin system. His new "foot" would be subdivided into 10 new inches. Despite prodding by both Jefferson and Washington, Congress adopted neither plan.

It is interesting to note that in the same year, 1790, the *metric system*, a decimal system, was developed in France. By 1840, metric measurement was compulsory in France. Shortly thereafter, many other nations adopted the metric system of measurement. By 1900, most of Europe and of South America used the system.

In 1866, metric measurements were legalized for official use in the United States, and English–metric equivalents were specified.

From 1870 to 1875, seventeen nations, including the United States, met in convention and formulated the Treaty of the Meter, which

1. reformulated the metric system and refined the accuracy of its standards
2. provided for construction and distribution of accurate standard copies to participating countries
3. set up the International Bureau of Weights and Measures

In 1893, the Secretary of the Treasury, by administrative order, declared the new metric standards to be the nation's "fundamental standards" of length and mass. Thus, indirectly, the United States *officially* became a metric nation in 1893. Even today, the English units are officially defined as fractions of the standard metric units.

In signing the Treaty of the Meter in 1875, the United States joined every other major nation in the world in endorsing the metric system as the internationally preferred system of weights and measures. This treaty made measurements *internationally* compatible at the highest level of accuracy. However, no effort was made to convert this nation practically and internally to the system that it had officially approved.

Throughout American history, many repeated and exhaustive efforts have been made without success to convert the United States to official use of the metric system. Then, in 1951, Japan decided to go metric. In 1957, Sputnik launched a scientific and technical revolution that staggered the imagination. As expenditures, interest, and research in the sciences surged, the U.S. government began to consider seriously the desirability of increasing the use of the metric system, the predominant measurement language of the sciences. In 1965, Great Britain decided to go metric. Thus, with most of our trading allies metric or going metric, we foresaw economic and balance-of-trade problems ahead.

In 1968, a three-year U.S. Metric Study was authorized by Congress. As an integral part of the study, hearings were conducted on the following:

1. labor
2. construction
3. education
4. consumer affairs
5. engineering-oriented industries
6. consumer-related industries
7. small business, state and local governments, health, transportation, and other services.

Supplemental investigations were done and conferences were held dealing with:

1. manufacturing industries
2. nonmanufacturing industries
3. education
4. consumers
5. international trade
6. engineering standards
7. international standards
8. Department of Defense
9. federal civilian agencies
10. commercial weights and measures
11. history of the metric system controversy in the United States.

The U.S. Metric Study provided the following recommendations through a broad, clear-cut consensus regarding three fundamental questions posed to the study group.

1. "Increased use of the metric system is in the best interests of the United States.
2. The nation should change to the metric system through a coordinated national program.
3. The transition period should be ten years, at the end of which the nation would be *predominantly* metric."

In July 1971, a 13-volume U.S. Metric Study Report was issued.

On December 23, 1975, President Gerald R. Ford signed into law the "Metric Conversion Act of 1975," which states in part that:

"The Congress finds as follows:

1. The United States was an original signatory to the 1875 Treaty of the Meter, which established the General Conference of Weights and Measures, the International Committee of Weights and Measures and the International Bureau of Weights and Measures.

2. Although the use of metric measurement standards in the United States has been authorized by law since 1866, this Nation today is the only industrially developed nation which has not established a national policy of committing itself and taking steps to facilitate conversion to the metric system.

"It is therefore declared that the policy of the United States shall be to coordinate and plan the increasing use of the metric system in the United States and to establish a United States Metric Board to coordinate the voluntary conversion to the metric system."

In some industries, the cost of conversion will be minimal, as in the bottling industry. Here the replacement of bottle molds with metric sizes could be done at minimal cost and would require only minor changes in the bottle-blowing machinery, especially if replacement were delayed until the molds wear out.

However, the cost of conversion will be expensive in industries that utilize machine tools and other complex equipment in making their products. Here more planning will be needed. Most machines can produce metric sizes and will need to be replaced only when they wear out. Others may require the conversion or replacement of gear trains, sizing dials, and lead screws. Alternatively, digital readout equipment, which makes the customary–metric conversions, can be used without changing the basic equipment. In many cases, the conversion timetable could be designed to coincide with the usual replacement schedule for a given industry so that the actual cost of conversion will be minimal. After a few years of increased exports, the cost of the conversion for many companies will be completely offset. Could the government perhaps make provisions whereby this cost could be fairly shared by all segments of our economy, as the overall increased exports will be shared by the country's economy as a whole?

In still other industries, change to the metric system offers an unparalleled opportunity for order and reason. One example is the fastener industry, in which an optimal system based on metric preferred numbers is being developed. This optimum series of screw thread dimensions and nut bolt sizes, in which each succeeding fastener has a definite proportional increase in strength over the next smaller one, would reduce the current worldwide production of hundreds of sizes to 25 general-use dimensions and an additional 14 dimensions for the aerospace industry. (How many different sizes of nuts and bolts do you see in your local hardware store? One source puts the number at a possible 116.) In this way, production would be more efficient and inventory could be significantly reduced.

Where we as a nation go from here will be determined by our industries and the many groups within and across all segments of our economy. They will need to study the technical problems, the revised standards, and the market's needs and demands. For the general public, the conversion process can be one of order and simple reason.

Because the United States is converting to the metric system, and at the urging and counsel of many colleagues, this book presents an all-metric version of technical physics. Since you, as students in this class, are the technicians of tomorrow, communicating and working with a worldwide technical and engineering community, you must be well versed in and very comfortable with the universal metric measuring system.

1.2 THE METRIC SYSTEM

The modern metric system is identified in all languages by the abbreviation SI (for Système International d'Unités—the international system of units of measurement, written in French).
The SI metric system has seven basic units:

Basic unit	SI abbreviation	Used for measuring
metre*	m	length
kilogram	kg	mass
second	s	time
ampere	A	electric current
kelvin	K	temperature
candela	cd	light intensity
mole	mol	molecular substance

All other SI units are called derived units; that is, they can be defined in terms of these seven basic units. For example, the newton (N) is defined as 1 kg m/s^2 (kilogram metre per second per second). Other commonly used derived SI units are:

Derived unit	SI abbreviation	Used for measuring
litre*	L	volume
cubic metre	m^3	volume
square metre	m^2	area
newton	N	force
metre per second	m/s	speed
joule	J	energy
watt	W	power

*At present, there is some difference of opinion in the United States on the spelling of metre and litre. We have chosen the "re" spellings for two reasons. First, this is the internationally accepted spelling for all English-speaking countries. Second, the word "meter" already has many different meanings—parking meter, electric meter, odometer, and so on. Many people feel that the metric unit of length should be distinctive and readily recognizable, which is what the "re" spelling is.

RELATIONSHIPS OF SI UNITS WITH NAMES

The chart below shows graphically how the 17 SI derived units with special names are derived in a coherent manner from the base and supplementary units. It was provided by the National Bureau of Standards.

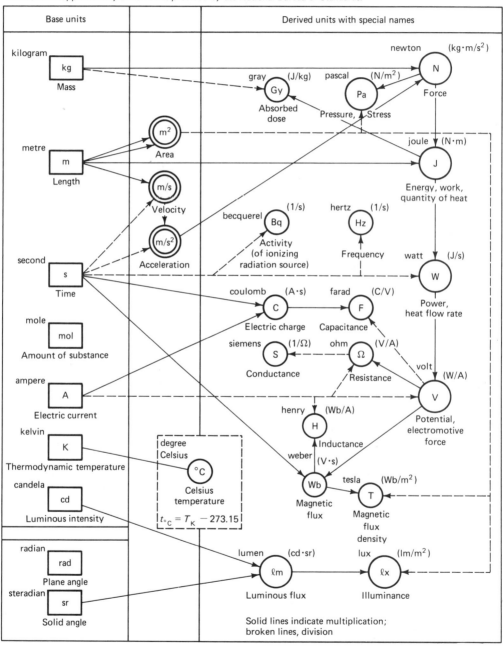

Since the metric system is a decimal or base 10 system, it is very similar to our decimal number system and our decimal money system. It is an easy system to use because calculations are based on the number 10 and its multiples. Special prefixes are used to name these multiples and submultiples which may be used with most all SI units. Since the same prefixes are repeatedly used, memorization of many conversions has been significantly reduced. The following table shows these prefixes and the corresponding symbols.

Prefixes for SI units

Multiple or submultiple[a] decimal form	Power of 10	Prefix	Prefix symbol	Pronun- ciation	Meaning
1,000,000,000,000	10^{12}	tera	T	tĕr′ă	one trillion times
1,000,000,000	10^{9}	giga	G	jĭg′ă	one billion times
1,000,000	10^{6}	mega	M	mĕg′ă	one million times
1,000	10^{3}	kilo[b]	k	kĭl′ō	one thousand times
100	10^{2}	hecto	h	hĕk′tō	one hundred times
10	10^{1}	deka	da	dĕk′ă	ten times
0.1	10^{-1}	deci	d	dĕs′ĭ	one tenth of
0.01	10^{-2}	centi[b]	c	sĕnt′ĭ	one hundredth of
0.001	10^{-3}	milli[b]	m	mĭl′ĭ	one thousandth of
0.000001	10^{-6}	micro	μ	mī′krō	one millionth of
0.000000001	10^{-9}	nano	n	năn′ō	one billionth of
0.000000000001	10^{-12}	pico	p	pē′kō	one trillionth of

[a]Factor by which the unit is multiplied.
[b]Most commonly used prefixes.

Length is one of the most common measurements. In the SI system, the *metre* (m) is the basic unit. Initially, in the eighteenth century, the metre was defined as 1/10,000,000 of the distance from the equator to the north pole along the meridian passing through Paris, France. However, you can imagine the error involved in measuring this tremendous distance through rugged mountains and over water using eighteenth-century equipment. Soon after, scientists found that a more precise definition of the metre was necessary. The metre was then defined as the distance between two marks on a special platinum–iridium bar which was kept in a vault near Paris. Duplicate bars were made and shipped to the various countries throughout the world.

Over a period of time, two problems arose using the bar as the standard. First, it was time consuming to check all bars against the one standard bar in Paris. Second, scientists were making measurements more precisely than they could measure the distance between the two scratches on the standard metre bar. Then in 1960, the standard metre was adjusted to a standard that was reproducible in the laboratory.

When an electric discharge is maintained in a sample of the gas krypton 86, the gas glows similarly to neon. The color produced by the atoms of krypton consists of several different wavelengths of light, one of which is orange-red. The standard metre of length is now based on the wavelength of the orange-red light emitted by krypton 86 atoms.

$$1 \text{ metre} = 1,650,763.73 \text{ wavelengths of orange-red light of krypton 86}$$

The number of wavelengths of light in a given distance can be counted using an optical interferometer. This instrument provides an accuracy of about 2 parts in 1 billion, or about 1 m of error in the distance between the earth and the moon.

Scientists are at work on an even more accurate standard. One such method of providing the metre standard of the future is by means of the laser beam. Accuracy of better than 1 part in 1 trillion (10^{12}) is theoretically possible with this method. (Light is discussed in detail in Chapter 28.)

The metric units for area and volume are derived from the metre. Units of area are based on the square and are called square units. For example, a *square metre* (m^2) is the amount of surface area contained within a square that measures 1 m on a side. The area of the rectangle shown below is 12 m^2.

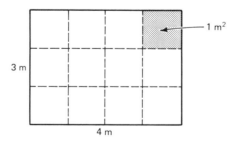

Units of volume are based on the cube and are called cubic units. For example, a *cubic metre* (m^3) is the amount of space contained in a cube that measures 1 m on an edge. The volume of the rectangular box shown on page 11 is 24 m^3.

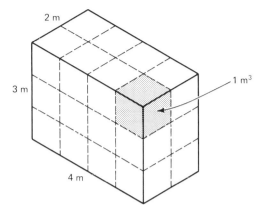

The litre (L) is a special volume unit which is equivalent to 1000 cm³ or a cubic decimetre. Since 1 L equals 1000 cm³, we have the volume equivalence

$$1 \text{ mL} = 1 \text{ cm}^3$$

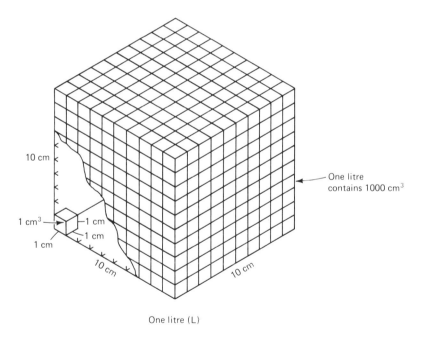

One litre (L)

The basic SI mass unit is the *kilogram* (kg). Originally, the kilogram was defined as the mass of 1 litre (1000 cm³) of water at 4° Celsius (°C), its temperature of maximum density. More accurate measurements showed that 1 kg of water at 4°C occupies 1000.027 cm³. Today, the standard kilogram mass is represented by a platinum–iridium cylinder kept near Paris. The *gram* (g) mass unit is 1/1000 kg. One millilitre (mL) of water has a mass of approximately 1 g.

The basic time unit is the *second* (s). Since

$$1 \text{ year} = 365\tfrac{1}{4} \text{ days} \quad (366 \text{ days every fourth year})$$

$$1 \text{ day} = 24 \text{ hours (h)}$$

$$1 \text{ h} = 60 \text{ minutes (min)}$$

$$1 \text{ min} = 60 \text{ seconds (s)}$$

a natural way of defining the second is

$$1 \text{ second} = 1 \text{ s} \times \frac{1 \text{ min}}{60 \text{ s}} \times \frac{1 \text{ h}}{60 \text{ min}} \times \frac{1 \text{ day}}{24 \text{ h}} = \frac{1}{86,400} \text{ day}$$

This is actually 1/86,400 of a mean solar day, which is the average time for the earth to make one complete rotation in relation to the sun. But again, careful measurements showed that the earth's rotation varies very slightly from year to year and even from day to day. So in 1956, the standard second was defined as 1/31,556,925.9747 of the year 1900. But this standard was not readily reproducible. In 1964, a new definition for the second was adopted which was based on the periodic time intervals in the vibration of the outermost electron of the cesium-133 atom. That is,

$$1 \text{ second} = 9,192,631,770 \text{ cesium-133 vibrations}$$

Other physical units of measurement will be developed as we progress through the text.

1.3   POWERS OF 10 AND SCIENTIFIC NOTATION

A command of powers of 10 and scientific notation is necessary to the study of physics.

First, let us review positive powers of 10.

$$10^1 = 10 \qquad\qquad = 10$$

$$10^2 = 10 \cdot 10 \qquad\qquad = 100$$

$$10^3 = 10 \cdot 10 \cdot 10 \qquad\qquad = 1000$$

$$10^4 = 10 \cdot 10 \cdot 10 \cdot 10 \qquad = 10,000$$

$$\vdots \qquad\qquad\qquad \vdots$$

$$10^n = \underbrace{10 \cdot 10 \cdot 10 \cdots 10}_{n \text{ factors}} = \underbrace{1000 \cdots 0}_{n \text{ zeros}}$$

Recall the following properties for negative powers of 10.

$$10^{-1} = \frac{1}{10} \qquad\qquad = \frac{1}{10} = 0.1$$

$$10^{-2} = \frac{1}{10 \cdot 10} \qquad\qquad = \frac{1}{10^2} = 0.01$$

$$10^{-3} = \frac{1}{10 \cdot 10 \cdot 10} \qquad\qquad = \frac{1}{10^3} = 0.001$$

$$10^{-4} = \frac{1}{10 \cdot 10 \cdot 10 \cdot 10} \qquad\qquad = \frac{1}{10^4} = 0.0001$$

$$\vdots \qquad\qquad\qquad\qquad \vdots \qquad \vdots$$

$$10^{-n} = \underbrace{\frac{1}{10 \cdot 10 \cdot 10 \cdots 10}}_{n \text{ factors}} = \frac{1}{10^n} = \underbrace{0.000 \cdots 01}_{(n-1) \text{ zeros}}$$

There are four basic laws of exponents involving powers of 10 that you should know.

> **LAW 1**
> $$10^m \cdot 10^n = 10^{m+n}$$

example 1 Multiply each of the following powers of 10.

(a) $10^3 \cdot 10^2 = 10^5$ (b) $10^6 \cdot 10^{-2} = 10^4$ (c) $10^{-3} \cdot 10^{-4} = 10^{-7}$

> **LAW 2**
> $$\frac{10^m}{10^n} = 10^{m-n}$$

example 2 Divide each of the following powers of 10.

(a) $\dfrac{10^8}{10^2} = 10^6$ (b) $10^{-2} \div 10^5 = 10^{-7}$ (c) $\dfrac{10^6}{10^{-3}} = 10^9$

> **LAW 3**
> $$(10^m)^n = 10^{mn}$$

example 3 Raise each of the following powers of 10 to the indicated power.

(a) $(10^2)^3 = 10^6$ (b) $(10^{-4})^2 = 10^{-8}$ (c) $(10^{-3})^{-2} = 10^6$

LAW 4

$$10^{-n} = \frac{1}{10^n} \quad \text{and} \quad \frac{1}{10^{-n}} = 10^n$$

---

example 4 Rewrite each of the following powers of 10 using positive exponents.

(a) $10^{-3} = \frac{1}{10^3}$ (b) $\frac{1}{10^{-4}} = 10^4$

Note that the second part of Law 4 can be shown by using the first part; that is,

$$\frac{1}{10^{-n}} = \frac{1}{\frac{1}{10^n}} = 1 \div \frac{1}{10^n} = 1 \times \frac{10^n}{1} = 10^n$$

The zero power of 10 is one; that is, $10^0 = 1$. To show this, we use the substitution principle, which states that if $a = b$ and $a = c$, then $b = c$.

$$\frac{10^n}{10^n} = 1 \qquad \text{(because any number other than zero divided by itself equals 1)}$$

and

$$\frac{10^n}{10^n} = 10^{n-n} \qquad \text{(by Law 2)}$$
$$= 10^0$$

Therefore,

$$10^0 = 1 \qquad \text{(by substitution)}$$

Scientific notation is especially useful for writing very large and very small numbers. To write a decimal number in scientific notation, write it as a product of a number between 1 and 10 and a power of 10; that is $N \times 10^m$, where $1 \leqslant N < 10$ and m is an integer.

example 5 Write each number in scientific notation.

(a) $4800 = 4.8 \times 10^3$
(b) $58{,}000{,}000 = 5.8 \times 10^7$
(c) $6{,}000{,}000{,}000 = 6 \times 10^9$
(d) $0.0045 = 4.5 \times 10^{-3}$
(e) $0.00000003 = 3 \times 10^{-8}$
(f) $0.0000000000057 = 5.7 \times 10^{-12}$

As you are probably already aware, many electronic calculators express very large and very small results in scientific notation, whereas others express all results in scientific notation.

example 6   Write each number in decimal form.

(a) $1.6 \times 10^4 = 16{,}000$
(b) $5.72 \times 10^6 = 5{,}720{,}000$
(c) $4.9 \times 10^{-5} = 0.000049$
(d) $1.5 \times 10^{-8} = 0.000000015$
(e) $2.998 \times 10^8 = 299{,}800{,}000$
(f) $5 \times 10^{-10} = 0.0000000005$

When a number in scientific notation is rounded to the nearest power of 10, the approximation is called the number's *order of magnitude*. For example,

Numbers	Order of magnitude
3.75×10^8	10^8
1.59×10^{-6}	10^{-6}
8.45×10^{12}	10^{13}
7.05×10^{-16}	10^{-15}

In science, comparisons between numbers are often made by comparing their orders of magnitude. For example, the masses of the earth, the moon, and the sun are given below. Compare their orders of magnitude.

Body	Mass	Order of magnitude
Earth	5.98×10^{24} kg	10^{25} kg
Moon	7×10^{22} kg	10^{23} kg
Sun	1.99×10^{30} kg	10^{30} kg

The sun is approximately $\dfrac{10^{30}}{10^{25}} = 10^5$, or 100,000, times the mass of Earth. Earth is approximately $\dfrac{10^{25}}{10^{23}} = 10^2$, or 100, times the mass of the moon.

EXERCISES

Simplify each power of 10 by using the laws of exponents.

1. $10^3 \times 10^5$ **2.** $10^8 \div 10^2$

3. $(10^4)^3$ **4.** $10^{-5} \cdot 10^8$

5. $\dfrac{10^{-6}}{10^3}$ **6.** $(10^{-3})^2$

7. $\dfrac{1}{10^{-5}}$

8. $\left(\dfrac{1}{10^{-6}}\right)^2$

9. $(10^3 \cdot 10^5)^2$

10. $\left(\dfrac{10^8}{10^{-3}}\right)^{-3}$

11. $\dfrac{10^3 \cdot 10^{-4} \cdot 10^8}{10^6 \cdot 10^{-8}}$

12. $\left(\dfrac{10^4 \times 10^4 \times 10^{-7} \times 10^{-9}}{10^8 \times 10^{-3} \times 10^{-2}}\right)^2$

Write each number in scientific notation.

13. 50,600

14. 4700

15. 0.0024

16. 0.0004

17. 8,500,000

18. 0.00000072

19. 0.0000000001

20. 50,000,000,000

Write each number in decimal form.

21. 4.1×10^3

22. 3.4×10^5

23. 6.32×10^{-4}

24. 5.37×10^{-3}

25. 1.05×10^9

26. 6.7×10^{-8}

27. 4×10^{-10}

28. 6×10^{12}

In Exercises 29–34, find the order of magnitude of each number.

29. 4.68×10^9

30. 8.63×10^{12}

31. 6.14×10^{-10}

32. 3.14×10^{-6}

33. 1.47×10^{-18}

34. 7.5×10^{19}

35. The masses of the sun and its planets are given below. Find the order of magnitude of the mass of each.

Body	Mass (kg)	Order of magnitude (kg)
(a) Sun	1.99×10^{30}	
(b) Mercury	3.3×10^{23}	
(c) Venus	4.87×10^{24}	
(d) Earth	5.98×10^{24}	
(e) Mars	6.4×10^{23}	
(f) Jupiter	1.9×10^{27}	
(g) Saturn	5.69×10^{26}	
(h) Uranus	8.7×10^{25}	
(i) Neptune	1.03×10^{26}	
(j) Pluto	6.6×10^{23}	

36. Comparing their masses, how many times larger is
   (a) Jupiter than Earth?
   (b) the sun than Jupiter?
   (c) Earth than Mars?

(d) Neptune than Uranus?

(e) Earth than Venus?

37. Comparing the orders of magnitude of their masses, how many times larger is

(a) Jupiter than Earth?

(b) the sun than Jupiter?

(c) Earth than Mars?

(d) Neptune than Uranus?

(e) Earth than Venus?

38. The masses of a proton, a neutron, and an electron are given below. Find the order of magnitude of the mass of each.

	Mass (kg)	Order of magnitude (kg)
(a) Proton	1.67252×10^{-27}	
(b) Neutron	1.67482×10^{-27}	
(c) Electron	9.10905×10^{-31}	

39. Comparing their masses, how many times larger is

(a) a proton than an electron?

(b) a neutron than an electron?

(c) a neutron than a proton?

(d) Earth than an electron?

40. Comparing the orders of magnitude of their masses, how many times larger is

(a) a proton than an electron?

(b) a neutron than an electron?

(c) a neutron than a proton?

(d) Earth than an electron?

1.4 UNIT ANALYSIS

One of the most basic and most useful concepts that you can learn in physics is often referred to as *unit analysis*. Unit analysis can be divided into three areas: (1) converting from one set of units to another; (2) determining and simplifying the units of a physical quantity obtained from the substitution of data into a formula; and (3) analyzing the derived units in terms of the basic units or other derived units.

To convert from one unit or set of units to another, we will use what is commonly called a *conversion factor*. We know that we can multiply any number or quantity by 1 (one) without changing the value of the original quantity. We also know that any fraction, whose numerator and denominator are equal, is equal to 1. For example, $\frac{3}{3}=1$, $\frac{15 \text{ m}}{15 \text{ m}} = 1$, and $\frac{4.5 \text{ kg}}{4.5 \text{ kg}} = 1$. In addition, since 1 m = 100 cm, $\frac{1 \text{ m}}{100 \text{ cm}} = 1$. Similarly, $\frac{100 \text{ cm}}{1 \text{ m}} = 1$, because the numerator equals the denominator.

We call this name for 1 a conversion factor. The information necessary for forming a conversion factor is usually found in tables.

As in the case 1 m = 100 cm, there are two conversion factors for each set of data:

$$\frac{1\text{ m}}{100\text{ cm}} \quad \text{and} \quad \frac{100\text{ cm}}{1\text{ m}}$$

The correct choice for a particular conversion is the one in which the old units are in the numerator of the original expression and in the denominator of the conversion factor, or in the denominator of the original expression and in the numerator of the conversion factor. That is, we want the old units to cancel each other out.

example 1 Express 245 cm in metres.
As we saw above, the two possible conversion factors are

$$\frac{1\text{ m}}{100\text{ cm}} \quad \text{and} \quad \frac{100\text{ cm}}{1\text{ m}}$$

We choose the conversion factor with centimetres in the *denominator* so that the cm units cancel each other out. That is,

$$245\text{ cm} \times \frac{1\text{ m}}{100\text{ cm}} = 2.45\text{ m}$$

example 2 Express 90 km/h in m/s.
This example involves two conversions: kilometres to metres and hours to seconds. To convert km to m (1 km = 1000 m), we have two possible conversion factors:

$$\frac{1\text{ km}}{1000\text{ m}} \quad \text{and} \quad \frac{1000\text{ m}}{1\text{ km}}$$

We choose the conversion factor with km in the *denominator* so that the km units cancel each other out.

To convert h to s (1 h = 3600 s), we have the following two possible conversion factors:

$$\frac{1\text{ h}}{3600\text{ s}} \quad \text{and} \quad \frac{3600\text{ s}}{1\text{ h}}$$

We choose the conversion factor with h in the *numerator* so that the h units cancel each other out. Thus,

$$90\,\frac{\cancel{\text{km}}}{\cancel{\text{h}}} \times \frac{1000\text{ m}}{1\,\cancel{\text{km}}} \times \frac{1\,\cancel{\text{h}}}{3600\text{ s}} = 25\text{ m/s}$$

A *formula* is an equation, written in letters, which expresses a relationship between two or among three or more physical quantities.

When using a formula to solve a given problem where all but the unknown quantity is known:

1. Solve the formula for the unknown quantity.
2. Substitute each known quantity with its units.

3. Use the conventional order of operations procedures to evaluate the resulting numerical quantity and to simplify the units.

A detailed discussion of problem solving is presented in Chapter 3.

example 3   Given the formula $d = rt$, where $r = 60$ km/h and $t = 3$ h, find d.

$$d = rt$$
$$= \left(60\,\frac{\text{km}}{\cancel{\text{h}}}\right)(3\cancel{h})$$
$$= 180 \text{ km}$$

example 4   When a satellite is in a circular orbit around the earth, the formula $M = gs^2/G$ expresses the relationship among M, the mass of the earth; g, the acceleration due to the force or pull of gravity of the earth; s, the distance of the satellite from the center of the earth; and G, a universal proportionality constant. Find the acceleration due to gravity (g) of a satellite in a circular orbit 800 km above the surface of the earth.
First, solve for g:

$$M = \frac{gs^2}{G}$$
$$g = \frac{GM}{s^2}$$

The mass of the earth $M = 5.98 \times 10^{24}$ kg. The average radius of the earth is 6370 km. Therefore, the distance from the center of the earth

$$s = 6370 \text{ km} + 800 \text{ km}$$
$$= 7170 \text{ km} \times \frac{1000 \text{ m}}{1 \text{ km}} = 7.17 \times 10^6 \text{ m}$$

The proportionality constant $G = 6.67 \times 10^{-11}$ m^3/kg s^2.
Now substitute for M, s, and G:

$$g = \frac{GM}{s^2}$$

$$= \frac{\left(6.67 \times 10^{-11}\,\dfrac{\text{m}^3}{\text{kg s}^2}\right)(5.98 \times 10^{24} \text{ kg})}{(7.17 \times 10^6 \text{ m})^2}$$

$$= 7.76 \text{ m/s}^2 \qquad \boxed{\dfrac{\left(\dfrac{\text{m}^3}{\text{kg s}^2}\right)\text{kg}}{\text{m}^2} = \dfrac{\text{m}}{\text{s}^2}}$$

(For comparison purposes, an average value of g on the earth's surface is 9.80 m/s^2.)

The analysis of derived units in terms of the basic units or other derived units will be an ongoing discussion as we proceed through the book.

1. Express 2500 m in km.
2. Express 150 cm in m.
3. Express 1300 mg in g.
4. Express 1.4 L in mL.
5. Express 0.65 km in m.
6. Express 14 mm in cm.
7. Express 60 km/s in m/s.
8. Express 400 m/s in km/s.
9. Express 5000 m/s in m/h.
10. Express 240 km/h in km/min.
11. Express 60 km/h in m/s.
12. Express 4 m/s in km/h.
13. Given the formula $V = lwh$, where $V = 2400$ cm^3, $l = 25$ cm, and $h = 8$ cm. Find w.
14. Given the formula $v = v_0 + at$, where $v = 60$ m/s, $a = 5$ m/s^2, and $t = 4$ s. Find v_0.
15. Given the formula $\dfrac{1}{R} = \dfrac{1}{R_1} + \dfrac{1}{R_2}$, where $R = 60$ ohms (Ω) and $R_2 = 240$ Ω. Find R_1.
16. Given the formula $v^2 = v_0^2 + 2a(\Delta s)$, where $v_0 = 15$ m/s, $a = 6$ m/s^2, and $\Delta s = 12$ m. Find v.

1.5 APPROXIMATE VERSUS EXACT NUMBERS

In your studies until now, probably all numbers and all measurements have been treated as exact numbers. An *exact number* is a number that has been determined as a result of counting, such as 24 students enrolled in this class, or by a definition, such as 1 h = 60 min or 1 in. = 2.54 cm, a conversion definition agreed to by the world governments' bureaus of standards. The treatment of the addition, subtraction, multiplication, and division of exact numbers normally comprises the content, or at least the emphasis, of courses in elementary mathematics.

However, nearly all data of a technical nature involve *approximate numbers*: they have been determined as a result of a measurement process—some direct, as with a ruler, and some indirect, as with a surveying transit. Before studying how to perform the calculations with approximate numbers (measurements), we first must determine the "correctness" of an approximate number. First, we realize that no measurement can be found exactly. The length of the cover of this book can be found using many instruments. The better the measuring device used, the better the measurement.

A measurement may be expressed in terms of its accuracy or its precision. The *accuracy* of a measurement refers to the number of digits, called *significant digits*, which indicate the number of units that we are reasonably sure of having counted when making a measurement. The greater the number of significant digits given in a measurement, the better the accuracy, and vice versa.

example 1   The average distance between the moon and the earth is 385,000 km. This measurement indicates measuring 385 thousands of kilometres; its accuracy is indicated by three significant digits.

example 2   A measurement of 0.035 cm indicates measuring 35 thousandths of a centimetre; its accuracy is indicated by two significant digits.

example 3   A measurement of 0.0200 mg indicates measuring 200 ten-thousandths of a milligram; its accuracy is indicated by three significant digits.

Notice that a zero is sometimes significant and sometimes not. To clarify this, we give the following rules for significant digits:

1. All nonzero digits are significant: 356.4 m has four significant digits (this measurement indicates 3564 tenths of metres).
2. All zeros between significant digits are significant: 406.02 km has five significant digits (this measurement indicates 40,602 hundredths of kilometres).
3. A zero in a number greater than 1 which is specially tagged, such as by a bar above it, is significant: $13\bar{0},000$ km has three significant digits (this measurement indicates $13\bar{0}$ thousands of kilometres).
4. All zeros to the right of a significant digit *and* a decimal point are significant: 36.10 cm has four significant digits (this measurement indicates $361\bar{0}$ hundredths of centimetres).
5. Zeros at the right in whole-number measurements which are not tagged are *not* significant: 2300 m has two significant digits (23 hundreds of metres).
6. Zeros at the left in measurements less than 1 are *not* significant: 0.00252 m has three significant digits (252 hundred-thousandths of a metre).

When a number is written in scientific notation, the decimal part indicates the number of significant digits. For example, $20\bar{0},000$ m would be written in scientific notation as 2.00×10^5 m.

The *precision* of a measurement refers to the smallest unit with which a measurement is made, that is, the position of the last significant digit.

example 4   The precision of the measurement 385,000 km is 1000 km. (The position of the last significant digit is in the thousands place.)

example 5   The precision of the measurement 0.035 cm is 0.001 cm. (The position of the last significant digit is in the thousandths place.)

example 6   The precision of the measurement 0.0200 mg is 0.0001 mg. (The position of the last significant digit is in the ten-thousandths place.)

Unfortunately, the terms "accuracy" and "precision" have several different common everyday meanings. Here, we will use each term consistently as we have defined them. A measurement of 0.0006 cm has good precision and poor accuracy when compared with the measurement 368.0 cm, which has much better accuracy (one versus four significant digits) and poorer precision (0.0001 cm versus 0.1 cm).

example 7 Determine the accuracy and the precision of each measurement.

Measurement	Accuracy (significant digits)	Precision
1. 3463 m	4	1 m
2. 3005 km	4	1 km
3. 10,809 kg	5	1 kg
4. 36,000 Ω	2	1000 Ω
5. 88$\overline{0}$0 V	3	10 V
6. 1,349,000 km	4	1000 km
7. 600$\overline{0}$ m	4	1 m
8. 0.00632 kg	3	0.00001 kg
9. 0.0401 m	3	0.0001 m
10. 0.0060 g	2	0.0001 g
11. 14.20 A	4	0.01 A
12. 30.00 cm	4	0.01 cm
13. 100.060 g	6	0.001 g

1.6 CALCULATIONS WITH APPROXIMATE NUMBERS

If one person measured the length of one of two parts of a shaft with a micrometer calibrated in 0.01 mm as 12.27 mm and another person measured the second part with a ruler calibrated in mm as 23 mm, would the total length be 35.27 mm? Note that the sum 35.27 mm indicates a precision of 0.01 mm. The precision of the ruler is 1 mm, which means that the measurement 23 mm with the ruler could actually be anywhere between 22.50 mm and 23.50 mm using the micrometer (which has a precision of 0.01 mm). That is, any measurement between 22.50 mm and 23.50 mm can only be read as 23 mm using the ruler. Of course, this means that the tenths and hundredths digits in the sum 35.27 mm are really meaningless. In other words, the sum or difference of measurements can be no more precise than the least precise measurement. That is:

TO ADD OR SUBTRACT MEASUREMENTS:

1. Make certain that all the measurements are expressed in the same units. If they are not, convert them all to the same units.
2. Next, round each measurement to the same precision as the least precise measurement.
3. Then, add or subtract.

---

example 1   Add the measurements: 1250 cm, 1562 mm, 2.963 m, and 9.71 m. First, convert all measurements to the same units, say m.

$$1250 \text{ cm} = 12.5 \text{ m}$$
$$1562 \text{ mm} = 1.562 \text{ m}$$

Next, round each measurement to the same precision as the least precise measurement, which is 12.5 m. Then add.

$$12.5 \text{ m} = 12.5 \text{ m}$$
$$1.562 \text{ m} = \ \ 1.6 \text{ m}$$
$$2.963 \text{ m} = \ \ 3.0 \text{ m}$$
$$\underline{9.71 \text{ m} = \ \ 9.7 \text{ m}}$$
$$26.8 \text{ m}$$

example 2   Subtract the measurements: 2567 g − 1.60 kg.
First, convert all measurements to the same units, say g.

$$1.60 \text{ kg} = 16\overline{0}0 \text{ g}$$

Next, round each measurement to the same precision as the least precise measurement, which is $16\overline{0}0$ g. Then subtract.

$$2567 \text{ g} = 2570 \text{ g}$$
$$\underline{16\overline{0}0 \text{ g} = 16\overline{0}0 \text{ g}}$$
$$970 \text{ g}$$

Suppose that you need to find the area of a rectangular room that measures 11.4 m by 15.6 m. If you multiply the numbers 11.4×15.6, the product 177.84 implies an accuracy of five significant digits. But note that each of the original measurements contains only three significant digits. To rectify this inconsistency, we say that the product or quotient of measurements can be no more accurate than the least accurate measurement. That is:

TO MULTIPLY OR DIVIDE MEASUREMENTS:

1.   First, multiply or divide the measurements as given.

2.   Then, round the result to the same number of significant digits as the measurement with the least number of significant digits.

---

example 3   Multiply the measurements: 11.4 m × 15.6 m.

$$11.4 \text{ m} \times 15.6 \text{ m} = 177.84 \text{ m}^2$$

Round this product to three significant digits, which is the accuracy of the least accurate measurement (and also the accuracy of each measurement in the example). That is,

$$11.4 \text{ m} \times 15.6 \text{ m} = 178 \text{ m}^2$$

example 4   Divide the measurements: 78,000 m² ÷ 654 m.

$$78,000 \text{ m}^2 \div 654 \text{ m} = 119.26605 \text{ m}$$

Round this quotient to two significant digits, which is the accuracy of the least accurate measurement (78,000 m²). That is,

$$78,000 \text{ m}^2 \div 654 \text{ m} = 120 \text{ m}$$

There are even more sophisticated methods for dealing with the calculations of measurements. The method one uses, and indeed if one should even follow any given procedure, depends on the number of measurements and the sophistication needed for a particular situation.

In this book, we will more generally follow the customary practice of expressing measurements in terms of three significant digits, which is the accuracy used in most engineering and design work.

When calculations involve a trigonometric function, we will follow a convenient rule of thumb:

Angle(s) expressed to the nearest:	Length(s) of side(s) of a triangle will contain:
1°	Two significant digits
0.1°	Three significant digits
0.01°	Four significant digits

EXERCISES

Determine the accuracy and the precision of each measurement.

Measurement	(a) Accuracy (significant digits)	(b) Precision
1. 368 km		
2. 208 kg		
3. 35,000 L		
4. 13$\bar{0}$0 m		
5. 4$\bar{0}$ mg		
6. 0.036 A		
7. 0.0050 V		
8. 10.00 m		
9. 0.0607 g		
10. 46.70 km		

Use the rules for addition of measurements to find the sum of each set of measurements.

11.	12.	13.	14.
19,200 m	735,000 V	14 V	13.8 m
8,940 m	490,000 V	1.005 V	140.2 cm
50,030 m	86,000 V	0.017 V	1.953 m
147 m	1,300,000 V	3.6 V	126.3 m
	20$\bar{0}$,000 V		

Use the rules for subtraction of measurements to find each difference.

15.	16.	17.	18.
9200 km	41.8 cm	1,800,000 V	160 mm
627 km	23.76 cm	655,000 V	10.5 cm

Use the rules for multiplication and/or division of measurements to evaluate each of the following.

19. $18.7 \text{ m} \times 48.2 \text{ m}$

20. $135 \text{ cm} \times 206 \text{ cm} \times 145 \text{ cm}$

21. $146\bar{0} \text{ cm} \times 123.5 \text{ cm} \times 118.6 \text{ cm}$

22. $5\bar{0}0 \text{ g} \times 136 \text{ cm}$

23. $360 \text{ m}^2 \div 12 \text{ m}$

24. $9180 \text{ m}^3 \div 36 \text{ m}^2$

25. $\dfrac{4800 \text{ V}}{14.2 \text{ A}}$

26. $\dfrac{27\bar{0}0 \text{ m}^2}{14.8 \text{ m}}$

27. $\dfrac{19,600 \text{ N} \times 4\bar{0} \text{ m}}{25 \text{ } s}$

28. $\dfrac{4750 \text{ N} \times 4.82 \text{ m}}{1.6 \text{ } s}$

29. $\dfrac{(120 \text{ V})^2}{47.6 \text{ } \Omega}$

30. $\dfrac{0.927 \text{ } N}{1.5 \text{ m} \times 6.0 \text{ m}}$

1.7   VARIATION

If as two quantities, y and x, change and their ratio remains constant ($y/x = k$), the quantities are said to *vary directly*, or y is *directly proportional* to x. In general, this is written in the form $y = kx$, where k is the *proportionality constant*.

example 1   Given the data:

y	4	10	6	20	14
x	6	15	9	30	21

Note that y varies directly with x because the ratio y/x is always $\frac{2}{3}$. Also note that $y = \frac{2}{3}x$, where $k = \frac{2}{3}$. Graphing the data above gives the graph on page 26. The graph of any direct variation relationship is a straight line.

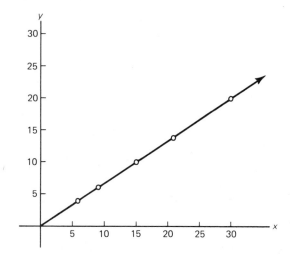

example 2   The circumference, C, of a circle varies directly with the radius, r. This is written $C = kr$, where $k = 2\pi$ is the proportionality constant.

example 3   The weight, w, of a body varies directly with its mass, m. This is written $w = km$, where $k = g$, the acceleration of gravity, which is the proportionality constant.

example 4   In a given wire the electrical resistance, R, varies directly with its length, L. This is written $R = kL$, where k is the proportionality constant.

example 5   Charles' law states that if the pressure on a gas is constant, its volume, V, varies directly with its absolute temperature, T. This is written $V = kT$, where k is the proportionality constant.

If as two quantities, y and x, change and their product remains constant ($yx = k$), the quantities are said to *vary inversely*, or y is *inversely proportional* to x. In general, this is written $y = \dfrac{k}{x}$, where k, again, is called the proportionality constant.

example 6   Given the data:

y	6	2	12	3	8	16
x	8	24	4	16	6	3

Note that y varies inversely with x because the product is always 48. Also note that $y = \dfrac{48}{x}$, where $k = 48$. Graphing the data above gives the graph on page 27. The graph of any inverse variation relationship is a hyperbola.

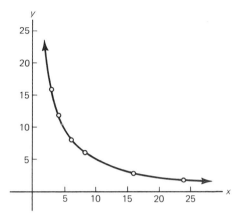

example 7 Boyle's law states that if the temperature of a gas is constant, its volume, V, varies inversely with its pressure, P. This is written $V = \dfrac{k}{P}$, where k is the proportionality constant.

example 8 The rate, r, at which an automobile covers a distance of 200 km varies inversely with the time, t. This is written $r = \dfrac{k}{t}$ or $r = \dfrac{200}{t}$, where $k = 200$ is the proportionality constant.

For many relationships, one quantity varies directly or inversely with a power of the other.

example 9 The area, A, of a circle varies directly with the square of its radius, r. This is written $A = kr^2$, where $k = \pi$, which is the proportionality constant.

example 10 In wire of a given length, the electrical resistance, R, varies inversely with the square of its diameter, D. This is written $R = \dfrac{k}{D^2}$, where k is the proportionality constant.

One quantity *varies jointly* with two or more quantities when it varies with the product of these quantities.

example 11 Coulomb's law for magnetism states that the force, F, between two magnetic poles varies jointly with the strengths, s_1 and s_2, of the poles and inversely with the square of their distance apart, d. This is written $F = \dfrac{ks_1s_2}{d^2}$, where k is the proportionality constant.

Once we are able to express a given variation sentence as an equation, the next step is to find the value of k, the proportionality constant. This value of k can then be used for all the different sets of data in a given problem or situation.

example 12   Given that y varies directly with x^2, and $x = 2$ when $y = 16$. Find y when $x = 3$ and when $x = \frac{1}{2}$.

First, write the variation equation:

$$y = kx^2$$

Second, to find k substitute the set of data that includes a value for each variable:

$$16 = k2^2$$
$$16 = 4k$$
$$4 = k$$

Therefore,

$$y = 4x^2$$

For $x = 3$,

$$y = 4(3)^2 = 36$$

For $x = \frac{1}{2}$,

$$y = 4\left(\frac{1}{2}\right)^2 = 1$$

example 13   At a given temperature the electrical resistance, R (in ohms, Ω), of a wire varies directly with its length, L, (in m) and inversely with the square of its diameter, D (in mm). If the resistance of 20.0 m of copper wire of diameter 0.81 mm is 0.67 Ω, what is the resistance of 40.0 m of copper wire 1.20 mm in diameter?

First, write the variation equation:

$$R = \frac{kL}{D^2}$$

Second, find k:

$$k = \frac{RD^2}{L}$$

$$k = \frac{(0.67 \ \Omega)(0.81 \ \text{mm})^2}{20.0 \ \text{m}}$$

$$= 0.022 \frac{\Omega \ \text{mm}^2}{\text{m}}$$

Therefore,

$$R = \frac{0.022L}{D^2}$$

For $L = 40.0$ m and $D = 1.20$ mm:

$$R = \frac{(0.022 \ \Omega \ \text{mm}^2/\text{m})(40.0 \ \text{m})}{(1.20 \ \text{mm})^2}$$

$$= 0.61 \ \Omega$$

EXERCISES

For each of the following sets of data, determine whether y varies directly with x or inversely with x. Also find k.

1.

y	6	15	54	1.5
x	8	20	72	2

2.

y	4	2	$\frac{1}{2}$	5
x	5	10	40	4

3.

y	$\frac{3}{4}$	$\frac{1}{4}$	$\frac{13}{20}$	$\frac{9}{16}$
x	2	6	$\frac{60}{26}$	$\frac{8}{3}$

4.

y	14	21	$\frac{7}{3}$	$\frac{14}{5}$
x	16	28	4	$\frac{16}{5}$

5.

y	12	18	0.75	1.5
x	20	30	2.5	5

6.

y	2	$\frac{1}{3}$	18	$\frac{2}{11}$
x	7	$\frac{7}{6}$	63	$\frac{7}{11}$

Write the variation equation for each of the following.

7. y varies directly with z.

8. p varies inversely with q.

9. a varies jointly with b and c.

10. m varies directly with the square of n.

11. r varies directly with s and inversely with the square root of t.

12. d varies jointly with e and the cube of f.

13. f varies jointly with g and h and inversely with the square of j.

14. m varies directly with the square root of n and inversely with the cube of p.

In Exercises 15–22, first find k; then find the given quantity.

15. y varies directly with x, and $y = 8$ when $x = 24$. Find y when $x = 36$.

16. m varies directly with n, and $m = 198$ when $n = 22$. Find m when $n = 35$.

17. y varies inversely with x, and $y = 9$ when $x = 6$. Find y when $x = 18$.

18. d varies inversely with e, and $d = \frac{4}{5}$ when $e = \frac{9}{16}$. Find d when $e = \frac{5}{3}$.

19. y varies directly with the square root of x, and $y = 24$ when $x = 16$. Find y when $x = 36$.

20. y varies jointly with x and the square of z, and $y = 150$ when $x = 3$ and $z = 5$. Find y when $x = 12$ and $z = 8$.

21. p varies directly with q and inversely with the square of r, and $p = 40$ when $q = 20$ and $r = 4$. Find p when $q = 24$ and $r = 6$.

22. m varies inversely with n and the square root of p, and $m = 18$ when $n = 2$ and $p = 36$. Find m when $n = 9$ and $p = 64$.

23. Use Charles' law in Example 5 (Section 1.7) to find the volume of oxygen when the absolute temperature is $3\overline{0}0$ K if the volume is 1500 cm^3 when the temperature is 250 K.

24. The electric power, P, generated by current varies jointly with the voltage drop, V, and the current, I. Find the power generated by 8.00 A of current on a 115-V circuit when 1320 W of power are generated by 6.00 A of current on a $22\overline{0}$-V circuit.

25. The electric power, P, used in a circuit varies directly with the square of the voltage drop, V, and inversely with the resistance, R. Find the power used in a circuit if the voltage drop is $12\overline{0}$ V and the resistance is $3\overline{0}$ Ω if 180 W is used by a $9\overline{0}$-V drop with a resistance of $4\overline{0}$ Ω.

VECTOR ANALYSIS

2.1 SCALAR VERSUS VECTOR QUANTITIES

A *scalar* quantity is a quantity that is completely described by its given magnitude —a number with suitable units. Physical quantities such as length, volume, time, mass, and temperature are scalars because each is completely described when its magnitude is given. For example, the mass of a certain steel beam is completely described as 158 kg. The temperature at 9:00 A.M. is completely described as 12°C.

We have other physical quantities in science and technology that require both a given magnitude and a given direction in order to be completely described. These are called *vector* quantities. That is, in addition to possessing a given magnitude, a vector always acts in a specific direction. For example, in order to completely describe wind velocity, both the speed and the direction must be known—15 km/h from the west. Examples of vector quantities include velocity, displacement, force, torque, and certain quantities from electricity.

Graphically, vectors are usually represented by means of an arrow. The length of the arrow corresponds to the magnitude of the vector, while the direction of the arrow shows the direction of the vector. For example, the wind velocity of 15 km/h from the west may be represented by the vector below.

15 km/h ———→

If *A* and *B* are the end points of a vector, the symbol **AB** represents the *vector from A to B*. Point *A* is called the *initial point* and point *B* is called the *terminal point*. Vectors may also be represented by a single lowercase letter, such as **u**, **v**, or **w**. The length of a vector **v** is written |**v**| and the length of a vector **AB** is written |**AB**|.

Two vectors are equal when they both have the same magnitude *and* the same direction.

Two vectors are opposites or negatives of each other when they have the same magnitude but opposite directions. Note that any vector **AB** is always the opposite of vector **BA**.

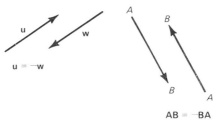

A given vector **v** may be placed in any position as long as its magnitude and direction are not changed.

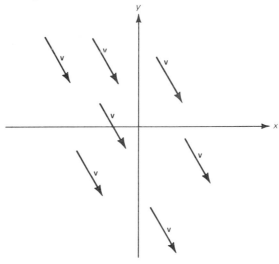

A vector is in *standard position* when its initial point is at the origin of the *xy*-coordinate system. A vector in standard position is expressed in terms of its length and its angle θ, where θ is measured counterclockwise from the positive *x*-axis to the vector. The vector below is in standard position.

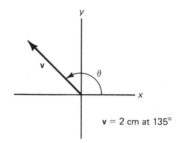

$\mathbf{v} = 2$ cm at $135°$

2.2 DISPLACEMENT—A VECTOR QUANTITY

Suppose that a friend who is a pilot offers to fly you from Landsville to a resort area, which is near an airfield. Suppose that he asks you how to get there and you reply, "It is 200 kilometres." Have you given him enough information to find the resort? Obviously not! You must also tell him in what direction to go. If you reply, "Go 200 kilometres due north," he can then find the resort. This change in position is represented by the vector at the left below.

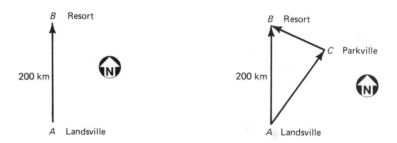

Perhaps the simplest vector to illustrate is one involving a *change of position*, which physicists call a *displacement*. As you saw above, displacement is a vector because it requires both a magnitude and a direction for its complete description.

Suppose that your friend needs to make a delivery to Parkville on the way and maps out the flight plan at right. Which is now the displacement vector?

The displacement, change of position, from one point to another is the net change of position; that is, the displacement from Landsville to the resort is the most direct route and is indicated by the vector from *A* to *B* above. The displacement vector is the same no matter which route is taken. That is, the displacement from Landsville may be expressed by the vector **AB**. If the plane stops in Parkville, the displacement from Landsville to Parkville may be expressed by **AC** and the displacement from Parkville to the resort may be expressed by **CB**. This gives the following vector result:

$$\mathbf{AC} + \mathbf{CB} = \mathbf{AB}$$

This result does not indicate that the distances are the same. In fact, we know that

$$|\mathbf{AC}| + |\mathbf{CB}| > |\mathbf{AB}|$$

Next, let us study the addition of vectors in more detail.

2.3   ADDITION OF VECTORS: GRAPHICAL METHODS

The sum of two or more vectors is called the *resultant*. This sum may be obtained graphically. To add two vectors, **v** and **w**, construct a parallelogram using **v** as one pair of parallel sides and **w** as the other pair. The diagonal of the parallelogram as shown below is the resultant or sum of the two vectors. (This is called the *parallelogram method* of adding vectors.)

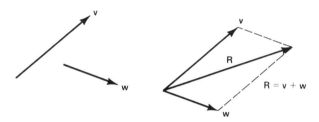

A second graphical method (*vector triangle method*) involves constructing the second vector with its initial point on the terminal point of the first vector. The resultant vector is the vector joining the initial point of the first vector to the terminal point of the second vector. (See figure at left below.)

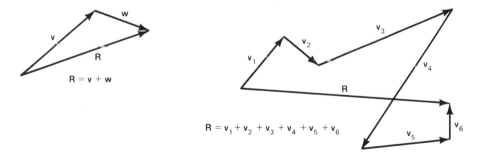

Addition of vectors is commutative; that is, the order in which the vectors are added does not affect the result.

$$\mathbf{v} + \mathbf{w} = \mathbf{w} + \mathbf{v}$$

This second method (right figure) is particularly useful when several vectors are to be added and is sometimes also called the *vector polygon method*.

A vector may be subtracted by adding its negative (a vector equal in magnitude but opposite in direction). That is, $\mathbf{v} - \mathbf{w} = \mathbf{v} + (-\mathbf{w})$. Construct $\mathbf{v}$ as usual; construct the negative of $\mathbf{w}$; find the resultant.

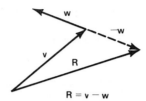

$$R = v - w$$

example 1 Using each graphic method, find the vector sum of the following two displacements.

$$\mathbf{v}: \quad 5.0 \text{ km at } 15° \text{ north of east } (15°)*$$

$$\mathbf{w}: \quad 7.5 \text{ km at } 6\bar{0}° \text{ north of east } (6\bar{0}°)$$

Choose a suitable scale.

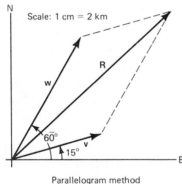

Parallelogram method

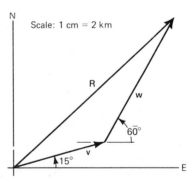

Vector triangle method

Using a ruler and protractor, we find that
$\mathbf{v} + \mathbf{w} = \mathbf{R} = 11.6 \text{ km at } 42° \text{ north of east}$

Using a ruler and protractor, we find that
$\mathbf{v} + \mathbf{w} = \mathbf{R} = 11.6 \text{ km at } 42° \text{ north of east}$

example 2 Find the vector sum of the following displacements using the vector polygon method. Choose a suitable scale.

$$\mathbf{v}_1 = 20\bar{0} \text{ km at } 3\bar{0}° \text{ north of west } (15\bar{0}°)$$

$$\mathbf{v}_2 = 5\bar{0} \text{ km due north} \qquad (\ 9\bar{0}°)$$

$$\mathbf{v}_3 = 15\bar{0} \text{ km at } 6\bar{0}° \text{ south of east } (30\bar{0}°)$$

*Angles in parentheses are expressed in standard position.

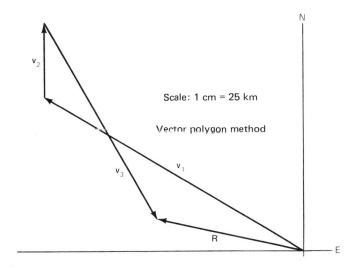

Using ruler and protractor, we find that

$$\mathbf{v}_1 + \mathbf{v}_2 + \mathbf{v}_3 = \mathbf{R} = 10\overline{0} \text{ km at } 12° \text{ north of west } (168°)$$

2.4   ADDITION OF VECTORS: TRIGONOMETRIC METHODS

Vector sums may be found using trigonometry.

example 1   Find the vector sum of the following displacements using trigonometric methods.

$$v = 8.50 \text{ km due west } (180.0°)$$
$$w = 4.60 \text{ km due south } (270.0°)$$

The vector triangle method gives us the following sketch.

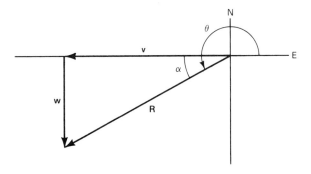

Using the Pythagorean theorem to find the length of R, we obtain

$$|\mathbf{R}| = \sqrt{|\mathbf{v}|^2 + |\mathbf{w}|^2}$$

$$= \sqrt{(8.50 \text{ km})^2 + (4.60 \text{ km})^2}$$

$$= 9.66 \text{ km}$$

$$\tan \alpha = \frac{\text{side opposite } \alpha}{\text{side adjacent } \alpha} = \frac{|\mathbf{w}|}{|\mathbf{v}|} \quad \text{(see Appendix C)}$$

$$\tan \alpha = \frac{4.60 \text{ km}}{8.50 \text{ km}} = 0.5412$$

$$\alpha = 28.4°$$

$$\theta = 180° + \alpha = 180.0° + 28.4° = 208.4°$$

Therefore, $\mathbf{R} = 9.66$ km at 28.4° south of west (208.4°).

Vector sums may also be found by using the Law of Sines and the Law of Cosines (see Appendix C).

example 2   Find the vector sum of the following displacements using trigonometric methods.

$$\mathbf{v} = 5.00 \text{ km at } 15.0° \text{ north of east } (15.0°)$$

$$\mathbf{w} = 7.50 \text{ km at } 60.0° \text{ north of east } (60.0°)$$

The vector triangle method gives us the following sketch. Using the Law of Cosines to find the length of $\mathbf{R}$, we obtain

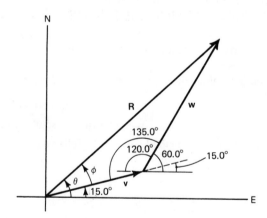

$$|\mathbf{R}|^2 = |\mathbf{v}|^2 + |\mathbf{w}|^2 - 2|\mathbf{v}||\mathbf{w}|\cos 135.0°$$

$$= (5.00 \text{ km})^2 + (7.50 \text{ km})^2 - 2(5.00 \text{ km})(7.50 \text{ km})(-0.7071)$$

$$= 134.3 \text{ km}^2$$

$$|\mathbf{R}| = 11.6 \text{ km}$$

Using the Law of Sines to find ϕ, we obtain

$$\frac{|\mathbf{R}|}{\sin 135.0°} = \frac{|\mathbf{w}|}{\sin \phi}$$

$$\frac{11.6 \text{ km}}{0.7071} = \frac{7.50 \text{ km}}{\sin \phi}$$

$$\sin \phi = 0.4572$$

$$\phi = 27.2°$$

$$\theta = \phi + 15.0° = 27.2° + 15.0° = 42.2°$$

Therefore, $\mathbf{R} = 11.6$ km at $42.2°$ north of east $(42.2°)$.

2.5   VECTOR COMPONENTS

We will find it desirable to express a given vector as the sum of two vectors, especially two vectors along the coordinate axes. The process of replacing one vector by two or more vectors is called the *resolution* of the original vector. In the figure below,

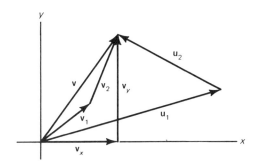

$$\mathbf{v} = \mathbf{v}_1 + \mathbf{v}_2$$
$$\mathbf{v} = \mathbf{u}_1 + \mathbf{u}_2$$

and

$$\mathbf{v} = \mathbf{v}_x + \mathbf{v}_y$$

The vectors $\mathbf{v}_1$, $\mathbf{v}_2$, $\mathbf{u}_1$, $\mathbf{u}_2$, $\mathbf{v}_x$, and $\mathbf{v}_y$ are called *components* of vector $\mathbf{v}$. That is, if two or more vectors are added and their sum is the resultant vector $\mathbf{v}$, each of these vectors is called a component of $\mathbf{v}$. We call $\mathbf{v}_x$ the horizontal component and $\mathbf{v}_y$ the vertical component.

In general, the horizontal and vertical components may be found using the definitions for sine and cosine:

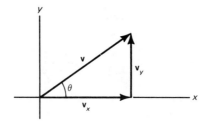

where θ is in standard position.

example 1 Find the horizontal and vertical components of the vector: 35.0 km at 60.0° north of east (60.0°).

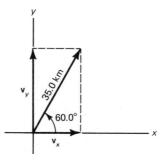

$\mathbf{v}_x = |\mathbf{v}|\cos\theta$

$\quad = (35.0\text{ km})(\cos 60.0°)$

$\quad = (35.0\text{ km})(0.5000)$

$\quad = 17.5\text{ km} \qquad$ (east or at 0° is understood)

$\mathbf{v}_y = |\mathbf{v}|\sin\theta$

$\quad = (35.0\text{ km})(\sin 60.0°)$

$\quad = (35.0\text{ km})(0.8660)$

$\quad = 30.3\text{ km} \qquad$ (north or at 90° is understood)

example 2 Find the horizontal and vertical components of the vector: 148 km at 25.0° north of west (155.0°).

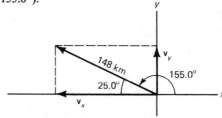

$$\mathbf{v}_x = |\mathbf{v}|\cos\theta$$
$$= (148\text{ km})(\cos 155.0°)$$
$$= (148\text{ km})(-0.9063)$$
$$= -134\text{ km} \qquad \text{(west or at }180°\text{ is understood)}$$

$$\mathbf{v}_y = |\mathbf{v}|\sin\theta$$
$$= (148\text{ km})(\sin 155.0°)$$
$$= (148\text{ km})(0.4226)$$
$$= 62.5\text{ km} \qquad \text{(north or at }90°\text{ is understood)}$$

Note: The minus sign in $\mathbf{v}_x = -134$ km indicates that the horizontal component is directed in the negative x-direction and does not indicate a negative distance.

2.6 ADDITION OF VECTORS: COMPONENT METHOD

Addition of vectors using graphic methods has a limited degree of accuracy. And, although accurate, addition of vectors using the Law of Sines and Law of Cosines is sometimes cumbersome. A component method of adding vectors is accurate and rather efficient.

To find the resultant vector **R** of two or more vectors using the component method:

1. Find the horizontal component, $\mathbf{R}_x$, of vector **R** by finding the algebraic sum of the horizontal components of each of the vectors being added.
2. Find the vertical component, $\mathbf{R}_y$, of vector **R** by finding the algebraic sum of the vertical components of each of the vectors being added.
3. Find the length of **R**: $|\mathbf{R}| = \sqrt{|\mathbf{R}_x|^2 + |\mathbf{R}_y|^2}$.
4. To find angle θ, first find angle α, the reference angle:

$$\tan\alpha = \frac{|\mathbf{R}_y|}{|\mathbf{R}_x|}$$

---

example 1 Find the vector sum of the following displacements using the component method.

$$\mathbf{v} = 5.00\text{ km at }15.0°\text{ north of east }(15.0°)$$
$$\mathbf{w} = 7.50\text{ km at }60.0°\text{ north of east }(60.0°)$$

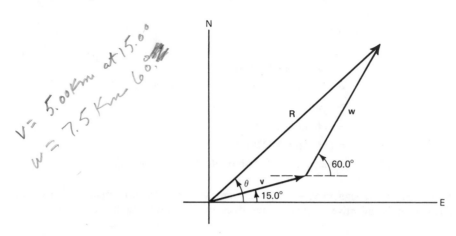

$$\mathbf{v}_x = |\mathbf{v}|\cos\theta$$

$$= (5.00 \text{ km})(\cos 15.0°)$$

$$= (5.00 \text{ km})(0.9659) \quad = 4.83 \text{ km}$$

$$\mathbf{w}_x = |\mathbf{w}|\cos\theta$$

$$= (7.50 \text{ km})(\cos 60.0°)$$

$$= (7.50 \text{ km})(0.5000) \quad = \underline{3.75 \text{ km}}$$

$\mathbf{R}_x$: sum of x-components $= 8.58$ km

$$\mathbf{v}_y = |\mathbf{v}|\sin\theta$$

$$= (5.00 \text{ km})(\sin 15.0°)$$

$$= (5.00 \text{ km})(0.2588) \quad = 1.29 \text{ km}$$

$$\mathbf{w}_y = |\mathbf{w}|\sin\theta$$

$$= (7.50 \text{ km})(\sin 60.0°)$$

$$= (7.50 \text{ km})(0.8660) \quad = \underline{6.50 \text{ km}}$$

$\mathbf{R}_y$: sum of y-components $= 7.79$ km

$$|\mathbf{R}| = \sqrt{|\mathbf{R}_x|^2 + |\mathbf{R}_y|^2}$$

$$= \sqrt{(8.58 \text{ km})^2 + (7.79 \text{ km})^2}$$

$$= 11.6 \text{ km}$$

$$\tan\theta = \frac{\mathbf{R}_y}{\mathbf{R}_x}$$

$$= \frac{7.79 \text{ km}}{8.58 \text{ km}} = 0.9079$$

$$\theta = 42.2°$$

Thus, $\mathbf{R} = 11.6$ km at $42.2°$ north of east ($42.2°$).

example 2   A ship travels $25\bar{0}$ km at $75.0°$ north of east, then $40\bar{0}$ km at $35.0°$ west of north, and then $15\bar{0}$ km at $25.0°$ west of south. Find the displacement; that is, find the net distance between the starting point and the stopping point and at what angle. Using the component method, let

$$\mathbf{u} = 25\bar{0} \text{ km at } 75.0° \text{ north of east } (75.0°)$$

$$\mathbf{v} = 40\bar{0} \text{ km at } 35.0° \text{ west of north } (125.0°)$$

$$\mathbf{w} = 15\bar{0} \text{ km at } 25.0° \text{ west of south } (245.0°)$$

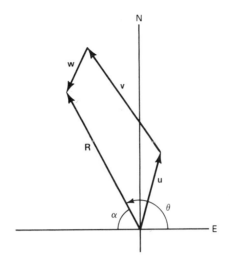

$$\mathbf{u}_x = |\mathbf{u}|\cos\theta$$

$$= (25\bar{0}\text{ km})(\cos 75.0°)$$

$$= (25\bar{0}\text{ km})(0.2588) \qquad = 65\text{ km}$$

$$\mathbf{v}_x = |\mathbf{v}|\cos\theta$$

$$= (40\bar{0}\text{ km})(\cos 125.0°)$$

$$= (40\bar{0}\text{ km})(-0.5736) \quad = -229\text{ km}$$

$$\mathbf{w}_x = |\mathbf{w}|\cos\theta$$

$$= (15\bar{0}\text{ km})(\cos 245.0°)$$

$$= (15\bar{0}\text{ km})(-0.4226) \quad = \underline{-63\text{ km}}$$

$\mathbf{R}_x$: sum of x-components $= -227$ km

$$\mathbf{u}_y = |\mathbf{u}|\sin\theta$$

$$= (25\bar{0}\text{ km})(\sin 75.0°)$$

$$= (25\bar{0}\text{ km})(0.9659) \qquad = 241\text{ km}$$

$$\mathbf{v}_y = |\mathbf{v}|\sin\theta$$

$$= (40\bar{0}\text{ km})(\sin 125.0°)$$

$$= (40\bar{0}\text{ km})(0.8192) \qquad = 328\text{ km}$$

$$\mathbf{w}_y = |\mathbf{w}|\sin\theta$$

$$= (15\bar{0}\text{ km})(\sin 245.0°)$$

$$= (15\bar{0}\text{ km})(-0.9063) \qquad = \underline{-136\text{ km}}$$

$\mathbf{R}_y$: sum of y-components $= 433$ km

$$|\mathbf{R}| = \sqrt{|\mathbf{R}_x|^2 + |\mathbf{R}_y|^2}$$

$$= \sqrt{(-227\text{ km})^2 + (433\text{ km})^2}$$

$$= 489\text{ km}$$

$$\tan\alpha = \frac{\mathbf{R}_y}{\mathbf{R}_x}$$

$$= \frac{433\text{ km}}{227\text{ km}} = 1.9075$$

$$\alpha = 62.3°$$

$$\theta = 180° - \alpha = 180.0° - 62.3° = 117.7°$$

Thus, **R** = 489 km at 62.3° north of west (117.7°). Or, the ship travels a net distance of 489 km at 62.3° north of west. (The angle in standard position is 117.7°.)

Although we have used displacement as the means for our initial discussion of vectors and vector operations, we will be presenting many other applications as we proceed through the developments of the various topics in physics.

EXERCISES *(Angles in parentheses are expressed in standard position.)*

Find the vector sum of each set of displacements by using a graphing method. Draw careful diagrams that are large enough for reasonably accurate results.

1. 15 km at $3\bar{0}°$ north of east $(3\bar{0}°)$
   45 km at $6\bar{0}°$ north of east $(6\bar{0}°)$

2. $14\bar{0}$ km at 55° north of west (125°)
   $18\bar{0}$ km at 35° east of south (305°)

3. $24\bar{0}$ km due west $(18\bar{0}°)$
   $10\bar{0}$ km due south $(27\bar{0}°)$

4. 25 km due east $(\bar{0}°)$
   45 km at $6\bar{0}°$ north of west $(12\bar{0}°)$
   $6\bar{0}$ km at $2\bar{0}°$ south of west $(20\bar{0}°)$

Find the vector sum of each set of displacements by using trigonometric methods.

5. 15.0 km at 30.0° north of east (30.0°)
   45.0 km at 60.0° north of east (60.0°)

6. 15.6 km at 25.0° south of west (205.0°)
   49.7 km at 16.0° north of west (164.0°)

7. $56\bar{0}$ km due south (270.0°)
   485 km due east (0.0°)

8. 145 km at 35.0° west of south (235.0°)
   365 km at 15.0° north of west (165.0°)
   275 km due north (90.0°)

In Exercises 9–15, find the vector sum of each set of displacements by using vector components.

9. 15.0 km at 30.0° north of east (30.0°)
   45.0 km at 60.0° north of east (60.0°)

10. 25.0 km at 40.0° south of west (220.0°)
    40.0 km at 50.0° east of north (40.0°)

11. 59.7 km due north (90.0°)
    85.6 km due west (180.0°)

12. 35.6 km at 25.0° north of west
    24.7 km at 65.0° west of north

13. 54.7 km at 40.0° west of north (130.0°)
    49.4 km at 10.0° west of south (260.0°)

14. 345 km at 15.0° north of west (165.0°)
    605 km at 25.0° east of south (295.0°)
    315 km at 18.0° south of west (198.0°)

15. $12\bar{0}$ km at $3\bar{0}$° south of east ($33\bar{0}$°)
    $15\bar{0}$ km due north ($9\bar{0}$°)
    75 km at 45° south of west (225°)
    25 km at $1\bar{0}$° south of west ($19\bar{0}$°)

16. An airplane takes off and flies $15\bar{0}$ km on a course 35.0° west of north, then changes course and flies 75.0 km due north to where it lands. Find the displacement from the starting point to the landing point.

17. A ship travels $25\bar{0}$ km on a course 35.0° north of east, then travels 125 km on a course 10.0° east of north to a point where it lands. Find the displacement from the starting point to the landing point.

18. An automobile is driven 16.0 km due east, then 4.00 km due north, then 6.00 km due east, then 10.0 km due south. Find the displacement from the starting point to the ending point.

19. In route between two cities, a car travels $5\bar{0}$ km due west, then 175 km southwest, then 75 km due south. Find the displacement from the starting point to the ending point.

Given two displacements with magnitudes of $4\bar{0}$ km and 25 km:

20. What is the magnitude of the maximum resultant displacement?

21. What is the magnitude of the minimum resultant displacement?

22. What is the angle between the two original displacements if the magnitude of the resultant displacement is 32 km?

23. What is the angle between the two original displacements if the magnitude of the resultant displacement is 15 km?

Given three displacements with magnitudes of $2\bar{0}$ km, 15 km, and $3\bar{0}$ km:

24. What is the magnitude of the maximum resultant displacement?

25. What is the magnitude of the minimum resultant displacement?

PROBLEM SOLVING

3.1 PROBLEM SOLVING

Problem solving in technical fields is more than substituting in formulas. It is necessary that you develop skill in taking data, analyzing the problems present, and finding the solution in an orderly manner.

Understanding the principle involved in a problem is more important than blindly substituting in a formula. By following an orderly procedure for problem solving, we hope to develop an approach to problem solving that you can use in your studies and on the job.

In all problems in the remainder of this book, the method described below will be applied to all problems where appropriate.

Problem-solving method

1. *Read the problem carefully*. This might appear obvious to you, but it is the most important step in solving a problem. As a matter of habit, you should read the problem at least twice.

    (a) The first time you should read the problem straight through from beginning to end. Do not stop this time to think about setting up an equation or formula. You are only trying to get a general overview of the problem during this first reading.

    (b) Next, read through a second time slowly and *completely*, beginning to think ahead to the following steps.

2. *Make a sketch.* All problems may not lend themselves to a sketch. However, make a sketch whenever it is possible. Often, seeing a sketch of the problem will show if you have forgotten important parts of the problem and may suggest the solution. This is a *very important* and often overlooked step in problem solving.

3. *Write down all given information.* This is necessary to get all essential facts in mind before looking for the solution. There are some common phrases that have understood physical meanings. For example, the term "from rest" means that the initial velocity equals zero, or $v_0 = 0$; the term "smooth surface" means to assume that no friction is present.

4. *Write down the unknown or quantity asked for in the problem.* Many students have difficulty solving problems because they do not know what they are looking for and solve for the wrong quantity.

5. *Write down the basic equation or formula that relates the known and unknown quantities.* We find the basic formula or equation to use by studying what we are given and what we are asked to find. Then look for a formula or equation that relates these quantities. Sometimes we may need to use more than one equation or formula in working a problem.

6. *Find a working equation by solving the basic equation or formula for the unknown quantity.*

7. *Substitute the data in the working equation, including the appropriate units.* It is important that you *carry the units all the way through the problem* as a check that you have solved the problem correctly. For example, if you are asked to find the weight of an object in newtons and the units of your answer work out to be metres, you need to review your solution for the error. (When the unit analysis is not obvious, we will go through it step by step in a box within the example.)

8. *Perform the indicated operations and work out the solution.* Although this will be your final written step in the solution, in every case you should ask yourself, "Is my answer reasonable?" Here and on the job you will be dealing with practical problems. A quick estimate will many times reveal an error in your calculation.

Sketch	Data	Basic Equation	Working Equation	Substitution
$12\ cm^2$, $b = 6.0\ cm$, $h = ?$	$A = 12\ cm^2$ $b = 6.0\ cm$ $h = ?$	$A = bh$	$h = \dfrac{A}{b}$	$h = \dfrac{12\ cm^2}{6.0\ cm}$ $= 2.0\ cm$

To help you recall the procedure detailed above with every problem set that follows, refer to the rectangular box. This box is not meant to be complete, simply an outline to assist you in remembering and following the procedure for solving problems. *You should follow this outline in solving all problems in this course.*

This problem-solving method will be demonstrated in terms of relationships and formulas with which you are probably familiar.

example 1   Find the volume of concrete needed to fill a bridge support whose dimensions are 2.00 m by 8.00 m by 28.0 m.

SKETCH:

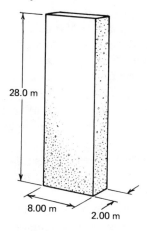

28.0 m

8.00 m

2.00 m

DATA:

$l = 8.00$ m
$w = 2.00$ m $\Big\}$ This is a listing of the information you know.
$h = 28.0$ m

$V = ?$ This identifies the unknown.

BASIC EQUATION: $V = lwh$

WORKING EQUATION: same

SUBSTITUTION: $V = (8.00 \text{ m})(2.00 \text{ m})(28.0 \text{ m})$

$= 448$ m^3 $\boxed{\text{m} \times \text{m} \times \text{m} = \text{m}^3}$

example 2   A cylindrical bin of radius 4.00 m contains 377 m^3 of storage space. Find its height.

SKETCH:

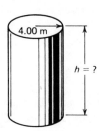

4.00 m

$h = ?$

DATA:

$$V = 377 \text{ m}^3$$

$$r = 4.00 \text{ m}$$

$$h = ?$$

BASIC EQUATION: $V = \pi r^2 h$ To find the working equation, solve the basic

WORKING EQUATION: $h = \dfrac{V}{\pi r^2}$ equation for the unknown letter.

SUBSTITUTION:

$$h = \dfrac{377 \text{ m}^3}{\pi (4.00 \text{ m})^2}$$

$$= 7.50 \text{ m} \qquad \boxed{\dfrac{\text{m}^3}{\text{m}^2} = \text{m}}$$

example 3 Cables are to be attached to a 95.0-m-tall television relay tower 20.0 m from the top. Each cable is anchored on the ground 45.0 m from its base. A cable of what length is needed?

SKETCH:

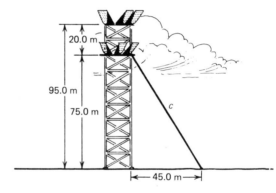

DATA:

$$a = 75.0 \text{ m}$$

$$b = 45.0 \text{ m}$$

$$c = ?$$

BASIC EQUATION: $c^2 = a^2 + b^2$

WORKING EQUATION: $c = \sqrt{a^2 + b^2}$

SUBSTITUTION: $c = \sqrt{(75.0 \text{ m})^2 + (45.0 \text{ m})^2}$

$$= \sqrt{7650 \text{ m}^2}$$

$$= 87.5 \text{ m}$$

example 4 A piece of conduit cuts across 4.50 m from a corner in a building at an angle of 25.0°. A conduit of what length is needed?

Sketch:

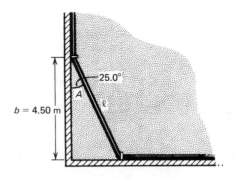

Data:
$$\angle A = 25.0°$$
$$b = 4.50 \text{ m}$$
$$\ell = ?$$

Basic Equation:
$$\cos A = \frac{\text{side adjacent } A}{\text{hypotenuse}} = \frac{b}{\ell}$$

Working Equation:
$$\ell = \frac{b}{\cos A}$$

Substitution:
$$\ell = \frac{4.50 \text{ m}}{\cos 25.0°} = \frac{4.50 \text{ m}}{0.9063}$$
$$= 4.97 \text{ m}$$

example 5 A trapezoidal plot of ground has an area of 3.40×10^4 m^2, a height of $12\bar{0}$ m, and one of the parallel bases has a length of 245 m. Find the length of the other parallel base.

Sketch:

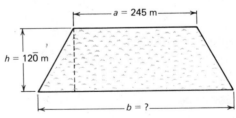

Data:
$$A = 3.40 \times 10^4 \text{ m}^2$$
$$h = 12\bar{0} \text{ m}$$
$$a = 245 \text{ m}$$
$$b = ?$$

Basic Equation:
$$A = \frac{h}{2}(a + b)$$

$$2A = h(a + b)$$
$$2A = ha + hb$$
$$2A - ha = hb$$
$$\frac{2A - ha}{h} = b$$

These steps will normally not be shown in the text.

WORKING EQUATION:
$$b = \frac{2A - ha}{h}$$

SUBSTITUTION:
$$b = \frac{2(3.40 \times 10^4 \ \text{m}^2) - (12\bar{0} \ \text{m})(245 \ \text{m})}{12\bar{0} \ \text{m}}$$

$$= 322 \ \text{m}$$

The problem-solving method helps you to solve problems by providing a framework with which to view, to understand, and to attack a problem. It also gives your work an order that is most important in eliminating careless errors, in checking your work and solution, and in communicating your solution to a fellow worker or to your supervisor. In industry and commerce, estimates of construction, requests for supplies for projects, solutions to engineering-type problems, and stocking a storeroom all require the communication of information that is easily followed and understood by another person (usually the boss!). This information must be presented in an orderly manner so that the data for each specific problem and its solution can be easily followed step by step as well as any overall data and general conclusions or implications.

EXERCISES

Use the problem-solving method to do each of the following. (Here, as throughout the text, follow the rules of calculations with measurements that were described in Section 1.6.)

Sketch	Data	Basic Equation	Working Equation	Substitution
$12 \ \text{cm}^2$ $h = ?$ $b = 6.0 \ \text{cm}$	$A = 12 \ \text{cm}^2$ $b = 6.0 \ \text{cm}$ $h = ?$	$A = bh$	$h = \dfrac{A}{b}$	$h = \dfrac{12 \ \text{cm}^2}{6.0 \ \text{cm}}$ $= 2.0 \ \text{cm}$

1. A rectangle has a length of 4.50 m and a width of 3.75 m.
   (a) Find its area.
   (b) Find its perimeter.

2. A rectangle has an area of 2700 cm². If its length is 75 cm, what is its width?

3. The dimensions of a rectangular box are 65.0 cm × 36.0 cm × 18.0 cm.
   (a) Find its volume.
   (b) Find its total surface area.

4. You are to build a rectangular bin that holds 1250 m³. If its length and width are to be 15.0 m × 12.0 m, what must its height be?

5. Each side of a square piece of sheet metal measures 64.0 cm on a side. A 10.0-cm square is then cut from each corner. The metal is then folded to form a box without a top. Find the volume of the box.

6. Find the cross-sectional area of a pipe whose outer diameter is 9.6 cm and whose inner diameter is 7.8 cm.

7. A piece of round stock of diameter 20.0 cm is milled into a square piece of stock of largest possible cross-sectional area. Find the cross-sectional area of this resulting square stock.

8. The centers of eight circular holes are equally spaced around a circle of diameter 15.0 cm. Find the (straight line) distance between the centers of any two adjacent holes.

9. An inverted conical tank 5.00 m in diameter holds 98.2 m³ of liquid.
   (a) What is its height?
   (b) If the tank is three-fourths full, what is the depth of the liquid?

10. A satellite is directly overhead one observation station at the same instant it is observed at an angle of 61° from the ground. If the two stations are 15.0 km apart, how high is the satellite above the level ground?

11. A bullet is found embedded in the wall of a room 2.5 m above the floor. The bullet entered the wall going upward at an angle of 76° with the wall. How far from the wall was the gun fired if the gun was held at a distance of 1.2 m above the floor?

12. A ship's navigator measures the angle from sea level to the beacon of a lighthouse as 9.80°. The navigator knows that this particular beacon is 225 m above sea level. How far is the ship from the lighthouse?

13. The total surface area of a right circular cylinder is given by the formula $S = 2\pi r(r + h)$. If the total surface area is 4280 cm² and the radius is 16.5 cm, find the height.

14. If the total surface area of a right circular cylinder is 4520 cm² and its height is 12.5 cm, find its radius.

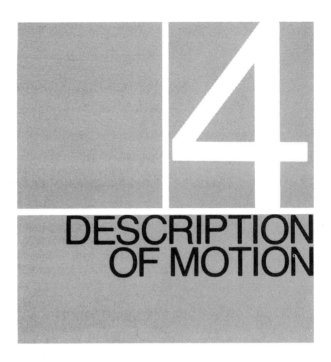

DESCRIPTION OF MOTION

4.1 MOTION

Motion is usually defined as a *continual change in position*. In Chapter 2, *displacement* was defined as a (net) *change in position*. Motion occurs while an object is undergoing a change in position. A displacement is the result of motion.

There are basically two types of motion: linear and rotary. *Linear* motion may be defined as motion along a straight line. *Rotary* motion may be defined as motion about an axis. More complex motions are essentially combinations of these two.

4.2 SPEED

An automobile has a speedometer which indicates the *speed*, or *the time rate of change of distance*, as you drive along the road. If you have "cruise control" or keep your foot pressure constant on the gas pedal, the speed is said to be uniform; that is, equal distances are covered during all equal time segments.

The graph on page 52 shows an automobile in uniform motion or moving at a uniform speed. Note that during any 2-s interval, the distance traveled is the same: 40 m.

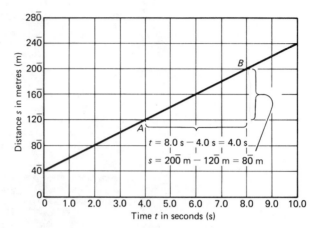

Graph of distance-time showing uniform or constant speed

The graphic representation of an object in uniform motion is a straight line. When you take a rather long automobile trip, you are more interested in your *average speed*.

$$\text{average speed} = \frac{\text{total distance traveled}}{\text{total elapsed time}}$$

That is, if you traveled $24\bar{0}$ km in 3.0 h, your average speed is

$$\bar{v} = \frac{s}{t} = \frac{24\bar{0} \text{ km}}{3.0 \text{ h}} = 8\bar{0} \text{ km/h}$$

The speed of the automobile as indicated in the graph above may be found from the graph by marking any two convenient points on the graph, say *A* and *B*. Find the distance traveled between *A* and *B* ($20\bar{0}$ m $- 12\bar{0}$ m $= 8\bar{0}$ m) and the corresponding time interval (8.0 s $-$ 4.0 s $=$ 4.0 s). The uniform speed is then

$$\bar{v} = \frac{s}{t} = \frac{8\bar{0} \text{ m}}{4.0 \text{ s}} = 2\bar{0} \text{ m/s}$$

Speed is usually measured in km/h or m/s.

Instantaneous speed is the speed at a specific instant of time. Instantaneous speed is measured by timing the motion over a very short time interval. The shorter the time interval and the more precise the clock, the more accurate the instantaneous speed. Note that both average speed and instantaneous speed are completely described in terms of a magnitude alone. Thus, speed is a scalar.

The instantaneous speed can also be found from a graph. The graph on page 53 shows the variable speed of an automobile traveling in heavy traffic.

The graphic representation of an object moving at a variable speed is a curved line. It can be shown mathematically that the instantaneous speed at any time t_1 is the slope of the tangent line to the curve at $t = t_1$. After drawing the tangent line to the curve at point *C*, we find that its slope is

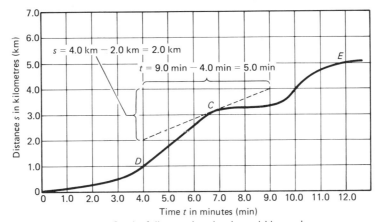

Graph of distance-time showing variable speed

$$\frac{2.0 \text{ km}}{5.0 \text{ min}} = 0.40 \text{ km/min}$$

which is the instantaneous speed at $t = 6.5$ min.

The average speed between any two points, say D and E, can be found by finding the distance traveled between D and E (5.0 km − 1.0 km = 4.0 km) and the corresponding time interval (12.0 min − 4.0 min = 8.0 min). The average speed between D and E is then

$$\bar{v} = \frac{s}{t} = \frac{4.0 \text{ km}}{8.0 \text{ min}} = 0.50 \text{ km/min}$$

Average speed is usually expressed by the following equation:

$$\boxed{\bar{v} = \frac{s}{t}}$$

where $\bar{v}$ is the average speed,

    t is the time interval, and

    s is the distance traveled during the time interval.

4.3 VELOCITY

When both speed and direction are needed to describe the motion of a object, the term "velocity" is used. *The velocity of an object is its time rate of change of its displacement.* Since both magnitude (speed) and direction are required to completely describe velocity, we must conclude that velocity is a vector quantity. Velocity may also be thought of as the rate of change of displacement. Recall that displacement is a vector quantity.

If plane A is flying east at $20\overline{0}$ km/h and plane B is flying north at $20\overline{0}$ km/h, we say that the planes are flying at the same speed (same magnitude) but that the velocities are different because the directions are different. (Recall that for two vector quantities to be equal, the magnitudes must be equal *and* the directions must be equal.)

The average velocity is found in essentially the same way as is the average speed except that displacement is used instead of distance. To find the average velocity, **v**, we use the equation

$$|\mathbf{v}| = \frac{|\mathbf{s}|}{t}$$

where $|\mathbf{v}|$ is the magnitude of the velocity vector,

t is the corresponding time interval, and

$|\mathbf{s}|$ is the magnitude of the displacement vector, that is, the net distance.

The direction of **v** is the direction of the displacement vector from the initial or starting point and the terminal or ending point.

example 1   An automobile travels 410 km in 5.5 h from Milwaukee to Grand Rapids around Lake Michigan. The air distance between the two cities is 195 km. Grand Rapids is due east of Milwaukee. Find
(a) the average speed.
(b) the average velocity of the automobile.

SKETCH:

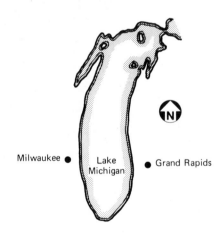

(a) DATA:

$$s = 410 \text{ km}$$

$$t = 5.5 \text{ h}$$

$$\bar{v} = ?$$

BASIC EQUATION:

$$\bar{v} = \frac{s}{t}$$

Working Equation:	same
Substitution:	$\bar{v} = \dfrac{410 \text{ km}}{5.5 \text{ h}} = 75 \text{ km/h}$
(b) Data:	$|\mathbf{s}| = 195 \text{ km}$
	$t = 5.5 \text{ h}$
	$|\mathbf{v}| = ?$
Basic Equation:	$|\mathbf{v}| = \dfrac{|\mathbf{s}|}{t}$
Working Equation:	same
Substitution:	$|\mathbf{v}| = \dfrac{195 \text{ km}}{5.5 \text{ h}} = 35 \text{ km/h}$

The direction of **v** is east. That is, **v** is 35 km/h due east.

Note: Since speed is the magnitude of velocity, it is common to use v for either speed or the magnitude of velocity and to use $\bar{v}$ for average speed. In addition, the terms *speed* and *velocity* are often used interchangeably because the direction of the motion is often understood. For example, velocity that results from motion in a straight line has the same direction as the direction of the motion. Now that you understand the difference between speed and velocity, you should be able to distinguish between them even by implication of the situation, if necessary.

Average velocity or average speed may also be found by

$$\boxed{v_{av} = \frac{s_f - s_i}{t_f - t_i}}$$

where v_{av} is the average velocity or average speed,
$\quad s_i$ is the initial position at time t_i, and
$\quad s_f$ is the final position at time t_f.
This relationship may also be written

$$\boxed{v_{av} = \frac{\Delta s}{\Delta t}}$$

where Δs* is the change in distance or the displacement, and
$\quad \Delta t$ is the change in time; that is, the corresponding time interval.
Since velocity is a vector quantity, we can use the vector methods of Chapter 2 to solve velocity problems.

---

*The Greek letter Δ is commonly used in mathematics and science to mean "the change in."

Suppose that a plane is flying in still air due east (0°) at $45\overline{0}$ km/h. Suddenly, there is a tail wind of $4\overline{0}$ km/h; that is, the velocity of the wind is $4\overline{0}$ km/h toward the east. What is the resultant velocity of the plane with respect to the ground; that is, what is the "ground speed"?

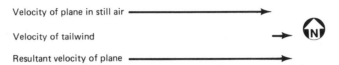

Since the two velocities are in the same direction, the magnitude of the resultant is the sum of the magnitudes of the component velocities. In other words, the resultant velocity is $49\overline{0}$ km/h ($45\overline{0}$ km/h + $4\overline{0}$ km/h) eastward. (*Note*: The plane's air speed remains $45\overline{0}$ km/h.)

Now, let us study the opposite case. Suppose that a plane is flying in still air due east (0°) at $45\overline{0}$ km/h. Suddenly, there is a head wind of $4\overline{0}$ km/h; that is, the velocity of the wind is $4\overline{0}$ km/h toward the west. What is the resultant velocity of the plane (its "ground speed")?

Since the two velocities are in opposite directions, the magnitude of the resultant is the difference of the magnitudes of the component velocities. In other words, the resultant velocity is $41\overline{0}$ km/h ($45\overline{0}$ km/h − $4\overline{0}$ km/h) eastward.

What happens if the plane is flying eastward at $45\overline{0}$ km/h in still air and suddenly there is a wind of $4\overline{0}$ km/h from the north? The resultant velocity is the sum of the two vectors.

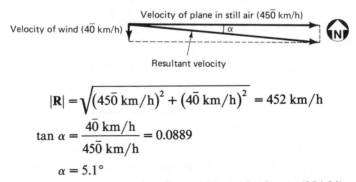

$$|\mathbf{R}| = \sqrt{\left(45\overline{0}\ \text{km/h}\right)^2 + \left(4\overline{0}\ \text{km/h}\right)^2} = 452\ \text{km/h}$$

$$\tan \alpha = \frac{4\overline{0}\ \text{km/h}}{45\overline{0}\ \text{km/h}} = 0.0889$$

$$\alpha = 5.1°$$

Thus, the resultant velocity is 452 km/h at 5.1° south of east (354.9°).

Of course, the pilot does not want to be $4\overline{0}$ km south of his course for each hour in flight. So, what does he or she do? Actually, the pilot wants the resultant velocity to be 45$\overline{0}$ km/h due east (0°). The problem is now one of finding one component knowing the resultant and the other component.

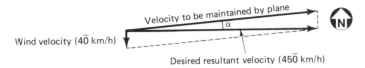

Velocity to be maintained by plane

Wind velocity (4$\overline{0}$ km/h)

Desired resultant velocity (45$\overline{0}$ km/h)

The magnitude of the velocity to be maintained by the plane is

$$\sqrt{\left(45\overline{0}\text{ km/h}\right)^2 + \left(4\overline{0}\text{ km/h}\right)^2} = 452\text{ km/h}$$

$$\tan\alpha = \frac{4\overline{0}\text{ km/h}}{45\overline{0}\text{ km/h}} = 0.0889$$

$$\alpha = 5.1°$$

Thus, the plane must fly at a heading of 5.1° north of east (5.1°) at 452 km/h in order to fly due east at a ground speed of 45$\overline{0}$ km/h with a 4$\overline{0}$ km/h wind from the north.

example 2   Suppose that a pilot wishes to fly a plane on a course 20.0° south of west (200.0°) at 40$\overline{0}$ km/h. The wind is 5$\overline{0}$ km/h from the south (90.0°). At what heading and at what speed must the plane fly to compensate for the wind?

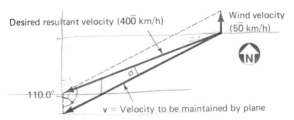

Desired resultant velocity (40$\overline{0}$ km/h)

Wind velocity (5$\overline{0}$ km/h)

110.0°

ϕ

v = Velocity to be maintained by plane

Using the Law of Cosines to find the speed, we obtain

$$|v|^2 = \left(40\overline{0}\text{ km/h}\right)^2 + \left(5\overline{0}\text{ km/h}\right)^2 - 2\left(40\overline{0}\text{ km/h}\right)\left(5\overline{0}\text{ km/h}\right)(\cos 110.0°)$$

$$|v| = 42\overline{0}\text{ km/h}$$

Using the Law of Sines to find angle ϕ, we obtain

$$\frac{5\overline{0}\text{ km/h}}{\sin\phi} = \frac{42\overline{0}\text{ km/h}}{\sin 110.0°}$$

$$\sin\phi = 0.1119$$

$$\phi = 6.4°$$

Thus, the plane must fly at a heading of 26.4° (20.0° + 6.4°) south of west (206.4°) at 42$\overline{0}$ km/h to compensate for the wind.

Up to now, we have given the usual standard position method of listing vectors. The navigational method also deserves mentioning. This reference plane is shown at the top of page 58.

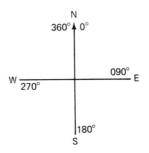

A plane flying due north is said to be flying on a heading of 0° or 360°. A plane flying due south is said to be flying on a heading of 180°. A plane flying northeast is said to be flying on a heading of 045°. Wind directions are given in terms of the direction *from* which it is blowing. For example, if a plane flying due east at 275 km/h with a tail wind of 45 km/h, the plane is flying on a heading of 090° at 275 km/h while the wind's heading is 270° at 45 km/h.

example 3   A pilot wishes to fly on a course of 285° at $35\bar{0}$ km/h. The wind is 75 km/h at 020°. At what navigational heading and at what speed must the plane fly to compensate for the wind?

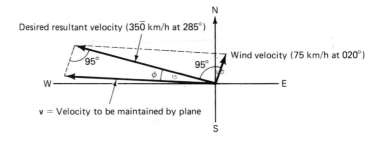

Using the Law of Cosines to find the speed, we obtain

$$|\mathbf{v}|^2 = \left(35\bar{0} \text{ km/h}\right)^2 + (75 \text{ km/h})^2 - 2(35\bar{0} \text{ km/h})(75 \text{ km/h})(\cos 95°)$$
$$|\mathbf{v}| = 364 \text{ km/h}$$

Using the Law of Sines to find angle ϕ, we obtain

$$\cdot \ \frac{364 \text{ km/h}}{\sin 95°} = \frac{75 \text{ km/h}}{\sin \phi}$$
$$\sin \phi = 0.2053$$
$$\phi = 12°$$

Thus, the plane must fly at a navigational heading of 273° (285° − 12°) at 364 km/h to compensate for the wind.

Note: We have not given any navigational method exercises so that the two methods are not confused.

4.4   ACCELERATION

The gas pedal of an automobile is commonly called the accelerator. When your foot pushes it down, the speed of the automobile increases. As the speed of the automobile increases, we say that the automobile is accelerating. *Acceleration* is the time rate of change of velocity. The faster the velocity increases, the greater the acceleration. Acceleration, being the rate of change of velocity, is also a vector. In this section we consider only acceleration that results from motion in a straight line; thus, the direction of the acceleration is the same as (or the opposite of) the direction of the motion. Note that since **v** is considered only in a straight line, it can be written as *v*. Acceleration resulting from a change in direction is discussed in Chapter 9.

Average acceleration may be found by the equation

$$a_{av} = \frac{v_f - v_i}{t_f - t_i}$$

where a_{av} is the average acceleration,
    v_i is the initial velocity at time t_i, and
    v_f is the final velocity at time t_f.
This relationship may also be written

$$a_{av} = \frac{\Delta v}{\Delta t}$$

where Δv is the change in velocity during the time interval Δt.

The units of acceleration deserve special attention. Acceleration is the rate of change of velocity:

$$a_{av} = \frac{\Delta v}{\Delta t} = \frac{\text{distance}/\text{time}}{\text{time}} = \frac{\text{distance}}{\text{time}^2}$$

The two most common metric units of acceleration are m/s/s or m/s² and km/h/s.

example 1   An automobile accelerates from 45 km/h to $8\bar{0}$ km/h in $1\bar{0}$ s. Find its average acceleration.

DATA: $\Delta v = 8\bar{0}$ km/h $- 45$ km/h $= 35$ km/h

                                           $\Delta t = 1\bar{0}$ s

                                           $a_{av} = ?$

BASIC EQUATION: $a_{av} = \dfrac{\Delta v}{\Delta t}$

WORKING EQUATION: same

SUBSTITUTION: $a_{av} = \dfrac{35 \text{ km/h}}{1\bar{0} \text{ s}}$

$$= 3.5 \text{ km/h/s}$$

example 2   An automobile starts from rest and accelerates to 25 m/s ($9\bar{0}$ km/h) in 8.0 s. Find its average acceleration.

DATA: $\Delta v = 25 \text{ m/s} - 0 \text{ m/s} = 25 \text{ m/s}$

$$\Delta t = 8.0 \text{ s}$$

$$a_{av} = ?$$

BASIC EQUATION: $a_{av} = \dfrac{\Delta v}{\Delta t}$

WORKING EQUATION: same

SUBSTITUTION: $a_{av} = \dfrac{25 \text{ m/s}}{8.0 \text{ s}}$

$$= 3.1 \text{ m/s/s} = 3.1 \text{ m/s}^2$$

A ball rolling down a plane is an example of uniform or constant acceleration.

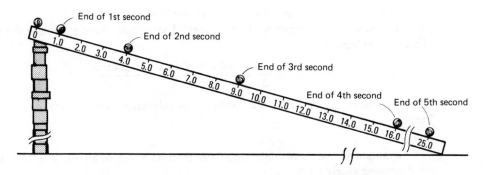

The graph of these data, shown on page 61, also shows that the ball is in uniform or constant acceleration. Note that during any 1-s interval, the velocity increases by the same quantity: 2.0 m/s.

Elapsed time, t (s)	Position at time, t (m)	Distance, Δs, covered (m)	Average velocity, v_{av}, for 1-s interval (m/s)	Acceleration during interval $\Delta v/\Delta t$ (m/s²)	Velocity, v, at time, t (m/s)
0.0	0.0				0.0
		1.0	1.0	2.0	
1.0	1.0				2.0
		3.0	3.0	2.0	
2.0	4.0				4.0
		5.0	5.0	2.0	
3.0	9.0				6.0
		7.0	7.0	2.0	
4.0	16.0				8.0
		9.0	9.0	2.0	
5.0	25.0				10.0

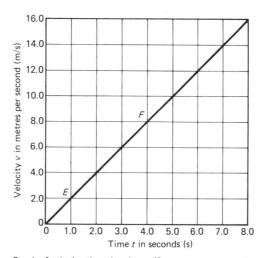

Graph of velocity-time showing uniform or constant acceleration

The graphic representation of the velocity of an object in uniform acceleration is a straight line. The acceleration of the ball can be found directly from the graph by marking any two convenient points on the graph, say E and F. Find the change in velocity between E and F (8.0 m/s − 2.0 m/s = 6.0 m/s) and the corresponding time interval (4.0 s − 1.0 s = 3.0 s). The uniform acceleration is then

$$a = \frac{\Delta v}{\Delta t} = \frac{6.0 \text{ m/s}}{3.0 \text{ s}} = 2.0 \text{ m/s}^2$$

Instantaneous acceleration is the acceleration at a specific instant of time. Instantaneous acceleration is measured by calculating the change in velocity over a very short time interval.

The graph below shows the velocity of an automobile which has variable acceleration while traveling in heavy traffic.

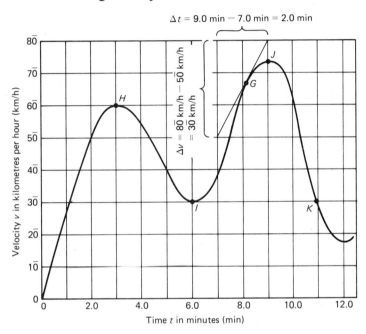

Graph of velocity-time showing variable acceleration

The graphic representation of an object in variable acceleration is a curved line. As in the case of variable velocity, it can be shown mathematically that the instantaneous acceleration at any time, t_1, is the slope of the tangent line to the velocity curve at $t = t_1$. Even though the variable velocity remains positive, the slope of the velocity curve does not always remain positive. That is, the velocity is *not* always increasing throughout all the time intervals. However, the displacement *is* always increasing throughout all the time intervals (see the figure on page 53).

After drawing the tangent line to the curve at point G, we find that its slope is

$$\frac{\Delta v}{\Delta t} = \frac{3\overline{0} \text{ km/h}}{2.0 \text{ min}} = 15 \text{ km/h/min}$$

which is the instantaneous acceleration at $t = 8.2$ min.

The average acceleration between two points, say H and I, can be found by finding the change in velocity between H and I ($3\overline{0}$ km/h $- 6\overline{0}$ km/h $= -3\overline{0}$ km/h) and the corresponding time interval (6.0 min $-$ 3.0 min $=$ 3.0 min).

The average acceleration between H and I is then

$$a_{av} = \frac{\Delta v}{\Delta t} = \frac{-3\bar{0} \text{ km/h}}{3.0 \text{ min}} = -1\bar{0} \text{ km/h/min}$$

Note here that the acceleration is negative, as is the slope of the line through points H and I. Negative acceleration is also commonly called deceleration.

Between points I and J, is the automobile accelerating or decelerating? Between points J and K? What can be said about the average acceleration or deceleration between points H and J?

4.5   ACCELERATION, VELOCITY, DISTANCE, AND TIME

Let us now study in detail the various relationships that exist between acceleration, velocity, distance or displacement, and time.

Acceleration, velocity, and time: *The final velocity of an object equals the sum of its initial velocity and the increase in velocity that is produced by the acceleration.* That is,

$$\boxed{v_f = v_i + a(\Delta t)} \qquad (1)$$

This result comes from the definition of acceleration:

$$a = \frac{v_f - v_i}{t_f - t_i} = \frac{v_f - v_i}{\Delta t}$$

$$a(\Delta t) = v_f - v_i$$

Thus,

$$v_f = v_i + a(\Delta t)$$

Of course, if the object starts at rest, $v_i = 0$ and

$$v_f = a(\Delta t)$$

example 1   An automobile traveling at an initial velocity of 8.00 m/s accelerates uniformly at 2.50 m/s². Find its final velocity (in km/h) after 6.00 s.

DATA: $v_i = 8.00$ m/s

$$a = 2.50 \text{ m/s}^2$$

$$\Delta t = 6.00 \text{ s}$$

$$v_f = ?$$

BASIC EQUATION: $v_f = v_i + a(\Delta t)$

WORKING EQUATION: same

SUBSTITUTION:
$$v_f = 8.00 \text{ m/s} + (2.50 \text{ m/s}^2)(6.00 \text{ s})$$

$$= 8.00 \text{ m/s} + 15.0 \text{ m/s}$$

$$= 23.0 \text{ m/s} \times \frac{1 \text{ km}}{1000 \text{ m}} \times \frac{3600 \text{ s}}{1 \text{ h}}$$

$$= 82.8 \text{ km/h}$$

The equation $v_f = v_i + a(\Delta t)$ is also applicable when the algebraic signs of v_i and a are negative. The initial velocity, v_i, is negative when the initial motion is opposite the specified direction; that is, the initial motion is backward (or opposite the specified direction). As we have already seen, the acceleration, a, is negative (deceleration) when the object is slowing down, as when an automobile is slowing down in a forward gear.

example 2   An automobile traveling at an initial velocity of 25.0 m/s accelerates uniformly at -4.00 m/s^2 (decelerates). Find its final velocity (in m/s) after 3.00 s.

DATA:
$$v_i = 25.0 \text{ m/s}$$
$$a = -4.00 \text{ m/s}^2$$
$$\Delta t = 3.00 \text{ s}$$
$$v_f = ?$$

BASIC EQUATION: $v_f = v_i + a(\Delta t)$

WORKING EQUATION: same

SUBSTITUTION:
$$v_f = 25.0 \text{ m/s} + (-4.00 \text{ m/s}^2)(3.00 \text{ s})$$

$$= 25.0 \text{ m/s} - 12.0 \text{ m/s}$$

$$= 13.0 \text{ m/s}$$

Initial, final, and average velocities: *The average velocity of an object in uniformly accelerated motion is the average of the initial and final velocities; that is, the average velocity is one-half the sum of the initial and final velocities.*

$$v_{av} = \frac{v_i + v_f}{2} \qquad (2)$$

Acceleration, velocity, distance, and time: *The distance traveled by an object in uniformly accelerated motion when the initial velocity, the acceleration, and* *the corresponding time interval are known is given by*

$$\Delta s = v_i(\Delta t) + \tfrac{1}{2}a(\Delta t)^2 \qquad\qquad (3)$$

Let us show mathematically how this result comes about.

Recall from Section 4.3 that

$$v_{av} = \frac{\Delta s}{\Delta t} \qquad\qquad (4)$$

or

$$\Delta s = v_{av}(\Delta t) \qquad\qquad (5)$$

Now substitute equation (1) into equation (2):

$$v_{av} = \frac{v_i + \left[v_i + a(\Delta t)\right]}{2}$$

$$v_{av} = v_i + \tfrac{1}{2}a(\Delta t) \qquad\qquad (6)$$

Next, substitute equation (4) into equation (6):

$$\frac{\Delta s}{\Delta t} = v_i + \tfrac{1}{2}a(\Delta t)$$

or

$$\Delta s = v_i(\Delta t) + \tfrac{1}{2}a(\Delta t)^2$$

which is equation (3). Of course, if the object starts from rest, $v_i = 0$, then

$$\Delta s = \tfrac{1}{2}a(\Delta t)^2$$

example 3   An automobile traveling at 78 km/h has its brakes applied at a uniformly decelerated rate of 6.0 m/s^2.

(a) How long (in s) will it take the automobile to stop?

(b) How far (in m) will the automobile travel before coming to a complete stop?

(a) DATA:

$$v_i = 78 \text{ km/h} \times \frac{1 \text{ h}}{3600 \text{ s}} \times \frac{1000 \text{ m}}{1 \text{ km}} = 22 \text{ m/s}$$

$$a = -6.0 \text{ m/s}^2$$

$$v_f = 0$$

$$\Delta t = ?$$

BASIC EQUATION:

$$a = \frac{v_f - v_i}{\Delta t}$$

WORKING EQUATION:

$$\Delta t = \frac{v_f - v_i}{a}$$

SUBSTITUTION:
$$\Delta t = \frac{(0 \text{ m/s}) - (22 \text{ m/s})}{-6.0 \text{ m/s}^2}$$

$$= 3.7 \text{ s}$$

(b) DATA:
$$v_i = 78 \text{ km/h} = 22 \text{ m/s}$$

$$\Delta t = 3.7 \text{ s}$$

$$a = -6.0 \text{ m/s}^2$$

$$\Delta s = ?$$

BASIC EQUATION:
$$\Delta s = v_i(\Delta t) + \tfrac{1}{2}a(\Delta t)^2$$

WORKING EQUATION: same

SUBSTITUTION:
$$\Delta s = (22 \text{ m/s})(3.7 \text{ s}) + \tfrac{1}{2}(-6.0 \text{ m/s}^2)(3.7 \text{ s})^2$$

$$= 81 \text{ m} - 41 \text{ m}$$

$$= 4\overline{0} \text{ m}$$

Acceleration, velocity, and distance: *When an object is uniformly accelerated, the square of the final velocity equals the sum of the square of the initial velocity and twice the product of the acceleration and distance.* That is,

$$\boxed{v_f^2 = v_i^2 + 2a(\Delta s)} \tag{7}$$

Solving equation (1) for Δt, we have

$$\Delta t = \frac{v_f - v_i}{a} \tag{8}$$

Now substitute equation (8) into equation (3):

$$\Delta s = v_i(\Delta t) + \tfrac{1}{2}a(\Delta t)^2$$

$$\Delta s = v_i\left(\frac{v_f - v_i}{a}\right) + \tfrac{1}{2}a\left(\frac{v_f - v_i}{a}\right)^2$$

$$\Delta s = \frac{v_i v_f - v_i^2}{a} + \frac{v_f^2 - 2v_i v_f + v_i^2}{2a}$$

Multiplying both sides by $2a$, we have

$$2a(\Delta s) = 2v_i v_f - 2v_i^2 + v_f^2 - 2v_i v_f + v_i^2$$

$$2a(\Delta s) = v_f^2 - v_i^2$$

or

$$v_f^2 = v_i^2 + 2a(\Delta s)$$

And, if $v_i = 0$,

$$v_f^2 = 2a(\Delta s)$$

Part (b) of Example 3 could also have been done using this formula without having to find Δt:

DATA:

$$v_f = 0$$

$$v_i = 78 \text{ km/h} = 22 \text{ m/s}$$

$$a = -6.0 \text{ m/s}^2$$

$$\Delta s = ?$$

BASIC EQUATION:

$$v_f^2 = v_i^2 + 2a(\Delta s)$$

WORKING EQUATION:

$$\Delta s = \frac{v_f^2 - v_i^2}{2a}$$

SUBSTITUTION:

$$\Delta s = \frac{0 - (22 \text{ m/s})^2}{2(-6.0 \text{ m/s}^2)}$$

$$= 4\bar{0} \text{ m}$$

4.6 FREELY FALLING BODIES

One of the most common examples of an object in uniformly accelerated motion is a freely falling body (when air resistance is neglected). Galileo (1564–1642) discovered that all bodies acted upon by gravity fall with the *same* acceleration.

All objects fall at the same rate in a vacuum, but many objects fall at different rates through air because of the air's resistance. However, two dense objects of different sizes with the same air resistance, such as a bowling ball and a marble, fall at the same rate.

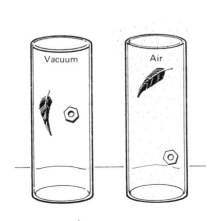

The value of the acceleration due to gravity varies with the mass of the planet and the distance of the object from the center of the planet. The value of the acceleration due to gravity for most locations in the United States is approximately 9.80 m/s². (This is commonly written $g = 9.80$ m/s².) This value varies from 9.782 m/s² at the equator to 9.832 m/s² at the poles. For comparison purposes, the value of g on the moon's surface is approximately 1.6 m/s².

Since freely falling bodies are objects in accelerated motion, all the equations from the last section apply. But since all freely falling bodies have the same acceleration, we replace a by g in these equations as follows:

Body has some initial velocity	Body starts from rest
$v_f = v_i + g(\Delta t)$	$v_f = g(\Delta t)$
$\Delta s = v_i(\Delta t) + \frac{1}{2}g(\Delta t)^2$	$\Delta s = \frac{1}{2}g(\Delta t)^2$
$v_f^2 = v_i^2 + 2g(\Delta s)$	$v_f^2 = 2g(\Delta s)$

Because g is an acceleration, g is actually a vector with its direction always toward the center of the earth. In addition, Δs, v_i, and v_f are also vector quantities; these quantities are commonly assigned a positive direction if they are directed downward and a negative direction if they are directed upward.

example 1 A hammer is dropped from a scaffold 125 m above the ground.

(a) How long does it take to hit the ground?
(b) How fast is it going as it strikes the ground?

(a) DATA: Note that the hammer is starting at rest; that is,

$$v_i = 0$$

$$\Delta s = 125 \text{ m}$$

$$g = 9.80 \text{ m/s}^2$$

$$\Delta t = ?$$

BASIC EQUATION: $\Delta s = \frac{1}{2}g(\Delta t)^2$

WORKING EQUATION: $\Delta t = \sqrt{\dfrac{2(\Delta s)}{g}}$

SUBSTITUTION: $\Delta t = \sqrt{\dfrac{2(125 \text{ m})}{9.80 \text{ m/s}^2}}$ $\boxed{\sqrt{\dfrac{\text{m}}{\text{m/s}^2}} = \sqrt{\text{s}^2} = \text{s}}$

$$= 5.05 \text{ s}$$

(b) DATA: $\Delta s = 125$ m

$$g = 9.80 \text{ m/s}^2$$

$$v_f = ?$$

BASIC EQUATION: $v_f^2 = 2g(\Delta s)$

WORKING EQUATION: $v_f = \sqrt{2g(\Delta s)}$

SUBSTITUTION: $v_f = \sqrt{2(9.80 \text{ m/s}^2)(125 \text{ m})}$ $\sqrt{\dfrac{m}{s^2} \cdot m} = \sqrt{\dfrac{m^2}{s^2}} = \dfrac{m}{s}$

$$= 49.5 \text{ m/s}$$

When any object is thrown vertically upward, its upward speed is uniformly decreased by gravity until it stops for an instant before falling back to the ground. And as it is falling to the ground, it is uniformly accelerated by gravity. If an object is thrown vertically upward and if the initial velocity is known, the previous acceleration–gravity equations may be used to find how high the object rises and how long it is in flight. (*Remember*: Upward direction is negative; downward direction is positive.)

example 2   A baseball is thrown vertically upward with an initial velocity of 21.0 m/s.

(a) How high does it go?
(b) How long will it take to reach its maximum height?
(c) How long is it in flight?

(a) DATA: $v_i = -21.0$ m/s   (v_i is negative because the initial velocity is directed opposite g)

$v_f = 0$ (at the moment of the ball's maximum height, its velocity is zero)

$g = 9.80 \text{ m/s}^2$

$\Delta s = ?$

BASIC EQUATION: $v_f^2 = v_i^2 + 2g(\Delta s)$

WORKING EQUATION: $\Delta s = \dfrac{v_f^2 - v_i^2}{2g}$

SUBSTITUTION: $\Delta s = \dfrac{0^2 - (-21.0 \text{ m/s})^2}{2(9.80 \text{ m/s}^2)}$ $\dfrac{(m/s)^2}{m/s^2} = \dfrac{m^2/s^2}{m/s^2} = m$

$$= -22.5 \text{ m}$$ (Δs being negative indicates an upward displacement)

(b) DATA:

$$v_i = -21.0 \text{ m/s}$$

$$v_f = 0$$

$$g = 9.80 \text{ m/s}^2$$

$$\Delta t = ?$$

BASIC EQUATION: $\qquad v_f = v_i + g(\Delta t)$

WORKING EQUATION: $\qquad \Delta t = \dfrac{v_f - v_i}{g}$

SUBSTITUTION: $\qquad \Delta t = \dfrac{0 - (-21.0 \text{ m/s})}{9.80 \text{ m/s}^2} \qquad \boxed{\dfrac{\text{m/s}}{\text{m/s}^2} = \dfrac{1}{1/\text{s}} = \text{s}}$

$$= 2.14 \text{ s}$$

(c) The ball decelerates on the way up and accelerates on the way down at the same rate because gravitational acceleration is constant (9.80 m/s^2). Therefore, the time for the ball to reach its peak is the same as the time for it to fall to the ground. Thus, its time in flight is $2 \times 2.14 \text{ s} = 4.28 \text{ s}$.

With what speed does the baseball in the example above hit the ground? The answer is 21.0 m/s. Can you explain why?

As we mentioned earlier, air resistance has some effect on freely falling bodies. What are the consequences of this air resistance? We saw that two dense and compact objects, such as a bowling ball and a marble, fall at the same rate in air. But we know that light objects, such as feathers and leaves, fall at much lesser rates because of the air's resistance.

The effects of air resistance can be discussed in terms of a sky diver. By using the air's resistance to his or her advantage, a skilled sky diver can control the speed, position, and direction of his or her body before the parachute opens. And an open parachute tremendously increases the air resistance, which reduces the diver's velocity to approximately 6 m/s (20 km/h).

What happens if the parachute does not open and the person forms his or her body into a ball? Does his or her velocity increase uniformly until he or she hits the ground? As the velocity increases, the air resistance increases. Since the gravitational pull and the air resistance are directed opposite each other, they tend to oppose or equalize each other. (Note that velocity and net acceleration are both directed downward while the air resistance is directed upward.) This equalization occurs when the friction of the air's resistance equals the force of gravity. When equalization occurs, the sky diver stops accelerating and continues falling at a constant velocity, which is called the *terminal velocity*. The terminal velocity of a person is approximately 50 m/s (180 km/h). In general, the terminal velocity of an object varies with its weight and its aerodynamic features. Aerodynamic features include the following:

1. the shape of the object. (A symmetrical object is more aerodynamic than a nonsymmetrical one.)
2. the orientation of the object as it is traveling. (The sky diver can slow the fall by spreading out his or her arms and legs and falling horizontally. The diver can increase speed by falling head or feet first.)
3. the "roughness" of the surface. (A body with a smooth outside surface provides less air resistance than a body with a rough surface and hence falls or flies faster.)

4.7   PROJECTILE MOTION

Let us now direct our attention to the motion of a projectile. Initially, we will assume that air resistance is negligible. In order to analyze the motion of a projectile, we need to separate the motion into its horizontal and vertical components. The horizontal component of the muzzle velocity is determined by the initial velocity and the angle of elevation of the projectile. The vertical component of the velocity is determined by the initial velocity, angle of elevation, and gravity.

The horizontal and vertical components are independent of each other, as shown in the figure below. In the figure both golf balls were released at the same

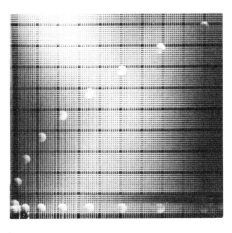

instant. The ball on the left simply drops while the ball on the right was given a horizontal velocity. Since both balls fall at the same rate, we conclude that the effect of gravity is unrelated to the horizontal velocity of the ball on the right.

This result leads to a most interesting situation, which many people, including hunters, seem to have initial difficulty in believing. Suppose that a rifle is fired horizontally with a velocity of $90\bar{0}$ m/s at a distance of 2.00 m above flat ground. Further, suppose that at the instant the bullet leaves the rifle, the casing simply falls from the rifle. Which hits the ground first, and how long does it take to hit the ground? Is this situation not identical to the golf ball experiment? (*Remember*: We

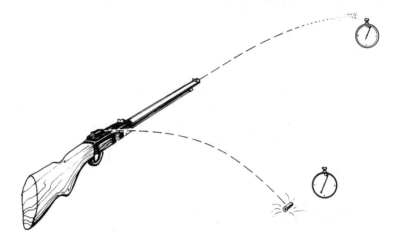

are assuming no air resistance.) Therefore, we must conclude that both bullet and casing hit the ground at the same instant. How long did it take? Since the initial vertical velocity was zero, we may use the equation

$$\Delta s = \tfrac{1}{2} g (\Delta t)^2$$

or

$$\Delta t = \sqrt{\frac{2(\Delta s)}{g}}$$

$$= \sqrt{\frac{2(2.00 \text{ m})}{9.80 \text{ m/s}^2}}$$

$$= 0.639 \text{ s} \qquad \text{for each to hit the ground}$$

After hunters think about this situation, they soon remember that to counteract the gravitational pull on the bullet, the rear sight on a rifle is adjusted so that the rifle is actually aimed a small distance above the target.

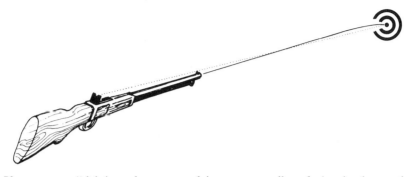

If you were "driving along a road in a vacuum" and simply dropped a ball out the "window," the ball would fall but travel along at the same speed as you were traveling until it hit the ground. As far as the ball is concerned, it is behaving as if it were shot from a gun; that is, the ball's initial or "muzzle" velocity is the speed of the "electric" car, and the ball would follow a path similar to the path of a bullet.

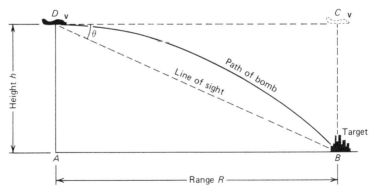

Theoretically, this is the same motion as when a plane drops a bomb. See the figure below.

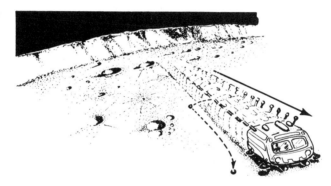

At the instant the plane simply drops the bomb, the bomb's horizontal velocity is the same as the plane's and its vertical velocity is 0. Since the vertical and horizontal components of the bomb's velocity are independent of each other, the time it takes for the bomb to hit the ground is the same as if the bomb were dropped vertically from a height of h with $v_i = 0$. Therefore,

$$h = \Delta s = \tfrac{1}{2}g(\Delta t)^2$$

or

$$\Delta t = \sqrt{\frac{2h}{g}}$$

The bomb and the plane are each traveling horizontally at velocity, v, which means the plane and the bomb each travel a horizontal distance of

$$R = \Delta s = v(\Delta t)$$

or

$$R = v\sqrt{\frac{2h}{g}}$$

This means that the bomb will hit and explode directly below the plane, assuming that the plane does not change course.

How is the correct angle θ determined? Look at right $\triangle\, BCD$ in the figure on page 73. Side $BC = h$ and side $DC = R = v\sqrt{2h/g}$.

$$\tan\theta = \frac{BC}{DC} = \frac{h}{R} = \frac{h}{v\sqrt{2h/g}}$$

which may be simplified to

$$\tan\theta = \frac{\sqrt{2hg}}{2v}$$

Again, we need to note that this discussion is all theoretical. Factors such as air resistance, wind and its direction, and the terminal velocity of the bomb must all be taken into account. All pertinent information is now collected by radar and other electronic equipment and fed into an on-board computer, which determines the correct release point.

example 1   A bomb is dropped from a plane flying at $30\bar{0}$ m/s at an altitude of $10,\bar{0}00$ m.

(a) How far ahead of the target should the bombardier release the bomb; that is, what is the range?

(b) How long will it take for the bomb to reach the ground; in other words, how much time does the pilot have to get away before the bomb hits?

(c) What is the value of angle θ?

(a) DATA:

$$v = 30\bar{0} \text{ m/s}$$

$$g = 9.80 \text{ m/s}^2$$

$$h = 10,\bar{0}00 \text{ m}$$

$$R = ?$$

BASIC EQUATION:

$$R = v\sqrt{\frac{2h}{g}}$$

WORKING EQUATION: same

SUBSTITUTION:

$$R = (30\bar{0} \text{ m/s})\sqrt{\frac{2(10,\bar{0}00 \text{ m})}{9.80 \text{ m/s}^2}}$$

$$= 13,600 \text{ m} \quad \text{or} \quad 13.6 \text{ km}$$

$$\boxed{\frac{\text{m}}{\text{s}}\sqrt{\frac{\text{m}}{\text{m/s}^2}} = \frac{\text{m}}{\text{s}}\sqrt{s^2} = \frac{\text{m}}{\text{s}}(\text{s}) = \text{m}}$$

(b) DATA:

$$h = 10,\bar{0}00 \text{ m}$$

$$g = 9.80 \text{ m/s}^2$$

$$\Delta t = ?$$

BASIC EQUATION:

$$\Delta t = \sqrt{\frac{2h}{g}}$$

WORKING EQUATION: same

SUBSTITUTION:

$$\Delta t = \sqrt{\frac{2(10,\bar{0}00 \text{ m})}{9.80 \text{ m/s}^2}}$$

$$= 45.2 \text{ s} \qquad \boxed{\sqrt{\frac{\text{m}}{\text{m/s}^2}} = \sqrt{s^2} = \text{s}}$$

(c) DATA:

$$h = 10,\bar{0}00 \text{ m}$$

$$v = 30\bar{0} \text{ m/s}$$

$$g = 9.80 \text{ m/s}^2$$

$$\theta = ?$$

BASIC EQUATION:

$$\tan \theta = \frac{\sqrt{2hg}}{2v}$$

WORKING EQUATION: same

SUBSTITUTION:

$$\tan \theta = \frac{\sqrt{2(10{,}0\overline{0}0 \text{ m})(9.80 \text{ m/s}^2)}}{2(30\overline{0} \text{ m/s})}$$

$$= 0.738 \qquad \boxed{\frac{\sqrt{(\text{m})(\text{m/s}^2)}}{\text{m/s}} = \frac{\text{m/s}}{\text{m/s}} = 1}$$

Thus, $\theta = 36.4°$

Now let us consider a gun being fired with an initial or muzzle velocity of $v_0 = 32\overline{0}$ m/s at an angle of $\theta = 30.0°$. How high and how far does the projectile go?

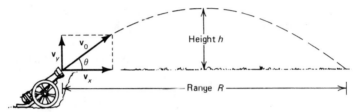

First, note that the initial velocity, v_0, can be resolved into its vertical and horizontal components, v_x and v_y. From trigonometry,

$$\cos \theta = \frac{v_x}{v_0} \quad \text{or} \quad v_x = v_0 \cos \theta = \left(32\overline{0} \text{ m/s}\right) \cos 30.0° = 277 \text{ m/s}$$

$$\sin \theta = \frac{v_y}{v_0} \quad \text{or} \quad v_y = v_0 \sin \theta = \left(32\overline{0} \text{ m/s}\right) \sin 30.0° = 16\overline{0} \text{ m/s}$$

To find how high the projectile goes is equivalent to finding how high the projectile would go if it were fired vertically with an initial velocity of $-16\overline{0}$ m/s.

$$v_f^2 = v_i^2 + 2g(\Delta s)$$

$$\Delta s = \frac{v_f^2 - v_i^2}{2g}$$

$$= \frac{0^2 - \left(-16\overline{0} \text{ m/s}\right)^2}{2(9.80 \text{ m/s}^2)}$$

$$= -1310 \text{ m or } 1310 \text{ m high}$$

To find how far the projectile goes, we first need to find how long the projectile is in flight; then multiply the time by the horizontal velocity ($s = vt$) to find the horizontal distance traveled.

To find the time in flight, let us find how long it takes for an object to fall 1310 m and then double it.

$$\Delta s = \tfrac{1}{2}g(\Delta t)^2$$

$$\Delta t = \sqrt{\frac{2(\Delta s)}{g}}$$

$$= \sqrt{\frac{2(1310 \text{ m})}{9.80 \text{ m/s}^2}}$$

$$= 16.4 \text{ s}$$

That is, 2(16.4 s) = 32.8 s is the total flight time.

$$s = v_x t$$

$$= (277 \text{ m/s})(32.8 \text{ s})$$

$$= 9090 \text{ m which is the horizontal distance traveled by the projectile.}$$

In general, knowing v_0 and θ, the maximum height is found as follows:

$$v_f^2 = v_i^2 + 2g(\Delta s)$$

$$\Delta s = \frac{v_f^2 - v_i^2}{2g}$$

$$= \frac{0^2 - v_y^2}{2g} = \frac{0 - (v_0 \sin \theta)^2}{2g}$$

Therefore, the maximum height is

$$\boxed{h = \frac{v_0{}^2 \sin^2 \theta}{2g}}$$

The total flight time is found as follows:

$$h = \Delta s = \tfrac{1}{2}g(\Delta t)^2$$

$$\Delta t = \sqrt{\frac{2h}{g}} \qquad \text{the time for the projectile to fall.}$$

or

$$t = 2\sqrt{\frac{2h}{g}} \qquad \text{the total flight time of the projectile.}$$

$$= 2\sqrt{\frac{2(v_0^2 \sin^2 \theta / 2g)}{g}}$$

The total flight time is then

$$t = \frac{2v_0 \sin \theta}{g}$$

Finally, the range is found as follows:

$$R = s = v_x t$$

$$= (v_0 \cos \theta)\left(\frac{2v_0 \sin \theta}{g}\right)$$

$$= \frac{v_0^2(2 \sin \theta \cos \theta)}{g}$$

$$= \frac{v_0^2 \sin 2\theta}{g} \qquad \text{(recall that } 2 \sin \theta \cos \theta = \sin 2\theta\text{)}$$

The range is

$$R = \frac{v_0^2 \sin 2\theta}{g}$$

What angle θ will produce the longest range for a given initial velocity v_0? Let us study the range equation

$$R = \frac{v_0^2 \sin 2\theta}{g}$$

Since v_0 and g are constants, the answer lies in what value of θ will make $\sin 2\theta$ a maximum. The sine function varies from -1 to 1. Therefore, R is a maximum when

$$\sin 2\theta = 1$$

Solving for 2θ, we have

$$2\theta = 90°$$

or

$$\theta = 45°$$

That is, an angle of 45° produces the greatest range for any given initial velocity. This fact is also shown in the figure on page 79, which shows the motion (flight paths) of projectiles fired at various angles at an initial velocity of $50\overline{0}$ m/s.

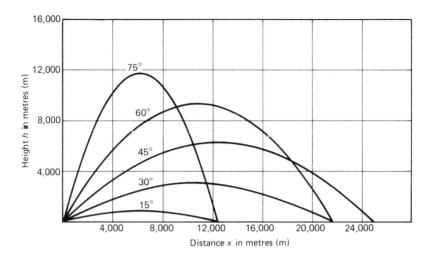

Angle	h (m)	R (m)
75.0°	11,900	12,800
60.0°	9,570	22,100
45.0°	6,380	25,500
30.0°	3,190	22,100
15.0°	854	12,800

(*Note*: When air resistance is neglected, the paths of projectiles are parabolas.)

Of course, these indicated heights and ranges are strictly theoretical and not realistic for this gun with this initial velocity because the air resistance reduces these values significantly.

EXERCISES

Sketch	Data	Basic Equation	Working Equation	Substitution
12 cm² $h = ?$ $b = 6.0$ cm	$A = 12$ cm² $b = 6.0$ cm $h = ?$	$A = bh$	$h = \dfrac{A}{b}$	$h = \dfrac{12 \text{ cm}^2}{6.0 \text{ cm}}$ $= 2.0$ cm

1. While driving at $9\bar{0}$ km/h, how far can one travel in 4.5 h?
2. While traveling at 90.0 km/h, how far (in m) does one travel in 10.0 s?
3. A road atlas lists the driving distance between Los Angeles and San Francisco as being 665 km with a driving time of 9.2 h. What is the average speed for this trip?

4. The speed at which light travels is 2.997×10^8 m/s.
   (a) How long (in min) does it take for light to travel from the sun to the earth, which is a distance of 1.5×10^{11} m?
   (b) How long (in s) does it take for light to travel from the moon to the earth, which is a distance of 3.8×10^8 m?

5. Plot carefully on graph paper the data in the table below.
   (a) Is the speed constant or variable?
   (b) If constant, what is the speed? If variable, what is the instantaneous speed at $t = 5.0$ s?

Time (s)	0.0	1.0	2.0	3.0	4.0	5.0	6.0	7.0	8.0	9.0	10.0
Distance (m)	0.0	6.4	12.9	19.8	26.0	32.1	39.5	45.5	51.7	58.3	65.1

6. Plot carefully on graph paper the data in the table below.
   (a) Is the speed constant or variable?
   (b) If constant, what is the speed? If variable, what is the instantaneous speed at $t = 5.0$ s?

Time (s)	0.0	1.0	2.0	3.0	4.0	5.0	6.0	7.0	8.0	9.0	10.0
Distance (m)	0.0	2.0	7.5	18.1	32.4	50.0	72.1	97.5	125.1	165.5	204.3

7. A given boat travels $2\bar{0}$ km/h in still water. The flow of a river is 4 km/h.
   (a) What is the speed of the boat (relative to the ground) going upstream?
   (b) What is the speed of the boat (relative to the ground) going downstream?

8. A plane is traveling 425 km/h in still air.
   (a) What would its ground speed be if the plane were flying into a 75-km/h head wind?
   (b) What would its ground speed be if the plane were flying with a 75-km/h tail wind?

9. A plane is traveling north at 425 km/h in still air. What is the velocity of the plane with a wind of 75 km/h blowing from the east?

10. A plane is traveling north (90.0°) at 425 km/h in still air. What is the velocity of the plane with a wind of 75 km/h blowing from 30.0° east of south (120.0°)?

11. A pilot wishes to fly a plane on a course due south (27$\bar{0}$°) at 45$\bar{0}$ km/h. The wind is 7$\bar{0}$ km/h from the west (0°). At what heading and at what speed (velocity) must the plane fly to compensate for the wind?

12. A pilot wishes to fly a plane on a course 45.0° north of west (135.0°) at 65$\bar{0}$ km/h. The wind is 55.0 km/h blowing from 30.0° south of west (30.0°). At what velocity must the plane fly to compensate for the wind?

13. A woman in a rowboat is crossing a river on a course to reach a point directly on the opposite side of a river.
   (a) If the woman can row at 5.0 km/h in still water and the speed of the current is 4.0 km/h, at what angle must she row to reach the opposite side?
   (b) If the river is 21$\bar{0}$ m wide, how long will it take to cross?

14. A man in a motorboat is to cross a river to a point on the opposite side and $3\bar{0}°$ upstream from where he begins.

    (a) If the boat travels 25 km/h in still water and the speed of the current is 5.0 km/h, at what angle must he steer to reach the point on the opposite side?

    (b) If the river is $65\bar{0}$ m wide, how long will it take to reach the destination point?

15. An automobile accelerates from $5\bar{0}$ km/h to 85 km/h in 8.0 s. Find its average acceleration (in m/s^2).

16. An automobile decelerates from $6\bar{0}$ km/h to a complete stop in 12 s. Find its average acceleration (in m/s^2).

17. An automobile is traveling uniformly at 8.2 m/s. It then accelerates at 6.00 m/s^2 for 4.00 s.

    (a) What is its speed after 4.00 s?

    (b) How far does it travel during this 4.00 s?

18. An automobile is traveling uniformly at 55 km/h. It then accelerates at 5.4 m/s^2 to 75 km/h.

    (a) How long does it take to reach the new speed?

    (b) How far does it travel while accelerating?

19. Plot carefully on graph paper the data in the table below.

    (a) Is the acceleration constant or variable?

    (b) If constant, what is the acceleration? If variable, what is the instantaneous acceleration at $t = 5.0$ s?

Time (s)	0.0	1.0	2.0	3.0	4.0	5.0	6.0	7.0	8.0	9.0	10.0
Velocity (m/s)	0.0	21.8	45.1	65.4	87.4	108.7	134.1	157.2	176.1	195.2	224.0

20. Plot carefully on graph paper the data in the table below.

    (a) Is the acceleration constant or variable?

    (b) If constant, what is the acceleration? If variable, what is the instantaneous acceleration at $t = 5.0$ s and at 2.5 s?

Time (s)	0.0	1.0	2.0	3.0	4.0	5.0	6.0	7.0	8.0	9.0	10.0
Velocity (m/s)	0.0	−4.0	−6.0	−6.0	−4.0	0	6.0	14.0	24.0	36.0	50.0

21. A large rock is dropped from a bridge to the water below.

    (a) If it takes 2.1 s for the rock to hit the water below, with what velocity (in m/s) does the rock hit the water?

    (b) What is the distance (in m) from the bridge to the water?

22. A bullet is fired vertically from a gun with an initial velocity of $25\bar{0}$ m/s.

    (a) How high does it go?

    (b) How long does it take to reach its maximum height?

    (c) How long is it in flight?

23. A bullet is fired vertically from a gun and reaches a height of $25\bar{0}0$ m.

    (a) What was its initial velocity?

(b) How long does it take to reach its maximum height?

(c) How long is it in flight?

24. Bill is standing on a steel beam 150.0 m above the ground. Jane is standing 30.0 m directly above Bill. Bill is to throw a hammer up to Jane. With what initial velocity must Bill throw the hammer in order for it to just reach Jane?

25. In Exercise 24, suppose that Bill throws the hammer straight up only 25.0 m.

   (a) How long does a child on the ground have to get out of the way from the time Bill lets go of the hammer?

   (b) How long does a child on the ground have to get out of the way from the time the hammer reaches its maximum height?

   (c) At what speed does the hammer hit the ground?

26. Suppose that in Exercise 24 Bill throws the hammer straight down with initial velocity of 15.0 m/s.

   (a) At what speed does the hammer hit the ground?

   (b) How long does a child on the ground have to get out of the way?

27. A baseball player throws a ball straight up with an initial velocity of 18.0 m/s.

   (a) How high does it go?

   (b) How long is it in flight?

28. Assume that the baseball player in Exercise 27 is on the moon, where the acceleration due to gravity is $\frac{1}{6}$ of the g-value on earth.

   (a) How high does the ball go?

   (b) How long is it in flight?

29. A bomb is dropped from a plane flying at $45\bar{0}$ m/s at an altitude of $90\bar{0}0$ m.

   (a) How far ahead of the target should the bombardier release the bomb?

   (b) How long will it take for the bomb to hit the ground?

   (c) At what vertical velocity will the bomb hit the ground?

   (d) At what horizontal velocity will the bomb hit the ground?

30. A gun fires a projectile with an initial velocity of 275 m/s at an angle 18° above the horizontal.

   (a) Find the horizontal component of the muzzle velocity.

   (b) Find the vertical component of the muzzle velocity.

   (c) Find the maximum height reached by the projectile.

   (d) How far does the projectile travel?

   (e) How long is the projectile in flight?

31. (a) At what angle should the projectile in Exercise 30 be fired so that the maximum range of the gun is attained?

   (b) What is the maximum range of the gun?

   (c) At what angle should the projectile be fired so that a maximum height is attained?

32. An airplane is flying at 375 m/s and diving at an angle of 15.0° below the horizontal. A bomb is released at an altitude of $60\bar{0}$ m.

   (a) How far ahead of the release point will the bomb hit the ground?

   (b) How long will it take for the bomb to hit the ground?

   (c) At what vertical velocity will the bomb hit the ground?

   (d) At what horizontal velocity will the bomb hit the ground?

5
LAWS
OF MOTION

5.1   INERTIA

In Chapter 4, we described motion. Next, we need to study the causes of various types of motion. But first, we must introduce the term force, which is central to the discussion of the causes of motion. A *force* is a push or a pull that tends to cause motion or tends to prevent motion. A force is a vector quantity and thus has both magnitude and direction. Examples of forces include:

1. The force produced by pushing your foot down on a bicycle pedal is transmitted through the chain and sprocket system to the ground whose reaction causes the bicycle to move forward.

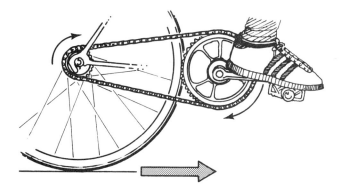

2. The reaction force produced by dragging your feet on the road when the bicycle is already moving causes the bicycle to slow down or stop its motion.

3. The force produced by a hot expanding gas on a piston of an engine causes it to move.

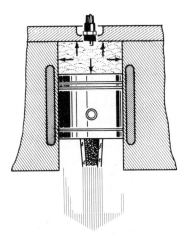

4. When a truck passes over a bridge, there are four vertical forces involved: the downward forces produced by the weights of the truck and the bridge, the upward forces produced by the bridge supports, and, of course, the upward forces that balance the downward forces. Structural engineers must design the supports of a bridge to allow for the weights of the traffic and of the bridge itself. Here, the upward forces prevent any downward motion—namely, the collapse of the bridge.

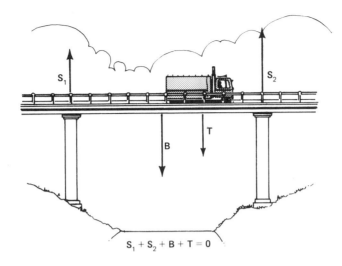

$$S_1 + S_2 + B + T = 0$$

Sir Isaac Newton (1642–1727) discovered that forces control the motion of objects. He described his discovery of the relationship between force and motion in three laws of motion, which are named in his honor. His laws are believed to be universally true; that is, they apply throughout the universe as well as on the earth.

If an automobile is at rest (stopped), it resists being set in motion. That is, it takes a tremendous push by a person to get it moving. Similarly, if an automobile is moving, even slowly, it takes a rather large force to stop it. This property of resisting a change in motion is called inertia. *Inertia* is the property of an object that causes it to remain at rest if it is at rest or to continue moving with a constant velocity. In other words, we say that the automobile possesses a certain amount of inertia which has to be overcome if the motion of the automobile is to be changed.

Assume that we have a coin resting on a card on a glass as shown below. If we quickly flick the card horizontally, the inertia of the coin tends to keep it at rest until gravity pulls it straight down into the glass.

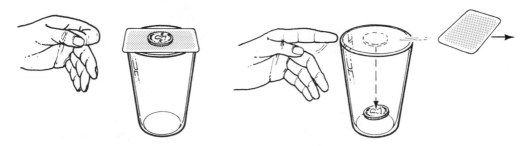

For those of you who are looking for a little excitement, try pulling the tablecloth out from under the dishes on the dinner table tonight. If you are fast enough, the inertia of the dishes will keep them on the table! If not , ...

NEWTON'S FIRST LAW OF MOTION (LAW OF INERTIA)

An object that is at rest will remain at rest and an object that is in motion will remain in motion at the same velocity (same speed and same direction) unless some unbalanced (outside) force acts upon it.

When the accelerating force of an automobile engine is no longer applied to a moving car, it will slow down. This is not a violation of the law of inertia because there are forces being applied to the car through air resistance, friction in the bearings, and the rolling resistance of the tires. If these forces could be removed, the auto would continue moving with a constant velocity. Anyone who has tried to stop quickly on ice knows the effect of the law of inertia when frictional forces are small.

If an automobile is traveling down the highway and its brakes are suddenly applied, what happens to the packages, the passengers, and any other loose objects? Because of their inertia, they tend to remain in motion at the same velocity until some other force acts upon them. Hopefully, the outside force acting upon the passengers is provided by the seatbelts. If not, all the objects will be stopped, or at least slowed, by the windshield!

Some objects tend to resist changes in their motion more than others. It is much easier to push a small automobile than to push a large truck into motion. *Mass is a measure of the resistance that a body has to change in its motion.*

The *kilogram* (kg) is the basic SI unit of mass. For very small masses, the gram (g), milligram (mg), or microgram (μg) are commonly used.

5.2   FORCE AND ACCELERATION

Newton's second law of motion describes the relationship between the force and the acceleration of an object.

NEWTON'S SECOND LAW OF MOTION (LAW OF ACCELERATION)

When an object is acted upon by an unbalanced (external) force, the object is given an acceleration that is in the same direction as the force and is directly proportional to the force and inversely proportional to the mass of the object. That is,

$$a \propto \frac{F}{m}$$

where a is the acceleration,
  F is the force, and
  m is the mass.

If we let k be the proportionality constant, we have

$$F = kma$$

It would be convenient if $k = 1$. What happens if $k = 1$? The force units would have to be defined in terms of the mass and acceleration units, which is what physicists have done. That is, if $k = 1$,

$$F = ma$$

In SI units, the mass unit is the kilogram (kg) and the acceleration unit is metre/second/second (m/s^2). That is, the force required to accelerate 1 kg of mass at the rate of 1 m/s^2 is

$$F = ma$$
$$= (1 \text{ kg})(1 \text{ m/s}^2)$$
$$= 1 \frac{\text{kg m}}{\text{s}^2}$$

The SI force unit is the newton (N), named in honor of Isaac Newton, and is defined as above; that is,

$$1 \text{ N} = 1 \text{ kg m/s}^2$$

Therefore,

Newton's second law of motion may be expressed as

$$F = ma$$

where m is the mass expressed in kilograms (kg),
      a is the acceleration expressed in metres/second/second (m/s^2), and
      F is the unbalanced force expressed in newtons (N).

The conversion factor

$$1 \text{ N} = 1 \frac{\text{kg m}}{\text{s}^2}$$

is most useful to recall and use in solving a variety of problems.

We should note here that there is one other metric unit of force, the dyne. One *dyne* is the force required to accelerate 1 g of mass at the rate of 1 cm/s^2. That is, 1 dyne $= 1$ g cm/s^2. However, since the dyne is not an SI unit, its use is becoming less and less common as the world more universally accepts and converts to the SI system of metric units. As a result, we will not use the dyne unit in this book.

example 1 What force is needed to produce an acceleration of 5.00 m/s² on a mass of 15.0 kg?

DATA:

$$a = 5.00 \text{ m/s}^2$$

$$m = 15.0 \text{ kg}$$

$$F = ?$$

BASIC EQUATION: $F = ma$

WORKING EQUATION: same

SUBSTITUTION:

$$F = (15.0 \text{ kg})(5.00 \text{ m/s}^2)$$

$$= 75.0 \text{ kg m/s}^2$$

$$= 75.0 \text{ N} \quad (1 \text{ kg m/s}^2 = 1 \text{ N})$$

example 2 What acceleration is produced by applying a force of 25$\overline{0}$ N to a mass of 20.0 kg?

DATA:

$$F = 25\overline{0} \text{ N}$$

$$m = 20.0 \text{ kg}$$

$$a = ?$$

BASIC EQUATION: $F = ma$

WORKING EQUATION: $a = \dfrac{F}{m}$

SUBSTITUTION:

$$a = \frac{25\overline{0} \text{ N}}{20.0 \text{ kg}}$$

$$= 12.5 \frac{\text{N}}{\text{kg}} \times \frac{1 \text{ kg m/s}^2}{1 \text{ N}}$$

$$= 12.5 \text{ m/s}^2$$

(*Note:* We must use a conversion factor to obtain acceleration units)

When the same force is applied to two different masses, these masses will have different accelerations. For example, a much smaller force is required to accelerate a baseball from rest to 3$\overline{0}$ m/s than to accelerate an automobile from rest to 3$\overline{0}$ m/s in the same time period. The reason for this is that the automobile has a much larger mass.

When the same amount of force is applied to two different masses, the smaller mass will be accelerated more than the larger mass. Compare samples 1 and 2 and samples 3 and 4 in the computer printout reproduced on pages 90–91, which illustrate these principles.

SAMPLE 1:

FORCE = 80.0 N
MASS = 4.00 kg
ACCELERATION = 20.0 m/s²

TIME (s)	POSITION (m)	VELOCITY (m/s)
0.000	0.00	0.00
0.100	0.10	2.00
0.200	0.40	4.00
0.300	0.90	6.00
0.400	1.60	8.00
0.500	2.50	10.0
0.600	3.60	12.0
0.700	4.90	14.0
0.800	6.40	16.0
0.900	8.10	18.0
1.00	10.0	20.0
1.10	12.1	22.0
1.20	14.4	24.0
1.30	16.9	26.0
1.40	19.6	28.0
1.50	22.5	30.0

SAMPLE 2:

FORCE = 80.0 N
MASS = 20$\bar{0}$ kg
ACCELERATION = 0.400 m/s²

TIME (s)	POSITION (m)	VELOCITY (m/s)
0.000	0.000	0.000
0.100	0.002	0.040
0.200	0.008	0.080
0.300	0.018	0.120
0.400	0.032	0.160
0.500	0.050	0.200
0.600	0.072	0.240
0.700	0.098	0.280
0.800	0.128	0.320
0.900	0.162	0.360
1.00	0.200	0.400
1.10	0.242	0.440
1.20	0.288	0.480
1.30	0.338	0.520
1.40	0.392	0.560
1.50	0.450	0.600

SAMPLE 3:

FORCE = 120̄0 N
MASS = 4.00 kg
ACCELERATION = 30̄0 m/s²

TIME (s)	POSITION (m)	VELOCITY (m/s)
0.000	0.00	00.0
0.100	1.50	30.0
0.200	6.00	60.0
0.300	13.5	90.0
0.400	24.0	120.
0.500	37.5	150.
0.600	54.0	180.
0.700	73.5	210.
0.800	96.0	240.
0.900	122.	270.
1.00	150.	300.
1.10	182.	330.
1.20	216.	360.
1.30	254.	390.
1.40	294.	420.

SAMPLE 4:

FORCE = 12̄00 N
MASS = 30̄00 kg
ACCELERATION = 0.400 m/s²

TIME (s)	POSITION (m)	VELOCITY (m/s)
0.000	0.00000	0.0000
0.100	0.00200	0.0400
0.200	0.00800	0.0800
0.300	0.0180	0.120
0.400	0.0320	0.160
0.500	0.0500	0.200
0.600	0.0720	0.240
0.700	0.0980	0.280
0.800	0.128	0.320
0.900	0.162	0.360
1.00	0.200	0.400
1.10	0.242	0.440
1.20	0.288	0.480
1.30	0.338	0.520
1.40	0.392	0.560
1.50	0.450	0.600

Newton's second law of motion may also be written in other forms. For example, we know that

$$a_{av} = \frac{\Delta v}{\Delta t}$$

which when substituted into $F = ma$ gives us

$$\boxed{F = \frac{m(\Delta v)}{\Delta t}}$$

example 3   What force is needed to accelerate a $15\bar{0}0$-kg car from 30.0 km/h to 85.0 km/h in 12.0 s?

DATA: $m = 15\bar{0}0$ kg

$\Delta v = 85.0$ km/h $- 30.0$ km/h $= 55.0$ km/h

$\Delta t = 12.0$ s

$F = ?$

BASIC EQUATION: $F = \dfrac{m(\Delta v)}{\Delta t}$

WORKING EQUATION: same

SUBSTITUTION: $F = \dfrac{(15\bar{0}0 \text{ kg})(55.0 \text{ km/h})}{12.0 \text{ s}}$

$= \dfrac{(15\bar{0}0 \text{ kg})(55.0 \text{ km/h})}{12.0 \text{ s}} \times \dfrac{1000 \text{ m}}{1 \text{ km}} \times \dfrac{1 \text{ h}}{3600 \text{ s}}$

$= 1910 \dfrac{\text{kg m}}{\text{s}^2}$

$= 1910$ N

Of course, the other acceleration equations from Chapter 4 may also be used in conjunction with Newton's second law of motion.

example 4   A $50\bar{0}0$-kg rocket sled is propelled from rest by an engine that produces a thrust of 625,000 N for 2.75 s.

(a) What is the sled's average acceleration?
(b) What is its velocity after 2.75 s?

(a) DATA: $F = 625,000$ N

$m = 50\bar{0}0$ kg

$a = ?$

BASIC EQUATION: $F = ma$

WORKING EQUATION: $a = \dfrac{F}{m}$

SUBSTITUTION: $a = \dfrac{625,000 \text{ N}}{50\bar{0}0 \text{ kg}}$

$$= 125 \frac{\text{N}}{\text{kg}} \times \frac{1 \text{ kg m}/s^2}{1 \text{ N}}$$

$$= 125 \text{ m}/s^2$$

(b) We first need to find Δs.

DATA:

$v_i = 0$

$\Delta t = 2.75 \text{ s}$

$a = 125 \text{ m}/s^2$

$\Delta s = ?$

BASIC EQUATION: $\Delta s = v_i(\Delta t) + \frac{1}{2}a(\Delta t)^2$

WORKING EQUATION: same

SUBSTITUTION: $\Delta s = 0 + \frac{1}{2}(125 \text{ m}/s^2)(2.75 \text{ s})^2$

$$= 473 \text{ m}$$

Then to find v_f:

DATA:

$v_i = 0$

$a = 125 \text{ m}/s^2$

$\Delta s = 473 \text{ m}$

$v_f = ?$

BASIC EQUATION: $v_f^2 = v_i^2 + 2a(\Delta s)$

WORKING EQUATION: $v_f = \sqrt{v_i^2 + 2a(\Delta s)}$

SUBSTITUTION: $v_f = \sqrt{0 + 2(125 \text{ m}/s^2)(473 \text{ m})}$

$$= 344 \text{ m}/s$$

5.3   MASS AND WEIGHT

The *mass* of an object is a measure of the amount of matter in the object. The *weight* of an object is a measure of the gravitational force or pull acting on the object. Mass is a scalar quantity, while weight, being a force, is a vector quantity whose direction is always toward the center of the earth.

The mass of an object remains constant while its weight varies according to its gravitational pull. For example, a $100\bar{0}$-kg spaceship on the launching pad also has a mass of $100\bar{0}$ kg out in space. But its weight varies according to its gravitational pull. Its weight will decrease as it moves away from the earth's gravitational field. If the spaceship is on the way to the moon, we say that the crew is "weightless" because they seem to float freely. Their mass remains the same while their weight is near zero. Since weight is a force and the acceleration due to gravity, g, is an acceleration, Newton's second law applies. If we let $F = F_w =$ the weight and $a = g =$ the acceleration due to gravity, then $F = ma$ becomes

$$\boxed{F_w = mg}$$

That is, the weight of an object is the product of its mass and its gravitational pull or acceleration.

The weight of the $100\bar{0}$-kg spaceship on the launching pad is

$$F_w = mg$$

$$= \left(100\bar{0} \text{ kg}\right)(9.80 \text{ m/s}^2)$$

$$= 98\bar{0}0 \text{ kg m/s}^2$$

$$= 98\bar{0}0 \text{ N}$$

Using the formula $F_w = mg$, we can show why the weight of the spaceship and its crew decreases. As they travel farther from the earth, the gravitational force of the earth decreases; that is, g decreases. The formula $F_w = mg$ says that the weight and the gravitational force are directly proportional. As g decreases, the weight decreases. And when $g = 0$, the weight is zero.

example 1   Find the weight of a person on earth whose mass is 80.0 kg.

$$F_w = mg$$

$$= (80.0 \text{ kg})(9.80 \text{ m/s}^2)$$

$$= 784 \text{ kg m/s}^2$$

$$= 784 \text{ N}$$

example 2   Find the weight of a person on the moon whose mass is 80.0 kg. The value of g on the moon is 1.63 m/s^2.

$$F_w = mg$$

$$= (80.0 \text{ kg})(1.63 \text{ m/s}^2)$$

$$= 13\bar{0} \text{ kg m/s}^2$$

$$= 13\bar{0} \text{ N}$$

As you are no doubt aware, the terms "mass" and "weight" are commonly used interchangeably by the general public. We have presented them as technical terms as they are commonly used in the scientific, engineering, and technical professions. Another distinction may be made as follows: mass is a property of an object; weight is the interaction of the object with its environment.

A simple, easy way to find a rather good approximation for g involves using a pulley and two masses. This, called *Atwood's machine*, is shown below. Assuming

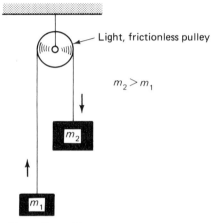

Atwood's machine

that $m_2 > m_1$, m_2 moves down and m_1 moves up when released from rest position. The net (unbalanced) force of the system is

$$F = F_{w_2} - F_{w_1} = m_2 g - m_1 g = (m_2 - m_1)g$$

By Newton's second law, this force may also be written

$$F = (m_2 + m_1)a$$

where $m_2 + m_1$ is the mass of the system. Therefore,

$$(m_2 - m_1)g = (m_2 + m_1)a$$

or

$$g = \frac{(m_2 + m_1)a}{m_2 - m_1}$$

For best results, choose m_2 only slightly larger than m_1 so that the masses move slowly and the acceleration, a, is much less than g. By using a stopwatch, the time for m_2 to fall a distance, s, can be measured so that the acceleration, a, can be found from the equation

$$s = \tfrac{1}{2}at^2$$

example 3   Atwood's machine is used to measure g as follows. The masses are 1.100 kg and 1.000 kg. When released, m_2 falls 46 cm in 1.4 s. Find g.
First, find a:

DATA: $\Delta s = 46 \text{ cm} = 0.46 \text{ m}$

$\Delta t = 1.4 \text{ s}$

$a = ?$

BASIC EQUATION: $\Delta s = \frac{1}{2} a (\Delta t)^2$

WORKING EQUATION: $a = \dfrac{2(\Delta s)}{(\Delta t)^2}$

SUBSTITUTION: $a = \dfrac{2(0.46 \text{ m})}{(1.4 \text{ s})^2}$

$= 0.47 \text{ m/s}^2$

Now find g:

DATA: $m_2 = 1.100 \text{ kg}$

$m_1 = 1.000 \text{ kg}$

$a = 0.47 \text{ m/s}^2$

$g = ?$

BASIC EQUATION: $g = \dfrac{(m_2 + m_1)a}{m_2 - m_1}$

WORKING EQUATION: same

SUBSTITUTION: $g = \dfrac{(1.100 \text{ kg} + 1.000 \text{ kg})(0.47 \text{ m/s}^2)}{1.100 \text{ kg} - 1.000 \text{ kg}}$

$= 9.9 \text{ m/s}^2$

example 4   A loaded elevator has a mass of $25\overline{0}0$ kg. Starting from rest, it accelerates to 3.00 m/s in 2.50 s.

(a) Find its acceleration.
(b) What force is needed for this acceleration?
(c) What is the tension in the cables?

(a) DATA: $v_f = 3.00 \text{ m/s}$

$v_i = 0$

$t = 2.50 \text{ s}$

$a = ?$

BASIC EQUATION: $v_f = v_i + at$

WORKING EQUATION:	$a = \dfrac{v_f - v_i}{t}$
SUBSTITUTION:	$a = \dfrac{3.00 \ m/s - 0}{2.50 \ s}$
	$= 1.20 \ \text{m/s}^2$
(b) DATA:	$m = 25\overline{0}0 \ \text{kg}$
	$a = 1.20 \ \text{m/s}^2$
	$F = ?$
BASIC EQUATION:	$F = ma$
WORKING EQUATION:	same
SUBSTITUTION:	$F = (25\overline{0}0 \ \text{kg})(1.20 \ \text{m/s}^2)$
	$= 30\overline{0}0 \ \text{N}$

(c) The cables must support the weight of the load plus the force due to the acceleration.

DATA:	$m = 25\overline{0}0 \ \text{kg}$
	$a = 1.20 \ \text{m/s}^2$
	$g = 9.80 \ \text{m/s}^2$
	$F = ?$
BASIC EQUATION:	$F = ma + mg$
WORKING EQUATION:	same
SUBSTITUTION:	$F = (25\overline{0}0 \ \text{kg})(1.20 \ \text{m/s}^2) + (25\overline{0}0 \ \text{kg})(9.80 \ \text{m/s}^2)$
	$= 30\overline{0}0 \ \text{N} \qquad\qquad + 24{,}500 \ \text{N}$
	$= 27{,}500 \ \text{N}$

5.4 ACTION AND REACTION

NEWTON'S THIRD LAW OF MOTION (LAW OF ACTION AND REACTION)

For every force applied by object A to object B (action), there is a force exerted by object B on object A (reaction) which has the same magnitude but is opposite in direction.

That is, for every action, there is an equal and opposite reaction.

Let us illustrate this law by studying the following examples.

One simple example of Newton's third law involves two spring balances. Attach one end of one spring balance to a fixed post. Then attach the two spring balances as shown below and pull with your hand. Note that as you pull with different amounts of force, both spring balances register larger but always equal readings. In each case the action force (F_1) is numerically equal to and opposite the reaction force (F_2).

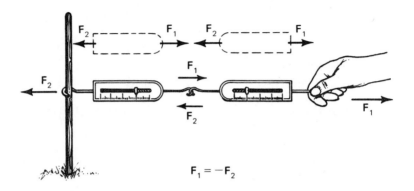

$$F_1 = -F_2$$

When you stand on firm ground, your body exerts a downward force on the ground equal to your weight. The ground, in turn, exerts an upward force on you equal to your weight.

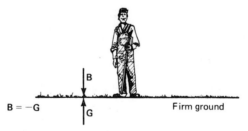

$B = -G$ Firm ground

What would happen if the ground would exert an upward force less than your weight? This actually happens when you stand in soft mud or quicksand. You sink!

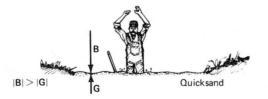

$|B| > |G|$ Quicksand

What would happen if the ground would exert an upward force more than your weight? You would be propelled upward by a force that was the difference between your weight and the upward force.

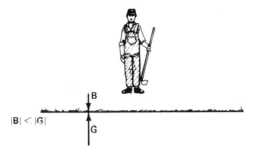

$|B| < |G|$

Suppose that two ice skaters were standing behind one another as shown below. If the man were to push the back of the woman, what would happen? The action F_1, would cause the woman to move away forward. The reaction, F_2, would cause the man to move away backward. Note that the action and reaction forces *never* act on the same object. The action force is on the woman; the reaction force is on the man.

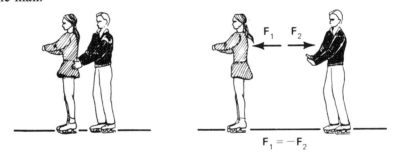

When a bullet is fired from a handgun (action), the recoil force felt on the hand is the reaction. Again, notice that the action and reaction forces are in opposite directions.

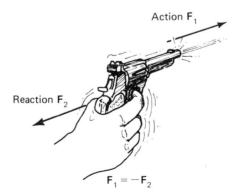

When an automobile accelerates, we know that a force is being applied to it. What applies this force? You may think that the tires exert this force on the auto.

This is not correct, since the tires move along with the auto. Actually, the force of the tires on the road is the action. And the force of the ground on the tires is the reaction.

F_2 = Force of tires on road (action)

F_1 = Force of road on tires (reaction)

$$F_1 = -F_2$$

Other examples of Newton's third law of motion include:

1. The force by the oar (action) of a rowboat against the water produces a force by the water (reaction) against the oar which causes the rowboat to move.

2. The force of the propellers of an airplane (action) against the air produces a force by the air (reaction) against the propellers which causes the airplane to move.

3. If a person were standing in the rowboat 1 m from shore and jumped from the boat to the shore, what would happen? Most probably, he or she would get wet! The force of the person jumping (action) against the boat produces a force by the boat (reaction) against the person. The forces are equal and opposite. But by Newton's second law of motion, we know that the acceleration of each object is inversely proportional

$$F = ma$$
$$\frac{F}{m} = a$$

to its mass. Because the boat and the person are both free objects and each has mass, the action–reaction forces will accelerate both boat and person away from each other. The person will not complete the jump because he or she has lost contact with the boat. In contrast, the person can jump with ease from an object of great mass such as a barge or the earth because it does not "leave the person" in the midst of the effort and let him or her fall short.

As you have seen from the preceding examples, Newton's third law of motion also states that forces cannot exist individually; forces must always exist in pairs.

Dynamics is a branch of physics that deals with forces and their relation to motion; dynamics is the study and application of the laws of motion.

EXERCISES

Sketch	Data	Basic Equation	Working Equation	Substitution
12 cm² $h = ?$ $b = 6.0$ cm	$A = 12$ cm² $b = 6.0$ cm $h = ?$	$A = bh$	$h = \dfrac{A}{b}$	$h = \dfrac{12 \text{ cm}^2}{6.0 \text{ cm}}$ $= 2.0$ cm

1. Find the total force (in newtons) necessary to give an automobile of mass 1750 kg an acceleration of 3.00 m/s².

2. Find the acceleration produced by a total force of 93.0 N on a mass of 6.00 kg.

3. Find the mass of an object that has an acceleration of 15.0 m/s² when an unbalanced force of 90.0 N acts on it.

4. Find the total force (in newtons) necessary to give a rocket of mass 365,000 kg an acceleration of 9.00 m/s².

5. Find the acceleration produced by a force of 5.72×10^6 N on a rocket whose mass is 5.45×10^5 kg.

6. An automobile has a mass of $145\bar{0}$ kg. Each passenger has a mass of 75 kg.
    (a) Find the acceleration of the auto and one passenger if the total force acting is 6650 N.
    (b) Find the acceleration of the auto and six passengers if the total force is again 6650 N.

7. A 1570-kg automobile is traveling at 90.0 km/h. A braking force of $85\bar{0}0$ N is applied.
    (a) How long will it take for the auto to come to a complete stop?
    (b) How far will the auto travel during its braking period?
    (c) What is its velocity after the braking force has been applied for 2.00 s?

8. A parachutist whose mass is 75.0 kg is free-falling at 65.0 m/s. The parachutist's speed decreases to 5.50 m/s in a vertical distance of 45.0 m as the chute opens. What force (in newtons) did the harness have to withstand?

9. An empty 2450-kg truck brakes from 90.0 km/h to 50.0 km/h in a distance of $15\bar{0}$ m.
    (a) What was its deceleration?
    (b) How much braking force was required?
    (c) How much braking force is required if the truck is carrying a load of 1450 kg?
    (d) What is the deceleration of the loaded truck if the braking force of the empty truck is used?

10. (a) What is your weight in newtons?
    (b) What would your weight be on the moon?

11. (a) Find the weight (in newtons) of a 1450-kg automobile.
    (b) How much would it weigh on the moon?

12. A model rocket produces a thrust of $20\bar{0}0$ N. The rocket weighs 833 N and is launched vertically. Find its acceleration at takeoff.

13. A loaded freight elevator weighs 35,0̄00 N. Starting from rest, it reaches its normal upward speed of 3.25 m/s in 2.15 s.

    (a) Find its acceleration.
    (b) Find the force needed for this acceleration.
    (c) Find the tension in the cables.

14. An 8.5-g bullet is fired from a rifle whose barrel is 75 cm long with a muzzle velocity of 6̄00 m/s.

    (a) Find its average acceleration.
    (b) Find the average force acting on the bullet while it is in the barrel.

15. (a) What is the mass of a locomotive whose weight is 1.8×10^5 N?
    (b) What force is needed to give it an acceleration of 0.75 m/s²?

16. Assume that the train in Exercise 15 has 100 freight cars, each having a mass of 6̄0 metric tons.

    (a) What force is needed to give the train an acceleration of 0.75 m/s²?
    (b) What distance is needed for the train (starting from rest) to reach a velocity of 3̄0 km/h?
    (c) How long would it take to reach this velocity of 3̄0 km/h if it started from rest?

17. Atwood's machine is used to measure g as follows. The masses used are 75̄0 g and 675 g. When released, m_2 falls 65 cm in 1.6 s.

    (a) Find the acceleration of the system.
    (b) Find g.

6.1 VECTOR PROPERTIES OF FORCES

In Chapter 5, we defined a force as a push or a pull that tends to cause motion or tends to prevent motion. When a force is applied, it must be applied in *some* direction. That is, forces are vector quantities. And being vector quantities, forces are analyzed mathematically as vectors were analyzed in Chapter 2. Concurrent forces are forces that are applied at the same point.

6.2 COMPONENTS OF FORCES

As we saw in Section 2.5, the components of a vector are usually two vectors whose sum equals the original vector. Most often we find it useful to choose the components to be perpendicular to each other. When a force or forces are acting in a horizontal plane, east-west and north-south components or x-y coordinate components are most useful. When gravity is involved, we find that horizontal and vertical components are most useful.

example 1 A person is pulling a sled on "frictionless" snow as shown on page 104 with a force of 80.0 N. What are the horizontal and vertical components of this force?

SKETCH:

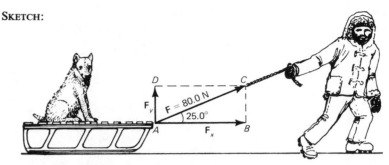

DATA: From $\triangle ABC$, $\angle CAB = 25.0°$

$$|\mathbf{F}| = 80.0 \text{ N}$$

$$\mathbf{F}_x = ?$$

BASIC EQUATION: $\cos CAB = \dfrac{\mathbf{F}_x}{|\mathbf{F}|}$

WORKING EQUATION: $\mathbf{F}_x = |\mathbf{F}|\cos CAB$

SUBSTITUTION: $\mathbf{F}_x = (80.0 \text{ N})(\cos 25.0°)$

$$= 72.5 \text{ N}$$

DATA: From $\triangle ADC$, $\angle ACD = 25.0°$ (since $AB \| CD, \angle CAB = \angle ACD$)

$$|\mathbf{F}| = 80.0 \text{ N}$$

$$\mathbf{F}_y = ?$$

BASIC EQUATION: $\sin ACD = \dfrac{\mathbf{F}_y}{|\mathbf{F}|}$

WORKING EQUATION: $\mathbf{F}_y = |\mathbf{F}|\sin ACD$

SUBSTITUTION: $\mathbf{F}_y = (80.0 \text{ N})(\sin 25.0°)$

$$= 33.8 \text{ N}$$

This means that the person is pulling straight ahead with an equivalent force of 72.5 N and actually lifting vertically with an equivalent force of 33.8 N.

In general,

$$\boxed{\begin{aligned} \mathbf{F}_x &= |\mathbf{F}|\cos\theta \\ \mathbf{F}_y &= |\mathbf{F}|\sin\theta \end{aligned}}$$

Note: Only "external forces" to the system are considered in the study of concurrent forces.

The figure on page 105 shows an object on a "frictionless" inclined plane. The gravitational attraction of the earth on the object causes the object to slide down the inclined plane. The weight of the object is indicated by $\mathbf{F}_w$. The object exerts a force perpendicular to the plane (called the *normal force*) which is indicated by $\mathbf{F}_N$. The plane, in turn, exerts an opposite force which prevents the object from "sinking into" the inclined plane. The force $\mathbf{F}$ acts parallel to the plane and indicates the force at which the object comes down the plane. Note that $\mathbf{F}$ and $\mathbf{F}_N$ are components of $\mathbf{F}_w$.

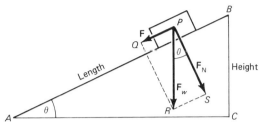

Next, we need to show that $\angle A = \angle RPS$. We will show this by first showing that $\triangle ABC$ and $\triangle PRQ$ are similar. The right angles are equal; that is, $\angle C = \angle Q$. $\overline{PQ} \| \overline{AB}$ and $\overline{PR} \| \overline{BC}$; therefore, $\angle QPR = \angle B$. Thus, $\triangle ABC \sim \triangle PRQ$. Then, $\angle A = \angle QRP$. Now since $PQRS$ is a rectangle, $QR \| SP$. Then $\angle QRP = \angle RPS$, because when two lines are parallel, the alternate interior angles are equal. Thus, we have

$$\angle A = \angle QRP \qquad \text{and}$$
$$\angle QRP = \angle RPS$$

Therefore,

$$\angle A = \angle RPS$$

From $\triangle RPS$,

$$\cos \theta = \frac{|\mathbf{F}_N|}{|\mathbf{F}_w|} \qquad \text{or} \qquad |\mathbf{F}_N| = |\mathbf{F}_w| \cos \theta$$

From $\triangle RPQ$,

$$\sin \theta = \frac{|\mathbf{F}|}{|\mathbf{F}_w|} \qquad \text{or} \qquad |\mathbf{F}| = |\mathbf{F}_w| \sin \theta$$

example 2 A block whose weight is 120 N is placed on the inclined plane below. Find the components $|\mathbf{F}_N|$ and $|\mathbf{F}|$.

SKETCH:

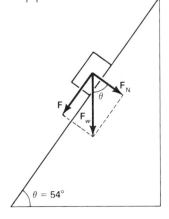

DATA:

$$|\mathbf{F}_w| = 120 \text{ N}$$

$$\theta = 54°$$

$$|\mathbf{F}_N| = ?$$

BASIC EQUATION:

$$|\mathbf{F}_N| = |\mathbf{F}_w|\cos\theta$$

WORKING EQUATION:

same

SUBSTITUTION:

$$|\mathbf{F}_N| = (120 \text{ N})(\cos 54°)$$

$$= 71 \text{ N}$$

DATA:

$$|\mathbf{F}_w| = 120 \text{ N}$$

$$\theta = 54°$$

$$|\mathbf{F}| = ?$$

BASIC EQUATION:

$$|\mathbf{F}| = |\mathbf{F}_w|\sin\theta$$

WORKING EQUATION:

same

SUBSTITUTION:

$$|\mathbf{F}| = (120 \text{ N})(\sin 54°)$$

$$= 97 \text{ N}$$

That is, a force of 71 N is pushing against the plane and a force of 97 N is pulling the block down the inclined plane.

At times we find it necessary to find the components of a vector which are not perpendicular to each other.

example 3   The load is exerting a force of 12,000 N on the crane boom below. What is the weight of the load, **W**, and the tension in the wire, T?

SKETCH:

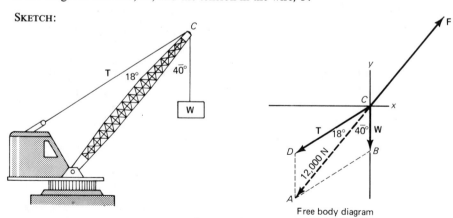

Free body diagram

Since $\overline{DC} \| \overline{AB}$, $\angle CAB = 18°$. Then $\angle ABC = 122°$. From $\triangle ABC$ and using the Law of Sines, we have:

DATA:

$$\angle CAB = 18°$$
$$\angle ABC = 122°$$
$$|\mathbf{AC}| = 12{,}000 \text{ N}$$
$$|\mathbf{W}| = ?$$

BASIC EQUATION:

$$\frac{|\mathbf{W}|}{\sin CAB} = \frac{|\mathbf{AC}|}{\sin ABC}$$

WORKING EQUATION:

$$|\mathbf{W}| = \frac{|\mathbf{AC}|(\sin CAB)}{\sin ABC}$$

SUBSTITUTION:

$$|\mathbf{W}| = \frac{(12{,}000 \, N)(\sin 18°)}{\sin 122°}$$
$$= 4400 \text{ N}$$

Similarly, from $\triangle ADC$, we have

DATA:

$$\angle DAC = 4\bar{0}°$$
$$\angle ADC = 122°$$
$$|\mathbf{AC}| = 12{,}000 \text{ N}$$
$$|\mathbf{T}| = ?$$

BASIC EQUATION:

$$\frac{|\mathbf{T}|}{\sin DAC} = \frac{|\mathbf{AC}|}{\sin ADC}$$

WORKING EQUATION:

$$|\mathbf{T}| = \frac{|\mathbf{AC}|(\sin DAC)}{\sin ADC}$$

SUBSTITUTION:

$$|\mathbf{T}| = \frac{(12{,}000 \, N)(\sin 4\bar{0}°)}{\sin 122°}$$
$$= 9100 \text{ N}$$

That is, the weight of the load is 4400 N and the tension in the wire is 9100 N.

6.3   ADDITION OF FORCES

When two or more forces act at a point, the resultant force is the sum of the forces or the single force applied at that same point that would produce the same result. Let us first consider forces in a straight line.

example 1   Suppose that we have a large crate on the floor which needs to be moved. Suppose that it has an eye screw to which a worker attaches a rope and pulls with a force of $80\bar{0}$ N to the right. A second worker attaches a rope to the same eye screw and pulls with a force of $95\bar{0}$ N to the right. What is the sum of the forces or the resultant force?

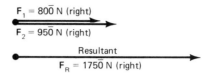

Note: The ropes and eye screw merely transmit force from the worker to the crate. This is equivalent to having the workers themselves apply the forces when in direct contact with the crate.

The resultant force of two or more forces acting in the same direction at the same point is the sum of the magnitudes of the forces and acts in the same direction.

example 2 Now suppose that the two workers attach the ropes on opposite sides of the crate above and pull with the same forces. The first worker pulls to the right. What is the sum of the forces now?

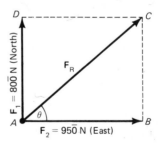

The resultant force of two forces acting in opposite directions at the same point is the difference of the magnitudes of the forces and acts in the direction of the greater force.

example 3 Next, suppose that the two workers attach their ropes to the crate at right angles and pull as shown below. What is the resultant force?

SKETCH:

```
D ----------------- C
|                  ↗|
|               ↗   |
|            ↗      |
F₁=800 N    ↗  F_R  |
(North)  ↗         |
|     ↗            |
|  ↗ θ             |
A ─────────────────→ B
    F₂ = 950 N (East)
```

First, find angle θ in $\triangle ABC$:

DATA:

$$|\mathbf{F}_1| = 80\overline{0} \text{ N} = \overline{BC}$$

$$|\mathbf{F}_2| = 95\overline{0} \text{ N}$$

$$\theta = ?$$

BASIC EQUATION:

$$\tan \theta = \frac{\overline{BC}}{|\mathbf{F}_2|}$$

WORKING EQUATION: same

SUBSTITUTION: $\tan \theta = \dfrac{80\bar{0}\ \text{N}}{95\bar{0}\ \text{N}} = 0.8421$

$$\theta = 40.1°$$

Now, find $|\mathbf{F}_R|$ in $\triangle ABC$:

DATA: $|\mathbf{F}_2| = 95\bar{0}\ \text{N}$

$$\theta = 40.1°$$

$$|\mathbf{F}_R| = \ ?$$

BASIC EQUATION: $\cos \theta = \dfrac{|\mathbf{F}_2|}{|\mathbf{F}_R|}$

WORKING EQUATION: $|\mathbf{F}_R| = \dfrac{|\mathbf{F}_2|}{\cos \theta}$

SUBSTITUTION: $|\mathbf{F}_R| = \dfrac{95\bar{0}\ \text{N}}{\cos 40.1°}$

$$= 1240\ \text{N}$$

That is, the crate moves at an angle 40.1° north of east as though one person were pulling with a force of 1240 N in a straight line along that angle.

example 4   Finally, suppose that two workers attach their ropes to the crate and pull as shown below. What is the resultant force?

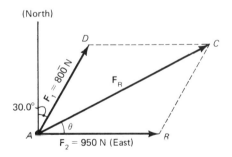

Solution I (using components):

Vector	Horizontal components	Vertical components
$\mathbf{F}_1$ $\theta = 60.0°$	$|\mathbf{F}_1| \cos \theta = (80\bar{0}\ \text{N})(\cos 60.0°) = \ 40\bar{0}\ \text{N}$	$|\mathbf{F}_1| \sin \theta = (80\bar{0}\ \text{N})(\sin 60.0°) = 693\ \text{N}$
$\mathbf{F}_2$ $\theta = 0.0°$	$|\mathbf{F}_2| \cos \theta = (95\bar{0}\ \text{N})(\cos 0.0°) = \ 95\bar{0}\ \text{N}$	$|\mathbf{F}_2| \sin \theta = (95\bar{0}\ \text{N})(\sin 0.0°) = \ \underline{\ \ 0\ \ }$
	Sum of horizontal components $= 135\bar{0}\ \text{N}$	Sum of vertical components $= 693\ \text{N}$

$$\tan \theta = \frac{\text{sum of vertical components}}{\text{sum of horizontal components}} = \frac{693 \text{ N}}{135\bar{0} \text{ N}} = 0.513$$

$$\theta = 27.2°$$

$$|\mathbf{F}_R| = \sqrt{(135\bar{0} \text{ N})^2 + (693 \text{ N})^2}$$

$$= 1520 \text{ N}$$

That is, the crate moves at an angle of 27.2° north of east as though one person were pulling with a force of 1520 N in a straight line along that angle.

Solution II (using the Law of Sines and Law of Cosines):

SKETCH:

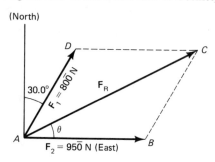

From $\triangle ABC$ and using the Law of Cosines, we have

DATA:

$$|\mathbf{F}_2| = \overline{AB} = 95\bar{0} \text{ N}$$

$$|\mathbf{F}_1| = \overline{BC} = 80\bar{0} \text{ N}$$

$$\angle DAB = 90° - 30.0° = 60.0°$$

$$\angle ABC = 180° - 60.0° = 120.0°$$ (the sum of any two consecutive angles of a parallelogram is 180°)

$$|\mathbf{F}_R| = \overline{AC} = ?$$

BASIC EQUATION: $(\overline{AC})^2 = (\overline{AB})^2 + (\overline{BC})^2 - 2(\overline{AB})(\overline{BC})(\cos ABC)$

WORKING EQUATION: $\overline{AC} = \sqrt{(\overline{AB})^2 + (\overline{BC})^2 - 2(\overline{AB})(\overline{BC})(\cos ABC)}$

SUBSTITUTION: $|\mathbf{F}_R| = \sqrt{(95\bar{0} \text{ } N)^2 + (80\bar{0} \text{ } N)^2 - 2(95\bar{0} \text{ } N)(80\bar{0} \text{ } N)(\cos 120.0°)}$

$$= 1520 \text{ N}$$

Next, find θ from $\triangle ABC$ using the Law of Sines:

DATA:

$$|\mathbf{F}_1| = \overline{BC} = 80\bar{0} \text{ N}$$

$$|\mathbf{F}_R| = \overline{AC} = 1520 \text{ N}$$

$$\angle ABC = 120.0°$$

$$\theta = ?$$

BASIC EQUATION: $$\frac{\overline{BC}}{\sin \theta} = \frac{\overline{AC}}{\sin ABC}$$

WORKING EQUATION: $$\sin \theta = \frac{(\overline{BC})(\sin ABC)}{\overline{AC}}$$

SUBSTITUTION: $$\sin \theta = \frac{(80\overline{0} \ N)(\sin 120.0°)}{1520 \ N} = 0.456$$

$$\theta = 27.1°$$

Note that the difference in the values of θ in the two solutions is due to rounding and is acceptable for our purposes. If greater precision were needed, each of the component forces would need to be measured with more precision.

When three or more forces act at a point, the resultant force can usually be found more easily using the component method (see Section 2.5).

6.4   CONCURRENT FORCES IN EQUILIBRIUM

An object is in equilibrium when there is no change in its motion. That is, an object in equilibrium is either at rest or moving at a constant speed in a straight line. There are two conditions for an object to be in equilibrium.

FIRST CONDITION OF EQUILIBRIUM

For an object in which concurrent forces are acting, the net force, or the sum of forces acting on the object, must be zero.

The second condition of equilibrium, which pertains to the rotation of an object, will be discussed in Section 10.1.

When two or more forces are acting at a point, the *equilibrant force* is that force which when applied at that same point produces equilibrium.

If a person is pushing on a box with a $12\overline{0}$-N force, it takes an equilibrant force of $12\overline{0}$ N pushing on the opposite side to get the box in a state of equilibrium.

Let us take another look at the two workers pulling on the crate on page 108. Recall that we found the resultant force to be 1240 N at 40.1° north of east. What single force would be required to bring the crate into a state of equilibrium?

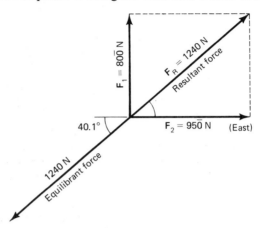

The required equilibrant force is one equal in magnitude but opposite in direction to the resultant force. In this case, the equilibrant force is 1240 N at 40.1° south of west.

Let us use the component method of adding vectors to solve equilibrium problems. The x and y components of the net force can be found by adding the x components and by adding the y components. The x component of the net force is

$$\mathbf{R}_x = \mathbf{A}_x + \mathbf{B}_x + \mathbf{C}_x + \cdots = \text{sum of } x \text{ components}$$

The y component of the net force is

$$\mathbf{R}_y = \mathbf{A}_y + \mathbf{B}_y + \mathbf{C}_y + \cdots = \text{sum of } y \text{ components}$$

If the object is in equilibrium, the net force, or the sum of the forces acting on the object, must be zero. That is,

$$\text{sum of } x \text{ components} = 0$$
$$\text{sum of } y \text{ components} = 0$$

Use the following procedure to solve equilibrium problems:

1. Draw a free-body diagram of the point at which the unknown forces act.
2. Find the x and y component of each force.
3. Substitute the components in the equations

$$\text{sum of } x \text{ components} = 0$$
$$\text{sum of } y \text{ components} = 0$$

4. Solve for the unknowns. This may involve two simultaneous linear equations.

In many problems we will be interested in finding the tension or compression in part of a structure, such as in a beam or a cable. *Tension* is a stretching force produced by forces pulling outward on the ends of the object. *Compression* is a compressing force produced by forces pushing inward on the ends of an object. A rubber band being stretched is an example of tension. A valve spring whose ends are pushed together is an example of compression.

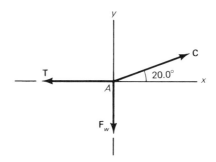

example 1   The crane shown below is supporting a beam that weighs $60\overline{0}0$ N. Find the tension in the horizontal supporting cable and the compression in the boom.

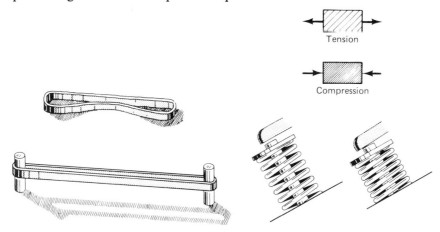

(a) First, draw a free-body diagram showing the forces acting *on* point A.
  $|\mathbf{T}|$ is the force exerted on A by the horizontal supporting cable.
  $|\mathbf{C}|$ is the force exerted by the boom on A.
  $|\mathbf{F}_w|$ is the force (weight of the beam) pulling straight down on A.

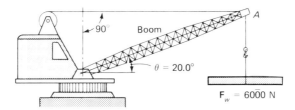

(b) Find the *x*- and *y*-components of each force:

Force	*x*-components	*y*-components
C	$\|C\|(\cos 20.0°)$	$\|C\|(\sin 20.0°)$
T	$-\|T\|$	0
F_w	0	$-\|F_w\|$

(c) Sum of *x*-components = 0   Sum of *y*-components = 0

$\|C\|(\cos 20.0°) + (-\|T\|) = 0$ $\|C\|(\sin 20.0°) + (-\|F_w\|) = 0$

(d) $\|T\| = \|C\|(\cos 20.0°)$ $\|C\| = \dfrac{\|F_w\|}{\sin 20.0°}$

$$= \frac{60\bar{0}0 \text{ N}}{\sin 20.0°}$$

$$= 17{,}500 \text{ N}$$

$= (17{,}500 \text{ N})(\cos 20.0°)$
$= 16{,}400 \text{ N}$

example 2 A 1600-N sign is supported by two cables as shown below. Find the tension in each rope.

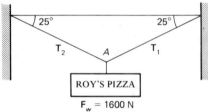

(a) First draw a free-body diagram showing the forces acting *on* point *A*. The weight, F_w, pulls straight down. T_1 and T_2 are the tension forces, which pull at 25° above the horizontal.

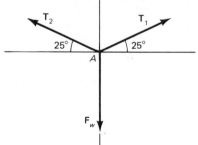

(b)

Force	*x*-components	*y*-components
F_w	0	-1600 N
T_1	$\|T_1\|(\cos 25°)$	$\|T_1\|(\sin 25°)$
T_2	$\|T_2\|(\cos 155°)$	$\|T_2\|(\sin 155°)$

(c) Sum of x-components = 0

$$|\mathbf{T}_1|(\cos 25°) + |\mathbf{T}_2|(\cos 155°) = 0$$
$$0.91|\mathbf{T}_1| - 0.91|\mathbf{T}_2| = 0 \qquad (1)$$

Sum of y-components = 0

$$-1600 + |\mathbf{T}_1|(\sin 25°) + |\mathbf{T}_2|(\sin 155°) = 0$$
$$0.42|\mathbf{T}_1| + 0.42|\mathbf{T}_2| = 1600 \qquad (2)$$

(d) Solve the system of equations (1) and (2) simultaneously. Dividing each side of equation (1) by 0.91, we have

$$|\mathbf{T}_1| - |\mathbf{T}_2| = 0 \qquad \text{or} \qquad |\mathbf{T}_1| = |\mathbf{T}_2|$$

Dividing each side of equation (2) by 0.42, we have

$$|\mathbf{T}_1| + |\mathbf{T}_2| = 3800$$

Substituting $|\mathbf{T}_1| = |\mathbf{T}_2|$, we have

$$2|\mathbf{T}_2| = 3800$$
$$|\mathbf{T}_2| = 1900$$

That is, the tension in each cable is 1900 N.

Note: We have used two-significant-digit accuracy.

6.5 FRICTION

When two objects slide across each other, a force that resists the motion is produced. This force is called *friction*. Friction is commonly caused by the irregularities of the two surfaces, which tend to catch on each other. Severe engine damage can be caused by friction if proper lubricants are not used. Friction makes it hard to push objects along the floor.

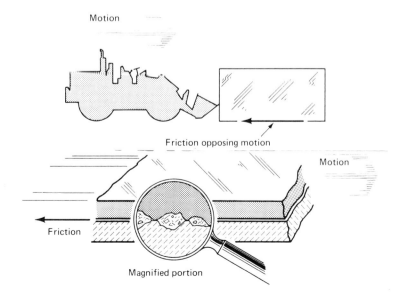

Friction is both a necessity and a hindrance to our everyday living. If there were no friction, how would you walk across the floor, start or stop your automobile on the highway, nail two boards together permanently, or keep the dishes and utensils on the table? Of course, friction is an enemy when it comes to engine wear, carpet wear, and rearranging the living room furniture!

Frictional relationships are difficult, at best, to describe in terms of simple laws or equations. Therefore, we shall discuss the following general principles of friction, which apply to most solid objects, so that we may have a better understanding of friction and its consequences. Fluid friction also exists, but as it is even more difficult to discuss, we will not go into it.

1. Friction is a force that acts parallel to the contact surfaces and always opposite to the motion or opposite to any force tending to produce motion.
2. Friction is dependent upon the kind of materials in contact and the smoothness of the contact surfaces. It is much easier to slide a crate on ice than on a wood floor. For certain materials, polishing the surfaces increases the friction; glass is an example.
3. Starting friction is greater than moving friction. If you have ever pushed an automobile by hand, you soon noticed that it took more force to start the car moving than it did to keep it moving.
4. Friction is practically independent of the surface area of contact. Sliding a box with dimensions 1.00 m × 2.00 m × 3.00 m on any side will result in practically the same frictional force.
5. Friction is directly proportional to the force pressing the two smooth unlubricated surfaces together. That is, the heavier the object, the more difficult to slide and vice versa. The relationship is usually written

$$F_f = \mu F_N$$

where F_f is the frictional force,
F_N is the normal force, and
μ is the coefficient of friction, which is the proportionality constant.

Some approximate values for the coefficients of friction of a few surfaces are given at the top of page 117.

In general, friction may be reduced by

1. Using a smoother surface.
2. Using a lubrication that provides a thin film between surfaces and thus reduces friction.
3. Using Teflon, which greatly reduces friction between surfaces when an oil lubricant is not desirable, such as in electric motors.
4. Substituting rolling friction for sliding friction. Using ball bearings and roller bearings greatly reduces friction.

Coefficient of friction μ between clean surfaces

Material	Starting	Sliding
Steel on steel	0.58	0.20
Glass on glass	0.95	0.40
Hardwood on hardwood	0.40	0.25
Steel on concrete	—	0.30
Aluminum on aluminum	1.9	—
Rubber on dry concrete	2.0	1.02
Rubber on wet concrete	1.5	0.97
Teflon on Teflon	0.04	0.04
Steel on steel (oiled)(approx.)	0.15	0.07
Aluminum on snow (wet)	0.4	0.02

example 1   It takes a force of 21 N to keep a 66-N wooden block sliding on a wood floor. What is the coefficient of friction?

DATA:

$$F_f = 21 \text{ N}$$

$$F_N = 66 \text{ N}$$

$$\mu = \,?$$

BASIC EQUATION:

$$F_f = \mu F_N$$

WORKING EQUATION:

$$\mu = \frac{F_f}{F_N}$$

SUBSTITUTION:

$$\mu = \frac{21 \text{ N}}{66 \text{ N}} = 0.32$$

example 2   A cart on wheels weighs 1500 N. The coefficient of rolling friction between the wheels and an inclined plane is 0.18. An inclined plane must be designed so that the cart rolls down the plane at a constant speed. Find angle θ.

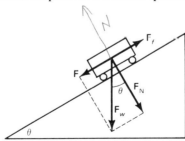

From the inclined plane,

$$|\mathbf{F}| = |\mathbf{F}_w| \sin \theta \quad \text{and} \quad |\mathbf{F}_N| = |\mathbf{F}_w| \cos \theta$$

Because of the friction,

$$|F_f| = \mu|F_N|$$

At a constant speed, the cart is in equilibrium; thus,

$$F + F_f = 0 \qquad \text{or} \qquad |F| = |F_f|$$

Therefore,

$$|F_w|\sin \theta = \mu|F_N|$$

and

$$|F_w|\sin \theta = \mu|F_w|\cos \theta$$

Thus,

$$\sin \theta = \mu \cos \theta$$

or

$$\mu = \frac{\sin \theta}{\cos \theta} = \tan \theta$$

Since $\mu = 0.18$,

$$\tan \theta = 0.18$$

$$\theta = 1\bar{0}°$$

That is, once started, the cart will travel at a uniform speed down an inclined plane whose angle of inclination is $1\bar{0}°$ and whose coefficient of friction is 0.18.

EXERCISES

Sketch	Data	Basic Equation	Working Equation	Substitution
(sketch: 12 cm², h = ?, b = 6.0 cm)	$A = 12 \text{ cm}^2$ $b = 6.0 \text{ cm}$ $h = ?$	$A = bh$	$h = \dfrac{A}{b}$	$h = \dfrac{12 \text{ cm}^2}{6.0 \text{ cm}}$ $= 2.0 \text{ cm}$

1. A gardener applies a 175-N force on the handle of a lawn mower that makes a 40.0° angle with the ground. Find the horizontal and vertical components of this force.

2. A worker pulls a loaded cart with a force of 475 N. The handle makes an angle of 22.0° with the ground. Find the horizontal and vertical components.

3. A block weighing 1200 N is placed on an inclined plane of 35°. Find the components F_N and F.

4. The load is exerting a compression force of 15,000 N on the crane boom below. Find the weight of the load, $|W|$, and the tension in the wire, $|T|$.

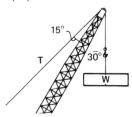

5. A force of 185 N and a second force of 115 N act on the same object.

    (a) What is the magnitude of the maximum force that they can exert together?

    (b) What is the magnitude of the minimum force that they can exert together?

6. Three forces, whose magnitudes are $15\overline{0}$ N, 75 N, and 125 N, act on the same object.

    (a) What is the magnitude of the maximum force that they can exert together?

    (b) What is the magnitude of the minimum force that they can exert together?

Find the sum of each set of vectors.

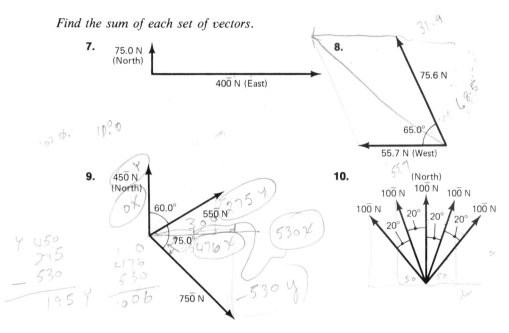

7. 75.0 N (North), $40\overline{0}$ N (East)

8. 75.6 N, 65.0°, 55.7 N (West)

9. $45\overline{0}$ N (North), 60.0°, $55\overline{0}$ N, 75.0°, $75\overline{0}$ N

10. (North), $10\overline{0}$ N, $10\overline{0}$ N, $10\overline{0}$ N, $10\overline{0}$ N, $10\overline{0}$ N, 20°, 20°, 20°, 20°

11.-14. *Find each equilibrant force in Exercises 7–10.*

Find the forces $|\mathbf{F}_1|$ and $|\mathbf{F}_2|$ necessary to produce equilibrium in the following free-body diagrams.

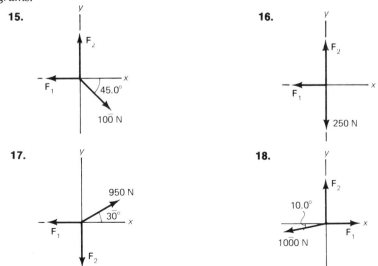

15. F_2, F_1, 45.0°, $10\overline{0}$ N

16. F_2, F_1, 250 N

17. 950 N, 30°, F_1, F_2

18. F_2, 10.0°, F_1, $10\overline{00}$ N

19.

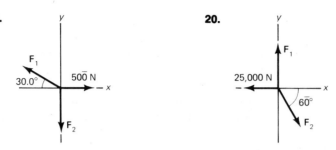

20.

21. Find the tension in the horizontal supporting cable and the compression in the boom of the crane below, which supports an 8900-N beam.

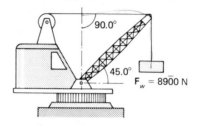

22. Find the tension in the horizontal supporting cable and the compression in the boom of the crane below, which supports a 45,000-N beam.

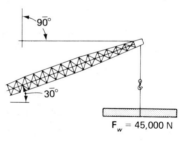

23. The rope shown below is attached to two buildings and supports a $120\overline{0}$-N sign. Find the tension in the two ropes, $|T_1|$ and $|T_2|$.

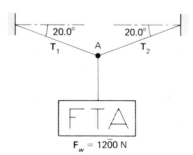

24. If the angle between the horizontal and the ropes in Exercise 23 is changed to 10.0°, what are the tensions in the two ropes, $|T_1|$ and $|T_2|$?

25. The frictional force in the mower below is $8\bar{0}$ N. What force must the man exert along the handle to push it at a constant velocity?

26. Find the tension in the cable and the compression in the support below.

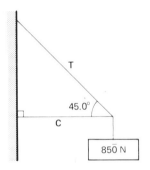

27. Find the compression in each support.

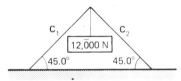

28. Find the compression in each support.

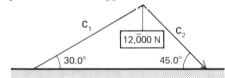

29. Find the tension in each cable.

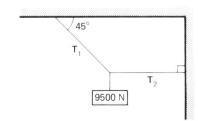

30. Find the compression in the boom and the tension in the cable of the crane.

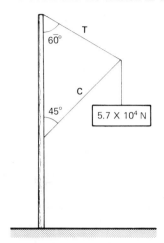

31. A wooden crate weighs 2150 N. What force is needed to start sliding the crate on a wood floor whose coefficient of friction is 0.45?

32. A wooden box weighs 1200 N. If a force of 480 N is needed to start sliding the box on a metal platform, what is the coefficient of starting friction?

33. What force is needed to push a 3000-N piano across a wooden floor if the coefficient of friction between the wheels and floor is 0.15?

34. (a) What force is needed to push the piano in Exercise 33 up a wooden ramp ($\mu = 0.15$) that makes a 30.0° incline?
(b) If the piano is allowed to roll down the incline, what is its acceleration?

35. At what angle of incline will the piano in Exercise 34 roll down the incline at a constant speed?

36. A metal block rests on a lubricated metal incline ($\mu = 0.15$) as shown below. What weight, **W**, is needed to keep the block from moving? (Assume that the pulley is frictionless.)

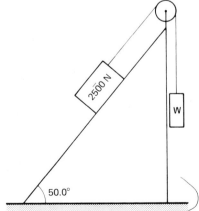

7.1 WORK

What is work? Are we doing work when we try to lift a large crate but it will not budge? Because we have exerted a force and feel tired, we would probably say that we had been working. But is it proper to say that we have done work *on* the crate? Is work done on the crate if we push the crate across the floor?

Work does have a technical meaning, however. When a stake is driven into the ground, work is done *by* the moving sledgehammer, and work is done *on* the stake. When a bulldozer pushes a boulder, work is done *by* the bulldozer, and work is done *on* the boulder.

These examples show a more limited meaning of work; that work is done when a force acts through a distance. The physical definition of work is even narrower: *work is the product of the force in the direction of the motion and the displacement* (or the product of the force and the displacement in the direction of the force); that is,

$$W = Fs$$

where W is the work,

F is the force *in the direction of the motion* (or the applied force), and

s is the displacement (or the displacement *in the direction of the force*).

Now, let us apply our physical definition of work to trying unsuccessfully to lift the crate. We applied a force by lifting on the crate. Have we done work? We were unable to move the crate. Therefore, the displacement was zero, and the product of the force and displacement must also be zero. Therefore, no work was done (on the crate). We should note that energy was expended in the form of body heat and chemical energy even though the energy of the crate was not increased.

In the SI system, the force is expressed in newtons (N) and the displacement is expressed in metres (m). By unit analysis, we see that the work unit is a newton metre (N m), which is called a *joule*, J (pronounced jool). That is, 1 joule of work is the result of a force of 1 newton acting through a distance of 1 metre. The joule unit is named in honor of James P. Joule, an English physicist who made important contributions in thermodynamics.

The *erg* (1 erg = 1 dyne centimetre) is another metric work unit. Since 1 J = 10^7 ergs, the erg is too small for most industrial uses. In addition, the erg is not included as a modern SI unit. As a result, the erg unit is gradually becoming extinct.

Work is not a vector quantity because it has no particular direction. So work is a scalar and has only magnitude.

example 1   A person lifting 220 N of bricks to a height of 1.5 m does how much work?

SKETCH:

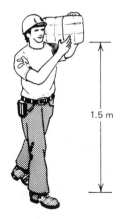

1.5 m

DATA:

$F = 220$ N

$s = 1.5$ m

$W = ?$

BASIC EQUATION: $W = Fs$

WORKING EQUATION: same

SUBSTITUTION: $W = (220 \text{ N})(1.5 \text{ m})$

$= 330$ N m

$= 330$ J

example 2 A person uniformly pushes a 2100-N loaded pallet (portable platform used in warehouses) a distance of $3\bar{0}$ m by exerting a constant horizontal force of $5\bar{0}$ N. How much work is done?

SKETCH:

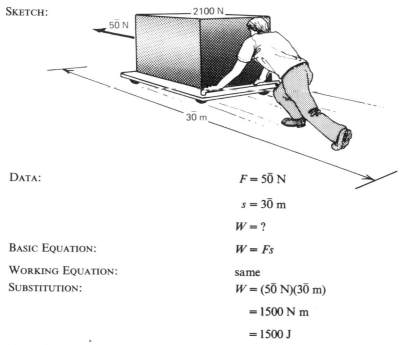

DATA:

$$F = 5\bar{0} \text{ N}$$

$$s = 3\bar{0} \text{ m}$$

$$W = ?$$

BASIC EQUATION:

$$W = Fs$$

WORKING EQUATION: same

SUBSTITUTION:

$$W = (5\bar{0} \text{ N})(3\bar{0} \text{ m})$$

$$= 1500 \text{ N m}$$

$$= 1500 \text{ J}$$

Note in Example 2 that the pallet weighs 2100 N. (Recall that the weight of an object is the measure of its gravitational attraction to the earth and is represented by a vertical vector pointing down to the center of the earth.) There is *no motion* in the direction that *this* force is exerted. Therefore, the weight of the box is not the force used to determine the work being done in the problem.

Work is being done by the person pushing the pallet. A force of $5\bar{0}$ N is exerted and there is a resulting displacement in the direction the force is applied. The work done is the product of this force ($5\bar{0}$ N) and the displacement ($3\bar{0}$ m) in the direction the force is applied.

How can we determine how much work is done when the applied force is *not* in the direction of the motion? Assume that a block is pulled along the ground with an applied force **F** by a rope that makes an angle θ with the ground. Recall that the

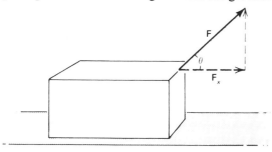

work done is the product of the force in the direction of the motion and the displacement. So, let us first find the component of the applied force in the direction of the motion.

$$\cos \theta = \frac{|\mathbf{F}_x|}{|\mathbf{F}|}$$

Therefore,

$$|\mathbf{F}_x| = |\mathbf{F}| \cos \theta$$

If the applied force is not in the direction of the motion, the work done is

$$\boxed{W = Fs \cos \theta}$$

where *W* is the work,
  F is the applied force,
  s is the displacement, and
  θ is the angle between the applied force and the direction of motion.

example 3   A worker now pulls the 2100-N loaded pallet from Example 2 with a rope that makes an angle of $3\overline{0}°$ with the floor. The worker exerts a constant pulling force of 58 N on the rope and moves the pallet $3\overline{0}$ m. How much work is done?

SKETCH:

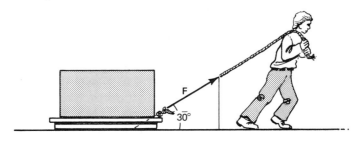

DATA:

$$F = 58 \text{ N}$$
$$s = 3\overline{0} \text{ m}$$
$$\theta = 3\overline{0}°$$
$$W = ?$$

BASIC EQUATION: $W = Fs \cos \theta$

WORKING EQUATION: same

SUBSTITUTION:

$$W = (58 \text{ N})(3\overline{0} \text{ m}) \cos 3\overline{0}°$$
$$= 1500 \text{ N m}$$
$$= 1500 \text{ J}$$

Note that the same amount of work was done on the pallet even though different amounts of force were exerted.

How much work is done in lifting an object of mass *m* vertically at constant speed? Here the force is equal to, but in opposite direction of, the weight of the object. The weight of the object is

$$F_w = mg$$

So the work done in lifting any object vertically is found by the equation

$$\boxed{W = Fs \quad \text{or} \quad W = mgh}$$

example 4 Find the amount of work done in vertically lifting at constant speed a beam of mass 650 kg a distance of $4\bar{0}$ m.

DATA:

$m = 650$ kg

$h = 4\bar{0}$ m

$g = 9.80$ m/s^2

$W = ?$

BASIC EQUATION: $W = mgh$

WORKING EQUATION: same

SUBSTITUTION:

$W = (650 \text{ kg})(9.80 \text{ m/s}^2)(4\bar{0} \text{ m})$

$= 2.5 \times 10^5 (\text{kg m/s}^2)(\text{m})$

$= 2.5 \times 10^5$ N m

$= 2.5 \times 10^5$ J

Note the unit analysis.

How much work is done by a uniform parallel force pushing a mass *m* up an inclined plane of length *s* which makes an angle θ with the horizontal when the coefficient of friction between the surfaces is μ?

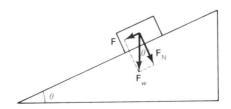

Recall that

$$\text{the frictional force is } F_f = \mu F_N$$

$$\text{the normal force is } F_N = F_w \cos \theta$$

$$\text{the weight is } F_w = mg$$

Therefore,

$$F_f = \mu mg \cos \theta$$

Also, recall that the force of the object that tends to pull it down the plane is

$$F = F_w \sin \theta = mg \sin \theta$$

To push an object *up* the plane, we must overcome both the frictional force, F_f, and the force that tends to pull the object down the plane. That is, the total force is

$$\mu mg \cos \theta + mg \sin \theta$$

and the work is

$$W = Fs = (\mu mg \cos \theta + mg \sin \theta)s$$

or

$$\boxed{W = mgs(\mu \cos \theta + \sin \theta)}$$

Note that if the object is pushed *down* the plane at constant speed, the force exerted is the difference between the frictional force, F_f, and the force that tends to pull the object down the plane. And the work is

$$W = Fs = (\mu mg \cos \theta - mg \sin \theta)s$$

or

$$\boxed{W = mgs(\mu \cos \theta - \sin \theta)}$$

example 5   A wooden crate of mass 250 kg is pushed uniformly up an 8.5-m metal ramp that makes an angle of $2\overline{0}°$ with the ground. How much work is done if the coefficient of friction between the crate and ramp is 0.35?

DATA: $m = 250 \text{ kg}$

$s = 8.5 \text{ m}$

$g = 9.80 \text{ m/s}^2$

$\mu = 0.35$

$\theta = 2\overline{0}°$

$W = ?$

BASIC EQUATION: $W = mgs(\mu \cos \theta + \sin \theta)$

WORKING EQUATION: same

SUBSTITUTION: $W = (250 \text{ kg})(9.80 \text{ m/s}^2)(8.5 \text{ m})[(0.35) \cos 2\bar{0}° + \sin 2\bar{0}°]$

$= 1.4 \times 10^4 \text{ (kg m/s}^2\text{)m}$

$= 1.4 \times 10^4 \text{ N m}$

$= 1.4 \times 10^4 \text{ J}$

7.2   POWER

Power is the rate of doing work; that is,

$$P = \frac{W}{t}$$

where P is the power,
     W is the work, and
      t is the time.

You do the same amount of work if you uniformly walk or run up a flight of stairs. However, you use more power when you run because your time is less.

The SI metric unit of power is the watt (W), which is named in honor of James Watt, who invented the first commercially successful steam engine. The watt is defined:

$$1 \text{ W} = 1 \text{ J/s}$$

Since the watt is a rather small unit of power, the kilowatt (kW, 1000 watts) and the megawatt (MW, 1 million watts) are more commonly used.

example 1   A 650-kg steel beam is uniformly raised a height of $4\bar{0}$ m in $3\bar{0}$ s. How many kilowatts of power are required?

DATA: $m = 650$ kg

$h = 4\bar{0}$ m

$t = 3\bar{0}$ s

$g = 9.80 \text{ m/s}^2$

$P = ?$

BASIC EQUATION: $P = \dfrac{W}{t}$ and $W = mgh$

WORKING EQUATION: $P = \dfrac{mgh}{t}$

SUBSTITUTION: $$P = \frac{(650 \text{ kg})(9.80 \text{ m/s}^2)(4\bar{0} \text{ m})}{3\bar{0} \text{ s}}$$

$$= 8500 \text{ W} \qquad \boxed{\frac{(\text{kg m/s}^2)\text{m}}{\text{s}} = \frac{\text{N m}}{\text{s}} = \frac{\text{J}}{\text{s}} = \text{W}}$$

$$= 8.5 \text{ kW}$$

example 2   A pump is needed to lift $10\bar{0}0$ L of water per minute (60.0 s) a distance of 20.0 m. What power, in kW, must the pump be able to deliver? (1 L of water has a mass of 1 kg.)

DATA: $$m = 10\bar{0}0 \text{ L} \times \frac{1 \text{ kg}}{1 \text{ L}} = 10\bar{0}0 \text{ kg}$$

$$h = 20.0 \text{ m}$$

$$t = 60.0 \text{ s}$$

$$g = 9.80 \text{ m/s}^2$$

$$P = ?$$

BASIC EQUATION: $$P = \frac{W}{t} \qquad \text{and} \qquad W = mgh$$

WORKING EQUATION: $$P = \frac{mgh}{t}$$

SUBSTITUTION: $$P = \frac{(10\bar{0}0 \text{ kg})(9.80 \text{ m/s}^2)(20.0 \text{ m})}{60.0 \text{ s}}$$

$$= 3270 \text{ W}$$

$$= 3.27 \text{ kW}$$

7.3   ENERGY

Physicists define mechanical energy as the ability to do work. Mechanical energy exists in two basic forms—kinetic and potential. *Kinetic energy* is the energy that an object contains due to its motion. The energy that an automobile possesses due to its motion going down the highway is called its kinetic energy. The energy that a hammer possesses due to its motion as it strikes a nail is called kinetic energy.

How do we measure the kinetic energy of an object of mass m? By Newton's second law, we know

$$F = ma \tag{1}$$

That is, F is the force necessary to give an object of mass m an acceleration a.

Multiplying both sides by s, we have

$$Fs = mas \tag{2}$$

Assuming that the object started at rest, its velocity at any time is given by

$$v^2 = 2as \tag{3}$$

or

$$\frac{v^2}{2} = as \tag{4}$$

Substituting equation (4) into equation (2), we have

$$Fs = \frac{mv^2}{2}$$

The product Fs represents the amount of work done on an object of mass m to accelerate it from rest to velocity v; that is, Fs represents the energy of the mass due to its motion. Therefore,

$$\boxed{KE = \frac{1}{2}mv^2}$$

where KE is the kinetic energy of the moving object,
$\quad\quad m$ is its mass, and
$\quad\quad v$ is its velocity.

Since energy is defined as the ability to do work, it is not very surprising that energy is expressed in the same unit as work—the joule.

example 1 A pile driver has a mass of $10,\overline{0}00$ kg and strikes the pile with a velocity of 10.0 m/s.

(a) What is the kinetic energy of the driver when it strikes the pile?
(b) If the pile is driven 0.200 m into the ground, what force is applied to the pile?

SKETCH:

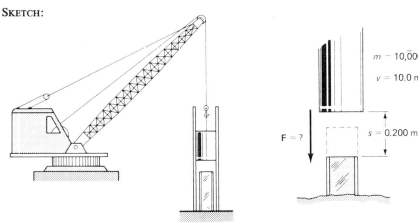

(a) DATA: $m = 1.00 \times 10^4$ kg

$v = 10.0$ m/s

KE = ?

BASIC EQUATION: $KE = \frac{1}{2}mv^2$

WORKING EQUATION: same

SUBSTITUTION: $KE = \frac{1}{2}(1.00 \times 10^4 \text{ kg})(10.0 \text{ m/s})^2$

$= 5.00 \times 10^5 \text{ (kg m/s}^2\text{)m}$

$= 5.00 \times 10^5 \text{ N m}$

$= 5.00 \times 10^5 \text{ J}$

(b) DATA: $KE = 5.00 \times 10^5$ J   (from part (a))

$s = 0.200$ m

$F = ?$

BASIC EQUATION: $KE = W = Fs$

WORKING EQUATION: $F = \dfrac{KE}{s}$

SUBSTITUTION: $F = \dfrac{5.00 \times 10^5 \text{ J}}{0.200 \text{ m}} \times \dfrac{\text{N m}}{\text{J}}$ (*Note*: conversion factor)

$= 2.50 \times 10^6 \text{ N}$

The other type of mechanical energy is due to an object's position or configuration and is called *potential energy*. One can think of potential energy as stored energy. A pile driver raised above a piling possesses a stored energy. For when it is released, it will fall and drive the piling into the ground. Note that if the pile driver is resting on the piling, it contains no potential, or stored, energy.

Internal potential energy is determined by the nature or condition of the substance, such as gasoline, a compressed spring, or a stretched rubber band. Gravitational potential energy is determined by the position of the object relative to a particular reference level.

How do we measure the gravitational potential energy of an object of mass *m*? We know that the amount of work needed to raise the mass to a height of *h* is

$$W = F_w s = mgh$$

Therefore, this same amount of work can be done *by* the object by dropping a height *h*. So we define the gravitational potential energy of an object as

$$\boxed{PE = mgh}$$

where PE is the potential energy of the object,
    m is its mass,
    g is the gravitational acceleration (9.80 m/s²), and
    h is the height above the reference level.

example 2   A wrecking ball of mass $20\bar{0}$ kg is poised 4.00 m above a concrete platform whose top is 2.00 m above the ground.

(a) With respect to the platform, what is the potential energy of the ball?
(b) With respect to the ground, what is the potential energy of the ball?

SKETCH:

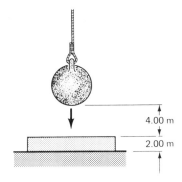

DATA: $m = 20\bar{0}$ kg

                                   $h_1 = 4.00$ m

                                   $h_2 = 2.00$ m

                                   PE $= ?$

BASIC EQUATION: PE $= mgh$

WORKING EQUATION: same

(a) SUBSTITUTION: PE $= (20\bar{0}$ kg$)(9.80$ m/s²$)(4.00$ m$)$

                                      $= 7840$ (kg m/s²)m

                                      $= 7840$ N m

                                      $= 7840$ J

(b) SUBSTITUTION: PE $= (20\bar{0}$ kg$)(9.80$ m/s²$)(6.00$ m$)$

                                      $= 11,800$ J

We have discussed only two kinds of energy—kinetic and potential. Keep in mind that energy exists in many forms—chemical, atomic, electrical, sound, and heat. These forms and the conversion of energy from one form to another will be studied later.

7.4 CONSERVATION OF MECHANICAL ENERGY

Now let us consider kinetic and potential energy together. Can you see how the two might be related? They are, in fact, related by the following important principle:

LAW OF CONSERVATION OF MECHANICAL ENERGY:

If no external forces act, the sum of the kinetic energy and potential energy in a system is constant.

Consider a swinging pendulum bob where there is no resistance involved. Pull the bob over to the right side so that the string makes an angle of 65° with the vertical. At this point, the bob contains its maximum potential energy and its minimum kinetic energy (zero). Note that a larger maximum potential energy is possible when an initial deflection of greater than 65° is made.

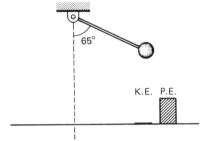

An instant later, the bob has lost some of its potential or stored energy, but it has gained in kinetic energy due to its motion.

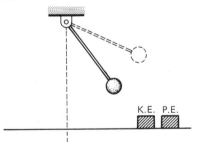

At the bottom of its arc of swing, its potential energy is zero and its kinetic energy is maximum (its velocity is maximum).

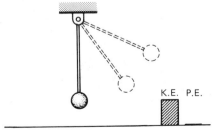

The kinetic energy of the bob then causes the bob to swing upward to the left. As it completes its swing, its kinetic energy is decreasing and its potential energy is increasing. That is, its kinetic energy is changing to potential energy.

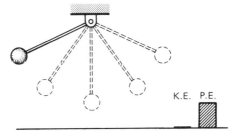

According to the law of conservation of mechanical energy, the sum of the kinetic energy and the potential energy of the bob at any instant is a constant. Assuming no resistant forces, such as friction or air resistance, the bob would swing uniformly "forever."

A pile driver also shows this energy conservation. When the driver is at its highest position, the potential energy is maximum and the kinetic energy is zero. Its potential energy is

$$PE = mgh$$

and its kinetic energy is

$$KE = \frac{1}{2}mv^2 = \frac{1}{2}m(0)^2 = 0$$

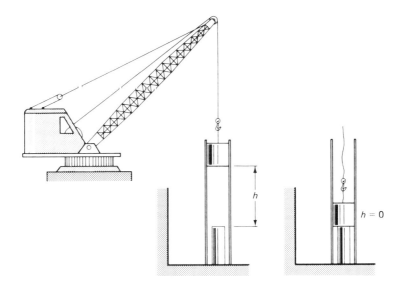

When the driver hits the top of the pile, it has its maximum kinetic energy and the potential energy is

$$PE = mgh = mg(0) = 0$$

Since the total energy in the system must remain constant, the maximum potential energy must equal the maximum kinetic energy.

$$PE_{max} = KE_{max}$$

$$mgh = \frac{1}{2}mv^2$$

Solving for the velocity of the driver just before it hits the pile, we get

$$v = \sqrt{2gh}$$

example   A pile driver falls freely from a height of 5.50 m above a pile. What is its velocity just before it hits the pile?

DATA:	$h = 5.50$ m
	$g = 9.80$ m/s^2
	$v = ?$
BASIC EQUATION:	$v = \sqrt{2gh}$
WORKING EQUATION:	same
SUBSTITUTION:	$v = \sqrt{2(9.80 \text{ m/s}^2)(5.50 \text{ m})}$
	$= 10.4$ m/s

EXERCISES

Sketch	Data	Basic Equation	Working Equation	Substitution
$h = ?$ $b = 6.0$ cm	$A = 12$ cm^2 $b = 6.0$ cm $h = ?$	$A = bh$	$h = \dfrac{A}{b}$	$h = \dfrac{12 \text{ cm}^2}{6.0 \text{ cm}}$ $= 2.0$ cm

1. How much work is required for a mechanical hoist to lift a 16,$\bar{0}$00-N automobile uniformly to a height of 2.00 m for repairs?
2. A hay wagon is used to move bales from the field to the barn. The tractor pulling the wagon exerts a constant force of 1500 N. The distance from field to barn is 0.50 km. How much work is done in moving one load of hay to the barn?
3. A worker lifts 75 concrete blocks at constant speed a distance of 1.50 m to the bed of a truck. Each block has a mass of 4.00 kg. How much work must he or she do to lift all the blocks to the truck bed?

4. The work required to lift eleven 5̄0-kg bags of cement at constant speed from the ground to the back of a truck is 8600 J. What is the distance from the ground to the bed of the truck?

5. A worker carries bricks at constant speed to a mason 2̄0 m away. If the amount of work required to carry one brick is 6.0 J, what force must the worker exert on each brick?

6. A gardener pushes a mower a distance of 90̄0 m in mowing a yard. The handle of the mower makes an angle of 4̄0° with the ground. The gardener exerts a force of 35.0 N along the handle of the mower. How much work does he do in mowing the lawn at constant speed?

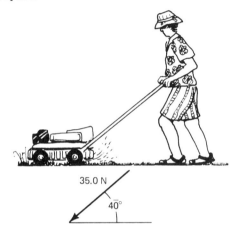

7. The handle of a vegetable wagon makes an angle of 25° with the horizontal. If the peddler exerts a force of 16̄0 N along the handle, how much work does he do in pulling the cart 1.00 km at constant speed?

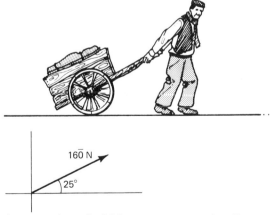

8. A 175-kg wooden crate is pushed 25 m across a wooden floor at constant speed. The coefficient of friction is 0.40.

   (a) What force is needed?

   (b) How much work is done?

9. A 175-kg wooden crate is pulled 25 m at constant speed by a rope that makes an angle of $4\bar{0}°$ with the horizontal. The coefficient of friction is 0.40.

   (a) What force is needed?
   (b) How much work is done?

10. A 175-kg wooden crate is pushed uniformly *up* a 25-m ramp that makes an angle of $2\bar{0}°$ with the horizontal. The coefficient of friction is 0.40.

   (a) What force is needed?
   (b) How much work is done?

11. A 175-kg wooden crate is pushed uniformly *down* a 25-m ramp that makes an angle of $2\bar{0}°$ with the horizontal. The coefficient of friction is 0.40.

   (a) What force is needed?
   (b) How much work is done?

12. A 175-kg wooden crate is pulled uniformly *up* a 25-m ramp that makes an angle of $2\bar{0}°$ with the horizontal by a rope that makes an angle of $4\bar{0}°$ with the ramp. The coefficient of friction is 0.40.

   (a) What force is needed?
   (b) How much work is done?

13. A 175-kg wooden crate is pulled uniformly *down* a 25-m ramp that makes an angle of $2\bar{0}°$ with the horizontal by a rope that makes an angle of $4\bar{0}°$ with the ramp. The coefficient of friction is 0.40.

   (a) What force is needed?
   (b) How much work is done?

14. A freight elevator with its operator weighs 2250 N. If it is raised at constant speed to a height of 20.0 m in 10.0 s, how many kilowatts of power are developed?

15. The mass of a large steel wrecking ball is $20\bar{0}0$ kg. What power (in kW) is used to raise it uniformly to a height of 40.0 m if the work is done in 20.0 s?

16. The work required to lift a crate is 36,500 J. If it can be lifted at constant speed in 25.0 s, what power is developed?

17. The power expended in lifting a $25\bar{0}$-kg girder uniformly to the top of a building 50.0 m high is 10.0 kW. What time is required to raise the girder?

18. A 1600-kg automobile is pushed at constant speed by its unhappy driver 0.50 km when it runs out of gas. To keep the car rolling he or she must exert a constant force of 750 N.

   (a) How much work does he or she do?
   (b) If it takes him or her 15 min, how much power does he or she exert?

19. An electric golf cart develops 1.25 kW of power while moving at a constant speed. If the cart travels $20\bar{0}$ m in 35.0 s, what force is exerted by the cart?

20. How many seconds would it take a 5.0-kW motor to raise a 215-kg boiler uniformly to a factory platform that is 12 m high?

21. How long would it take a $95\bar{0}$-W motor to raise a $36\bar{0}$-kg mass to a height of 16.0 m at constant speed?

22. A 1650-kg casting is raised 8.5 m at constant speed in 1.5 min. Find the required power in kW.

23. An escalator is needed to carry 125 passengers per minute at constant speed a vertical distance of 12 m. Assume that the mass of the average passenger is $7\bar{0}$ kg.

(a) What is the power (in kW) of the motor needed?

(b) What is the power (in kW) of the motor needed if $4\bar{0}\%$ of the power is lost to friction and heat loss?

24. A pump is needed to lift 1500 L of oil per minute at constant speed a distance of 25 m.

   (a) What power, in kW, must the pump be able to deliver, assuming no power loss?

   (b) What power, in kW, is needed if 45% of the power is lost to friction and heat loss? (1 L of oil has a mass of 0.68 kg.)

25. A truck is going along a highway with a uniform velocity of 90 km/h. The mass of the truck is 13,500 kg. What is the kinetic energy of the truck?

26. A bullet travels at 415 m/s. If it has a mass of 12.0 g, what is its kinetic energy?

27. An electron of mass 9.1×10^{-31} kg is moving at a uniform velocity of 1.0×10^7 m/s. What is its kinetic energy?

28. A 25.0-g bullet is fired from a gun and possesses 1530 J of kinetic energy. What is its velocity?

29. A crate of mass 475 kg is raised uniformly to a height of 17.0 m from the floor. What potential energy has it acquired with respect to the floor?

30. The potential energy possessed by a girder after being lifted uniformly to the top of a new building is 1.6×10^5 J. If the mass of the girder is 355 kg, how high is the girder?

31. Water is pumped uniformly at the rate of $30\bar{0}$ m^3/min from a lake into a tank that is 50.0 m above the lake.

   (a) What power is delivered by the pump?

   (b) What is the increase in potential energy of the water? (1.00 L of water has a mass of 1.00 kg.)

32. A 2.00-kg projectile is fired vertically with an initial velocity of 98.0 m/s. Find the kinetic energy, the potential energy, and the sum of kinetic and potential energies at

   (a) the instant of firing;

   (b) $t = 1.00$ s;

   (c) $t = 2.00$ s;

   (d) $t = 5.00$ s;

   (e) $t = 10.0$ s;

   (f) $t = 12.0$ s;

   (g) $t = 15.0$ s;

   (h) $t = 20.0$ s.

33. A pile driver falls freely through a distance of 4.00 m before hitting a pile. What is the velocity of the driver just before it hits the pile?

34. A sky diver jumps out of a plane at a height of 1500 m. If the parachute does not open until he or she reaches $30\bar{0}$ m, what is the diver's velocity if air resistance can be neglected?

35. A piece of shattered glass falls from the eighty-second floor of a skyscraper, $27\bar{0}$ m above the ground. What is the velocity of the glass when it hits the ground, if air resistance can be neglected?

MOMENTUM

8.1 MOMENTUM

We all know that if two automobiles are moving with the same speed and one is heavier than the other, the heavier auto would cause more damage in a head-on collision. The lighter auto can cause as much or more damage if its velocity is greater than that of the heavier auto.

A bullet fired from a gun penetrates a target more than a bullet of the same mass that is thrown by hand.

Momentum is a concept and a physical quantity that is related to Newton's laws of motion. The *momentum* of an object is defined as the product of its mass and its velocity. That is,

$$p = mv$$

where p is the momentum,

m is the mass, and
v is the velocity.

A dictionary definition of momentum is "a property of a moving body that determines the length of time required to bring it to rest when under the action of a constant force." Momentum is a vector quantity because it is the product of a scalar (mass) and a vector (velocity). It is in the same direction as the velocity.

example 1 Find the momentum of an auto that has a mass of 1450 kg and a velocity of 24.0 m/s.

DATA:

$m = 1450$ kg

$v = 24.0$ m/s

$p = ?$

BASIC EQUATION: $p = mv$

WORKING EQUATION: same

SUBSTITUTION: $p = (1450 \text{ kg})(24.0 \text{ m/s})$

$= 34{,}800$ kg m/s

(*Note*: There is no special unit for momentum.)

example 2 Find the velocity of a bullet of mass 1.00×10^{-2} kg if it is to have the same momentum as a bullet of mass 1.80×10^{-3} kg and a velocity of $30\bar{0}$ m/s.

DATA:

heavier bullet	*lighter bullet*
$m_1 = 1.00 \times 10^{-2}$ kg	$m_2 = 1.80 \times 10^{-3}$ kg
$v_1 = ?$	$v_2 = 30\bar{0}$ m/s

BASIC EQUATIONS:

$p_1 = m_1 v_1$

$p_2 = m_2 v_2$

we want:

$p_1 = p_2$

or

$m_1 v_1 = m_2 v_2$

WORKING EQUATION: $v_1 = \dfrac{m_2}{m_1} v_2$

SUBSTITUTION: $v_1 = \dfrac{1.80 \times 10^{-3} \text{ kg}}{1.00 \times 10^{-2} \text{ kg}} \times 30\bar{0} \text{ m/s}$

$= 54.0$ m/s

The *impulse* of an object is the product of the average force and the time interval during which the force acts on the object. That is,

$$\boxed{\text{impulse} = F(\Delta t)}$$

where F is the average force in newtons and
     Δt is the time.
     How are impulse and momentum related? Let us start with Newton's second law of motion.

$$F = ma$$

Since

$$v = a(\Delta t)$$

then

$$\frac{v}{\Delta t} = a$$

and substituting, we have

$$F = \frac{mv}{\Delta t}$$

Multiplying both sides by Δt, we have

$$F(\Delta t) = mv$$

or

$$F(\Delta t) = m(\Delta v)$$

That is,

$$\text{impulse} = \text{change in momentum}$$

A common example which illustrates the relationship between impulse and change in momentum is a golf club hitting a golf ball. During the time that the club and ball are in contact, the force of the swinging club is transmitting most of its

momentum to the ball. This impulse transmitted to the ball is directly proportional to the force with which the ball was hit and the length of time that the club and ball were in contact. After the ball leaves the club, the ball acquires a momentum equal to its mass times its velocity.

example 3   A 60.0-g bullet is fired with a muzzle velocity of $55\bar{0}$ m/s from a gun whose mass is 6.50 kg and whose barrel is 84.7 cm long.

(a) What is the force on the bullet as it leaves the barrel?
(b) What is the impulse on the bullet while it is in the barrel?
(c) What is the momentum of the bullet as it leaves the barrel?

(a) We first need to find how long the bullet is in the barrel.

DATA:

Δs = length of barrel = 84.7 cm = 0.847 m

$\bar{v}$ = average velocity of the bullet while in the barrel

$$= \frac{55\bar{0} \text{ m/s} + 0 \text{ m/s}}{2} = 275 \text{ m/s}$$

$\Delta t = ?$

BASIC EQUATION:

$\Delta s = \bar{v}(\Delta t)$

WORKING EQUATION:

$\Delta t = \dfrac{\Delta s}{\bar{v}}$

SUBSTITUTION:

$\Delta t = \dfrac{0.847 \text{ m}}{275 \text{ m/s}} = 0.00308 \text{ s}$

Now, to find the force:

DATA:

$\Delta t = 0.00308 \text{ s}$

$m = 60.0 \text{ g} = 0.0600 \text{ kg}$

$\Delta v = 55\bar{0} \text{ m/s} - 0 \text{ m/s} = 55\bar{0} \text{ m/s}$

$F = ?$

BASIC EQUATION:

$F(\Delta t) = m(\Delta v)$

WORKING EQUATION:

$F = \dfrac{m(\Delta v)}{\Delta t}$

SUBSTITUTION:

$F = \dfrac{(0.0600 \text{ kg})(55\bar{0} \text{ m/s})}{0.00308 \text{ s}}$

$= 10,700 \text{ kg m/s}^2$

$= 10,700 \text{ N}$

(b) DATA:

$\Delta t = 0.00308 \text{ s}$

$F = 10,700 \text{ N}$

impulse = ?

BASIC EQUATION:	impulse $= F(\Delta t)$
WORKING EQUATION:	same
SUBSTITUTION:	impulse $= (10{,}700 \text{ N})(0.00308 \text{ s})$

$$= 33.0 \text{ N s}$$
$$= 33.0 \text{ (kg m/s}^2\text{)s} \qquad (1 \text{ N} = 1 \text{ kg m/s}^2)$$
$$= 33.0 \text{ kg m/s}$$

(c) DATA:	$v = 55\bar{0} \text{ m/s}$
	$m = 60.0 \text{ g} = 0.0600 \text{ kg}$
	$p = ?$
BASIC EQUATION:	$p = mv$
WORKING EQUATION:	same
SUBSTITUTION:	$p = (0.0600 \text{ kg})(55\bar{0} \text{ m/s})$

$$= 33.0 \text{ kg m/s}$$

(*Note*: The impulse equals the change in momentum.)

8.2   LAW OF CONSERVATION OF MOMENTUM

One of the most important principles or laws of physics involves the momentum of objects in motion. This law has many practical applications.

LAW OF CONSERVATION OF MOMENTUM

When no outside forces are acting on a system of moving objects, the vector sum of the momentums of the system remains constant.

Let us look at some examples. Consider a 35-kg boy and a 75-kg man standing next to each other on ice skates on "frictionless ice." The man pushes on

$V_b = 0.40 \text{ m/s} \qquad V_m = 0.19 \text{ m/s}$

the boy, which gives the boy a velocity of 0.40 m/s. What happens to the man? Initially, the total momentum was zero because the initial velocity of each was zero. Because of the law of conservation of momentum, the total momentum must still be zero. That is,

$$m_{boy}v_{boy} + m_{man}v_{man} = 0$$
$$(35 \text{ kg})(0.40 \text{ m/s}) + (75 \text{ kg})v_{man} = 0$$
$$v_{man} = -0.19 \text{ m/s}$$

(*Note*: The minus sign indicates that the man's velocity and the boy's velocity are in opposite directions.)

Rocket propulsion is another illustration of conservation of momentum. Like the skaters, the total momentum of a rocket on the launch pad is zero. When the rocket engines are fired, hot exhaust gases (actually gas particles) are expelled downward through the rocket nozzle at tremendous speeds. As the rocket takes off, the sum of the total momentums of the rocket and the gas particles must remain zero. The total momentum of the gas particles is the sum of the products of each mass and its corresponding velocity and is directed down. The momentum of the rocket is the product of its mass and its velocity and is directed up.

When the rocket is in space, its propulsion works in the same manner. The conservation of momentum is still valid except that when the rocket engines are fired, the total momentum is a nonzero constant. This is because the rocket has velocity.

Actually, repair in space is more difficult than on earth because of the conservation of momentum and the "weightlessness" of objects in orbit. On earth, when a hammer is swung, the person is coupled to the earth by frictional forces so that the magnitude of the momentum of the person and earth *combined* equals the magnitude of the momentum of the hammer. (The hammer moves in one direction while the earth and person move immeasurably in the opposite direction.) In space orbit, because the person is weightless, there is no friction to couple him or her to the spaceship, so that the magnitude of the person's momentum *alone* must equal the magnitude of the momentum of the hammer. (The hammer moves in one direction and the person moves in the opposite direction.) The person in space has roughly the same problem driving a nail as a person on earth would have wearing a pair of "frictionless" roller skates.

8.3 COLLISIONS

When two objects collide, the resulting motion can be predicted in terms of the law of conservation of momentum. The types of collisions vary on a scale that ranges from perfectly elastic to perfectly inelastic. In a perfectly elastic collision, all the kinetic energy is conserved. That is, the total kinetic energy of the bodies before impact equals the total kinetic energy after impact. No kinetic energy is lost due to the collision. In a perfectly inelastic collision, the bodies stick together after impact and the loss in kinetic energy is greatest, although the momentum is conserved. In general, there is still some kinetic energy after impact in perfectly inelastic collisions.

Collision Continuum

Perfectly elastic collision					Perfectly inelastic collision

All KE is conserved.   Some KE is lost.   KE loss is greatest.

KE (before impact) =   KE (before impact) >   Bodies stick together after impact.
KE (after impact)   KE (after impact)

Let us consider collisions in one dimension. Assume that we have two carts of equal mass running toward each other at the same speed (2.00 m/s) on a "frictionless" track. The ends of the carts are made of a very hard or "perfectly elastic" rubber. In such a perfectly elastic collision, no kinetic energy is lost due to the collision. By the law of conservation of momentum, the total momentum before the collision equals the total momentum after the collision. Thus, after the collision, the carts are traveling at the same speed as before the collision but in opposite directions.

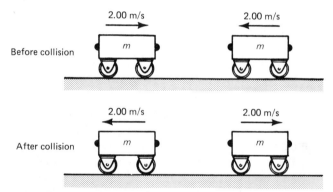

example 1 Cart *A* of mass 5.00 kg is traveling at 6.00 m/s on a "frictionless" track in the same direction (to the right) as cart *B* of mass 3.00 kg traveling at 2.00 m/s. After the collision, cart *A* is traveling at 4.00 m/s to the right. What is the velocity of cart *B*?

SKETCH:

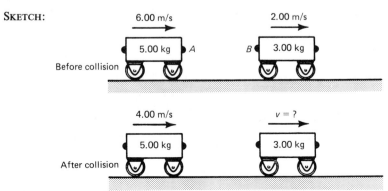

DATA:

$$m_A = 5.00 \text{ kg} \qquad\qquad m_B = 3.00 \text{ kg}$$

$$v_A = 6.00 \text{ m/s} \qquad\qquad v_B = 2.00 \text{ m/s}$$

$$v'_A = 4.00 \text{ m/s} \qquad\qquad v'_B = ?$$

By the law of conservation of momentum:

total momentum before impact = total momentum after impact

BASIC EQUATION:
$$m_A v_A + m_B v_B = m_A v'_A + m_B v'_B$$

WORKING EQUATION:
$$v'_B = \frac{m_A v_A + m_B v_B - m_A v'_A}{m_B}$$

SUBSTITUTION:

$$v'_B = \frac{(5.00 \text{ kg})(6.00 \text{ m/s}) + (3.00 \text{ kg})(2.00 \text{ m/s}) - (5.00 \text{ kg})(4.00 \text{ m/s})}{3.00 \text{ kg}}$$

$$= 5.33 \text{ m/s to the right}$$

example 2 Cart A of mass 2.00 kg is traveling on a "frictionless" track at 8.00 m/s to the right. Cart B of mass 6.00 kg is traveling at 4.00 m/s to the left. If the collision is perfectly elastic, what is the velocity of each cart after they collide?

SKETCH:

DATA:

$$m_A = 2.00 \text{ kg} \qquad\qquad m_B = 6.00 \text{ kg}$$

$$v_A = 8.00 \text{ m/s} \qquad\qquad v_B = -4.00 \text{ m/s*}$$

$$v'_A = ? \qquad\qquad\qquad v'_B = ?$$

By the law of conservation of momentum:

total momentum before impact = total momentum after impact

BASIC EQUATION:
$$m_A v_A + m_B v_B = m_A v'_A + m_B v'_B$$

$$(2.00 \text{ kg})(8.00 \text{ m/s}) + (6.00 \text{ kg})(-4.00 \text{ m/s}) = (2.00 \text{ kg})v'_A + (6.00 \text{ kg})v'_B$$

After simplifying and dropping significant zero digits and units, we have the equation

*We have designated direction to the right as positive and direction to the left as negative.

WORKING EQUATION: $\qquad\qquad -4 = v'_A + 3v'_B$ (1)

Since the equation was perfectly elastic:

total kinetic energy before impact = total kinetic energy after impact

BASIC EQUATION: $\qquad\qquad \frac{1}{2}m_A v_A^2 + \frac{1}{2}m_B v_B^2 = \frac{1}{2}m_A v_A'^2 + \frac{1}{2}m_B v_B'^2$

$$\frac{1}{2}(2.00\text{ kg})(8.00\text{ m/s})^2 + \frac{1}{2}(6.00\text{ kg})(-4.00\text{ m/s})^2 = \frac{1}{2}(2.00\text{ kg})v_A'^2 + \frac{1}{2}(6.00\text{ kg})v_B'^2$$

After simplifying and dropping significant zero digits and units, we have the equation

WORKING EQUATION: $\qquad\qquad 112 = v_A'^2 + 3v_B'^2$ (2)

Solve the system of equations (1) and (2) by substituting v'_A from equation (1) into equation (2):

$$112 = (-4 - 3v'_B)^2 + 3v_B'^2$$
$$112 = 16 + 24v'_B + 9v_B'^2 + 3v_B'^2$$
$$0 = 12v_B'^2 + 24v'_B - 96$$
$$0 = v_B'^2 + 2v'_B - 8$$
$$0 = (v'_B - 2)(v'_B + 4)$$
$$v'_B = 2 \qquad \text{or} \qquad v'_B = -4$$

Substituting $v'_B = 2$ into $-4 = v'_A + 3v'_B$, we have

$$-4 = v'_A + 3(2)$$
$$-10 = v'_A$$

That is, after impact the cart on the right is traveling at 2.00 m/s to the right while the cart on the left is traveling at 10.0 m/s to the left.
Substituting $v'_B = -4$ into $-4 = v'_A + 3v'_B$, we have

$$-4 = v'_A + 3(-4)$$
$$8 = v'_A$$

This result is trivial because it represents the original data.

In an inelastic collision, the total energy of the system is conserved while the total kinetic energy is not. Here most of the kinetic energy is lost in the form of heat energy. However, as in any collision, the momentum of the system is conserved.

example 3   A cart of mass 2.00 kg is traveling on a "frictionless" track at 5.00 m/s to the right. A 0.250-kg lump of soft clay, which is traveling at 25.0 m/s to the left, hits and sticks to the front of the cart. The clay and cart then travel together. What is the velocity of the cart and clay after impact?

SKETCH:

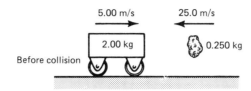

Before collision

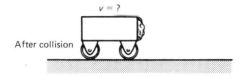

After collision

DATA:

$$m_1 = 2.00 \text{ kg}$$

$$v_1 = 5.00 \text{ m/s}$$

$$m_2 = 0.250 \text{ kg}$$

$$v_2 = -25.0 \text{ m/s}$$

$$v' = ?$$

By the law of conservation of momentum:

total momentum before impact = total momentum after impact

BASIC EQUATION: $$m_1 v_1 + m_2 v_2 = (m_1 + m_2)v'$$

WORKING EQUATION: $$v' = \frac{m_1 v_1 + m_2 v_2}{m_1 + m_2}$$

SUBSTITUTION: $$v' = \frac{(2.00 \text{ kg})(5.00 \text{ m/s}) + (0.250 \text{ kg})(-25.0 \text{ m/s})}{2.00 \text{ kg} + 0.250 \text{ kg}}$$

$$= 1.67 \text{ m/s to the right}$$

example 4 Police, after investigating an accident between two cars involved in a head-on collision, reported that car A, 25$\overline{0}$0 kg, and car B, 20$\overline{0}$0 kg, had skid marks of 24.0 m and 14.0 m, respectively. Car A remained at the point of impact while car B recoiled 8.0 m from the point of impact. The coefficient of friction between the tires of the cars and the road is taken to be 0.62. In order to make a low limit for the estimate of the speeds, it was assumed that car B had completely come to rest before the collision.

(a) What was the speed of car B after impact?
(b) What was the speed of car A at impact?
(c) What was the speed of car B at the beginning of its skid mark?
(d) What was the speed of car A at the beginning of its skid mark?

SKETCH:

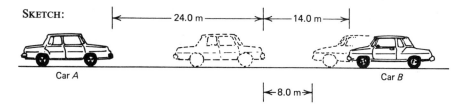

Car A   Car B

Note: We will let

$$v_0 = \text{velocity before brakes are applied,}$$

$$v = \text{velocity at instant of impact, and}$$

$$v' = \text{velocity immediately after impact}$$

(a) DATA:

$$m_A = 25\overline{0}0 \text{ kg}$$

$$m_B = 20\overline{0}0 \text{ kg}$$

$$d = 8.0 \text{ m}$$

$$\mu = 0.62$$

$$g = 9.80 \text{ m/s}^2$$

$$v_B' = ?$$

Car B immediately after impact had a kinetic energy of $\frac{1}{2}m_B v_B'^2$, which was dissipated by friction between the tires and the road.

BASIC EQUATION:

$$\tfrac{1}{2}m_B v_B'^2 = \mu m_B g d$$

WORKING EQUATION:

$$v_B' = \sqrt{2\mu g d}$$

SUBSTITUTION:

$$v_B' = \sqrt{2(0.62)(9.80 \text{ m/s}^2)(8.0 \text{ m})}$$

$$= 9.9 \text{ m/s}$$

(b) DATA:

$$m_A = 25\overline{0}0 \text{ kg}$$

$$m_B = 20\overline{0}0 \text{ kg}$$

$$v_B = 0$$

$$v_B' = 9.9 \text{ m/s} \qquad \text{[from part (a)]}$$

$$v_A' = 0$$

$$v_A = ?$$

The momentum of the system, car A and car B, must be conserved.

momentum before impact = momentum after impact

BASIC EQUATION:

$$m_A v_A + m_B v_B = m_A v_A' + m_B v_B'$$

WORKING EQUATION:
$$v_A = \frac{m_B}{m_A} v_B'$$

SUBSTITUTION:
$$v_A = \frac{20\bar{0}0 \text{ kg}}{25\bar{0}0 \text{ kg}} (9.9 \text{ m/s})$$

$$= 7.9 \text{ m/s}$$

(c) DATA:
$$m_B = 20\bar{0}0 \text{ kg}$$

$$d_B = 14.0 \text{ m}$$

$$v_B = 0$$

$$v_{B_0} = ?$$

The change in the kinetic energy of car B is equal to the work done on it by the road.

BASIC EQUATION:
$$0 - \tfrac{1}{2} m_B v_{B_0}^2 = -\mu m_B g d_B$$

WORKING EQUATION:
$$v_{B_0} = \sqrt{2\mu g d_B}$$

SUBSTITUTION:
$$v_{B_0} = \sqrt{2(0.62)(9.80 \text{ m/s}^2)(14.0 \text{ m})}$$

$$= 13 \text{ m/s} \quad \text{(about 47 km/h)}$$

(d) DATA:
$$m_A = 25\bar{0}0 \text{ kg}$$

$$d_A = 24.0 \text{ m}$$

$$v_A = 7.9 \text{ m/s} \quad \text{[from part (b)]}$$

$$v_{A_0} = ?$$

The change in the kinetic energy of car A is equal to the work done on it by the road.

BASIC EQUATION: $\tfrac{1}{2} m_A v_A^2 - \tfrac{1}{2} m_A v_{A_0}^2 = -\mu m_A g d_A$

WORKING EQUATION:
$$v_{A_0} = \sqrt{v_A^2 + 2\mu g d_A}$$

SUBSTITUTION:
$$v_{A_0} = \sqrt{(7.9 \text{ m/s})^2 + 2(0.62)(9.80 \text{ m/s}^2)(24.0 \text{ m})}$$

$$= 19 \text{ m/s} \quad \text{(about 68 km/h)}$$

The estimate that the respective speeds of cars A and B were at least 68 km/h and 47 km/h is based on the assumption that car B had stopped just before impact.

example 5 An alternate assumption in Example 4 might be that car B was traveling at the legal residential speed limit of 55 km/h when cars A and B first saw each other. Based on this assumption:

(a) What was the speed of car B at impact?
(b) What was the speed of car A at impact?
(c) What was the speed of car A at the beginning of its skid mark?

(a) DATA: $v_{B_0} = 55$ km/h or 15 m/s

$m_B = 20\bar{0}0$ kg

$\mu = 0.62$

$g = 9.80$ m/s^2

$v_B = ?$

The change in the kinetic energy of car B is equal to the work done on it by the road.

BASIC EQUATION: $\frac{1}{2}m_B v_B^2 - \frac{1}{2}m_B v_{B_0}^2 = -\mu m_B g d_B$

WORKING EQUATION: $v_B = \sqrt{v_{B_0}^2 - 2\mu g d_B}$

SUBSTITUTION: $v_B = \sqrt{(15 \text{ m/s})^2 - 2(0.62)(9.80 \text{ m/s}^2)(14.0 \text{ m})}$

$= 7.4$ m/s

(b) DATA: $m_B = 20\bar{0}0$ kg

$v_B = 7.4$ m/s   [from part (a)]

$v_B' = 9.9$ m/s   [from part (a) Example 4]

$m_A = 25\bar{0}0$ kg

$v_A' = 0$

$v_A = ?$

The law of the conservation of momentum requires that the total momentum of the system, cars A and B, be the same after as before the collision.

BASIC EQUATION: $m_A v_A + m_B v_B = m_A v_A' + m_B v_B'$

(*Note*: Momentum is a vector quantity and cars A and B were moving in opposite directions before impact.)

WORKING EQUATION: $v_A = \dfrac{m_B}{m_A}(v_B' + v_B)$

SUBSTITUTION: $v_A = \dfrac{20\bar{0}0 \text{ kg}}{25\bar{0}0 \text{ kg}}(9.9 \text{ m/s} + 7.4 \text{ m/s})$

$= 14$ m/s or $5\bar{0}$ km/h

(c) DATA: $v_A = 14$ m/s   [from part (b)]

$m_A = 25\bar{0}0$ kg

$\mu = 0.62$

$g = 9.80$ m/s^2

$d_A = 24.0$ m

$v_{A_0} = ?$

The change in the kinetic energy of car A from the beginning to the end of its skid is equal to the work done on it by the road.

BASIC EQUATION: $\frac{1}{2}m_A v_A^2 - \frac{1}{2}m_A v_{A_0}^2 = -\mu m_A g d_A$

WORKING EQUATION: $v_{A_0} = \sqrt{v_A^2 + 2\mu g d_A}$

SUBSTITUTION: $v_{A_0} = \sqrt{(14.0 \text{ m/s})^2 + 2(0.62)(9.80 \text{ m/s}^2)(24.0 \text{ m})}$

$= 22 \text{ m/s or } 79 \text{ km/h}$

The conclusion that car A was initially traveling at a speed of 79 km/h is based on the assumption that car B was initially traveling at 55 km/h.

These two calculations in Examples 4 and 5 place the speeds of car A between 68 and 79 km/h and car B between 47 and 55 km/h.

The law of conservation of momentum is also valid for collisions in two dimensions—when objects collide at angles other than in a straight line. A common example is billiard balls. Since momentum is a vector quantity, collisions in two dimensions may be treated and analyzed in terms of vector diagrams. However, because of the tedious mathematics involved, we shall not present a discussion of collisions in two (or three) dimensions here.

EXERCISES

Sketch	Data	Basic Equation	Working Equation	Substitution
	$A = 12 \text{ cm}^2$	$A = bh$	$h = \dfrac{A}{b}$	$h = \dfrac{12 \text{ cm}^2}{6.0 \text{ cm}}$
$b = 6.0 \text{ cm}$	$b = 6.0 \text{ cm}$			$= 2.0 \text{ cm}$
	$h = ?$			

1. (a) Find the momentum of an automobile of mass 2630 kg traveling at $25\overline{0}$ m/s.
   (b) Find the velocity of an automobile of mass 1170 kg if it is to have the same momentum as the automobile in part (a).

2. (a) Find the momentum of a bullet of mass $15\overline{0}$ g traveling at $25\overline{0}$ m/s.
   (b) Find the velocity of a bullet of mass $50\overline{0}$ g if it is to have the same momentum as the bullet in part (a).

3. A cannon is mounted on a railroad car. The cannon shoots a 1.50-kg ball with a muzzle velocity of 275 m/s. The cannon and the railroad car together have a mass of $40\overline{0}0$ kg. If the cannon is shot horizontally and is initially at rest, what is its recoil velocity?

4. A $15\overline{0}$-g bullet is fired at a muzzle velocity of $85\overline{0}$ m/s from a gun whose mass is 7.50 kg and whose barrel is 67.0 cm.

   (a) What is the force of the bullet coming out of the barrel?
   (b) What is the impulse on the bullet while in the barrel?

(c) What is the momentum of the bullet as it leaves the barrel?

(d) What is the average acceleration of the bullet while in the barrel?

5. A 14$\overline{0}$-kg pile driver falls the height of its boom, which is 12.5 m.

(a) What is its velocity as it hits the piling?

(b) With what momentum does it hit the piling?

(c) What is its kinetic energy as it hits the piling?

6. Cart *A* of mass 4.00 kg is traveling to the right on a "frictionless" track at 5.00 m/s in the same direction as cart *B* of mass 4.00 kg traveling at 2.00 m/s. After the collision, cart *A* continues in the same direction at 2.50 m/s. What is the velocity of cart *B*?

7. Cart *A* of mass 4.00 kg is traveling on a "frictionless" track at 5.00 m/s to the left. Cart *B* of mass 4.00 kg is traveling at 2.00 m/s to the right. After the collision, cart *A* is traveling at 1.50 m/s to the right. What is the velocity of cart *B*?

8. Cart *A* of mass 4.00 kg is traveling on a "frictionless" track at 5.00 m/s to the right and collides with cart *B* of mass 4.00 kg, which is initially at rest. After the collision, cart *A* is traveling at 2.00 m/s to the right. What is the velocity of cart *B*?

9. Cart *A* of mass 6.00 kg is traveling to the right on a "frictionless" track at 10.0 m/s in the same direction as cart *B* of mass 4.00 kg traveling at 3.00 m/s. After the collision, cart *A* continues in the same direction at 5.50 m/s. What is the velocity of cart *B*?

10. Cart *A* of mass 12.0 kg is traveling on a "frictionless" track at 3.00 m/s to the right. Cart *B* of mass 4.00 kg is traveling at 10.0 m/s to the left. After the collision, cart *A* is traveling at 2.00 m/s to the left. What is the velocity of cart *B*?

11. Cart *A* of mass 8.00 kg is traveling on a "frictionless" track at 6.00 m/s to the right and collides with cart *B* of mass 5.00 kg, which is initially at rest. After the collision, cart *A* is traveling at 3.00 m/s to the right. What is the velocity of cart *B*?

12. In Exercise 11, what is the velocity of cart *B* if cart *A* is traveling at 3.00 m/s to the left after the collision?

13. Cart *A* of mass 4.00 kg is traveling on a "frictionless" track at 10.0 m/s to the right. Cart *B* of mass 12.0 kg is traveling at 5.00 m/s to the left. If the collision is perfectly elastic, what is the velocity of each cart after they collide?

14. Cart *A* of mass 3.00 kg is traveling on a "frictionless" track at 6.00 m/s to the right. Cart *B* of mass 9.00 kg is traveling at 6.00 m/s to the left. If the collision is perfectly elastic, what is the velocity of each cart after they collide?

15. Cart *A* of mass 5.00 kg is traveling on a "frictionless" track at 8.00 m/s to the right and collides with cart *B* of mass 10.0 kg, which is initially at rest. If the collision is perfectly elastic, what is the velocity of each cart after they collide?

16. A cart of mass 4.00 kg is traveling on a "frictionless" track at 6.00 m/s to the right. A 0.200-kg lump of soft clay, which is traveling at 30.0 m/s to the left, hits and sticks to the front of the cart. The clay and cart then travel together. What is the velocity of the cart and clay after impact?

17. A cart of mass 4.00 kg is traveling on a "frictionless" track at 6.00 m/s to the right. A 0.200-kg lump of soft clay, which is traveling at 30.0 m/s to the right, hits and sticks to the back of the cart. The clay and cart then travel together. What is the velocity of the cart and clay after impact?

18. A freight car of mass 2.5×10^5 kg is traveling east at 2.0 m/s. It collides with another freight car of mass 1.5×10^5 kg, which is at rest. The cars link together and roll after impact on a horizontal track. What is the velocity of the linked-up freight cars?

19. A 1500-kg automobile is traveling east at $4\bar{0}$ km/h on an ice-covered road. It hits a $2\bar{0}00$-kg pickup truck traveling at $2\bar{0}$ km/h from the rear. After impact, the bumpers hook together. What is their velocity after impact?

20. In Exercise 19, what is their velocity after impact if the collision is head-on (the truck is headed west)?

21. Two cars of $18\bar{0}0$ kg and $15\bar{0}0$ kg collide head-on and stop at the point of impact locked together.

(a) If the $15\bar{0}0$ kg car had been traveling at the speed limit of $8\bar{0}$ km/h, what would have been the speed of the $18\bar{0}0$-kg car?

(b) If the $18\bar{0}0$-kg car had been traveling at the legal speed limit, what speed was the $15\bar{0}0$-kg car traveling?

22. Two $17\bar{0}0$-kg cars are involved in a head-on collision. Police measure skid marks of 25.0 m for car *A* and 30.0 m for car *B*. Car *A* stopped at the point of impact and car *B* recoiled 5.0 m. (Use the value of 0.70 as the coefficient of friction between the tires and the road.)

(a) Find the speed of car *B* immediately after impact.

(b) Assume that car *B* was traveling at the legal speed limit of 85 km/h at the beginning of the skid. Find the speed of car *B* just before impact.

(c) Find the speed of car *A* just before impact.

(d) Find the speed of car *A* at the beginning of its skid mark.

23. Two $17\bar{0}0$-kg cars are involved in a head-on collision. Police measure skid marks of 25.0 m for car *A* and 30.0 m for car *B*. Car *A* stopped at the point of impact and car *B* recoiled 5.0 m. (Use the value of 0.70 as the coefficient of friction between the tires and the road.)

(a) Find the speed of car *B* immediately after impact.

(b) Find the speed of car *A* just before impact, assuming that its speed at the beginning of its skid was the legal speed limit of 85 km/h.

(c) Find the speed of car *B* just before impact.

(d) Find the speed of car *B* at the beginning of its skid mark.

ROTATIONAL AND CIRCULAR MOTION

9.1 ANGLE MEASUREMENT

So far, we have studied rectilinear motion—motion in a straight line, and curvilinear motion—motion along a curved path. The third type of motion we need to study is rotational motion or rotary motion—motion of a rigid body when the body itself is spinning about an axis. Examples of rotational motion include the earth spinning or rotating on its axis, a wheel spinning, a turning driveshaft, and the turning shaft of an electric motor.

The first prerequisite of quantitatively studying rotational motion is defining a system of angle measurement. There are three basic systems: revolutions, degrees, and radians.

One unit of angle measurement in rotational motion is the number of rotations—the number of times the object completely turns around. This unit is called a *revolution* (rev). This is the unit used most often in industry.

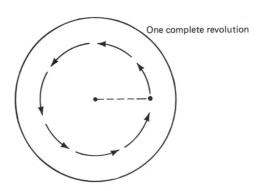

One complete revolution

Another system of angle measurement divides the circle of rotation in 360 equal parts called *degrees* (°); that is, 1 rev = 360°. This system allows a more precise measure of the angle.

Scientists and engineers use a third system to measure angles, which is based on the radian. A *radian* (rad) is defined as the measure of an angle with its vertex at the center of a circle whose intercepted arc is equal to the radius of the circle. In other words, when the length of arc *PQ* equals the length of the radius, the central angle θ equals 1 radian.

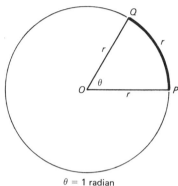

$\theta = 1$ radian

In general, any angle θ in radians is defined as the ratio of the length of arc *PQ*, *s*, to the length of the radius, *r*; that is, $\theta = s/r$. As a result, the radian unit has no physical dimensions, since it is the ratio of two lengths.

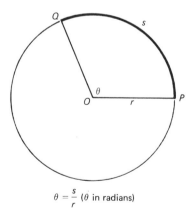

$\theta = \dfrac{s}{r}$ (θ in radians)

The circumference of any circle is given by the formula $C = 2\pi r$. The ratio of the circumference of a circle to its radius is $C/r = 2\pi r/r = 2\pi$ (6.28 approx.). That is, 2π rad is the measure of the central angle of any complete circle (one complete revolution).

The radian is also the standard SI metric unit of a plane angle. Many calculations require that either the angle be expressed in radians or converted to radians. Of course, some common everyday measurements will continue to be made in degrees. Probably, some trades will even continue to subdivide the degree into the traditional

$$1° = 60 \text{ minutes } (') \qquad \text{and} \qquad 1' = 60 \text{ seconds } ('')$$

But, most will find that tenths and hundredths of degrees are easier to use for calculations.

For conversion purposes, you should find it convenient to remember the following relationship:

$$\boxed{1 \text{ rev} = 360° = 2\pi \text{ rad}}$$

Since

$$360° = 2\pi \text{ rad}$$

we find that

$$1° = \left(\frac{2\pi}{360}\right) \text{rad} = \left(\frac{\pi}{180}\right) \text{rad} = 0.01745 \text{ rad (approx.)}$$

and

$$1 \text{ rad} = \left(\frac{360}{2\pi}\right)° = \left(\frac{180}{\pi}\right)° = 57.30° \text{ (approx.)}$$

On a rigid body in rotary motion about an axis, all points move or rotate through the same angle. We define angular displacement to be this change of angular position; that is, *angular displacement* is the angle through which any point on the body in rotary motion is rotated. As the record below makes one revolution (rotates 360°), each point rotates 360°. The angular displacement of each point is 360°.

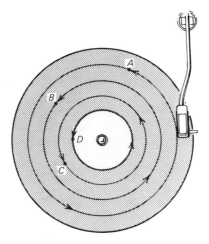

To find the total linear distance traveled by a point on a body that is rotated about an axis, we use a variation of the equation $\theta = s/r$, namely,

$$s = r\theta$$

where s is the arc length or linear distance traveled by the point,
   r is the radius or distance from the axis to the point, and
   θ is the angle of rotation (*in radians*).

example 1   A flywheel of radius 80.0 cm is rotating on its axis.

(a) Find the linear distance traveled by a point (A) on the rim in one complete revolution.
(b) Find the linear distance traveled by a point (B) 40.0 cm from the axis in one complete revolution.

SKETCH:

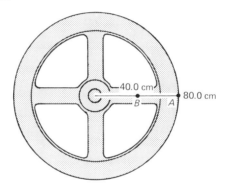

(a) DATA:

$r = 80.0$ cm

$\theta = 2\pi$ rad

$s = ?$

BASIC EQUATION: $s = r\theta$

WORKING EQUATION: same

SUBSTITUTION: $s = (80.0 \text{ cm})(2\pi \text{ rad})$

$= 503$ cm

Note: As we saw earlier, the radian unit has no physical dimension. Some refer to it as a "phantom unit" which "fades in and out" of calculations.

(b) DATA:

$r = 40.0$ cm

$\theta = 2\pi$ rad

$s = ?$

BASIC EQUATION: $s = r\theta$

WORKING EQUATION: same

SUBSTITUTION: $s = (40.0 \text{ cm})(2\pi \text{ rad})$

$\qquad\qquad\qquad\qquad\qquad\quad = 251 \text{ cm}$

example 2   Find the length of arc of a circle of radius 16.0 cm with a central angle of $24\bar{0}°$.

SKETCH:

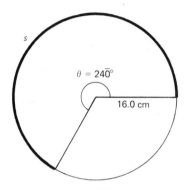

DATA: $r = 16.0 \text{ cm}$

$\qquad\qquad\qquad\qquad \theta = 24\bar{0}° \times \dfrac{\pi \text{ rad}}{180°} = \dfrac{4\pi}{3} \text{ rad}$

$\qquad\qquad\qquad\qquad s = ?$

BASIC EQUATION: $s = r\theta$

WORKING EQUATION: same

SUBSTITUTION: $s = (16.0 \text{ cm})\left(\dfrac{4\pi}{3} \text{ rad}\right)$

$\qquad\qquad\qquad\qquad\qquad\quad = 67.0 \text{ cm}$

9.2   ANGULAR VELOCITY

In uniform linear motion, we defined uniform or average velocity as the rate of change of displacement; that is,

$$\bar{v} = \frac{\Delta s}{\Delta t}$$

where $\bar{v}$ is the uniform or average velocity,
$\qquad \Delta s$ is the change in displacement, and
$\qquad \Delta t$ is the change in time or the time interval.
Similarly, for uniform rotational motion, we define angular velocity as the rate of

change of angular displacement; that is,

$$\omega = \frac{\Delta \theta}{\Delta t}$$

where ω is the uniform angular velocity,
  $\Delta \theta$ is the angular displacement, and
  Δt is the time interval.
Angular velocity can be measured in revolutions per second or in degrees per second, but it is more commonly measured in radians per second (rad/s) or revolutions per minute (rev/min or rpm).

To find the linear velocity of a point traveling on a body that is rotated about an axis, we substitute the linear velocity equation

$$\Delta s = v(\Delta t)$$

into the equation from Section 9.1:

$$\Delta \theta = \frac{\Delta s}{r}$$

This gives us

$$\Delta \theta = \frac{v(\Delta t)}{r}$$

which we substitute into the angular velocity equation

$$\omega = \frac{\Delta \theta}{\Delta t}$$

which gives us

$$\omega = \frac{\frac{v(\Delta t)}{r}}{\Delta t}$$

or

$$\omega = \frac{v}{r}$$

or

$$v = \omega r$$

where v is the linear velocity of the point,
  ω is the angular velocity of the point, in rad/time, and
  r is the radius or distance from the axis to the point.

There are some comparisons that can be made between rotational motion and circular motion—the motion of an object in a circular path or orbit. Let us consider a model airplane attached by a wire to a fixed post and traveling in a circular path at a constant speed. Actually, the magnitude of the plane's velocity remains constant while its direction changes constantly, but uniformly.

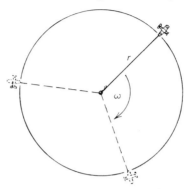

Assume that the plane is traveling at uniform angular velocity, ω. What happens if the wire breaks?

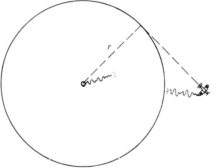

The plane travels in a straight line that is tangent to the circle (circular path) at the point where the plane was at the instant the break occurred. This tangential velocity corresponds to and is equal to the linear velocity, ωr. (This tangential velocity at any point, P, also corresponds to the instantaneous velocity at point P.)

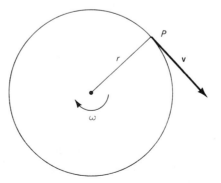

example 1   The flywheel of an engine rotates at $30\overline{0}0$ rpm. Find its angular displacement (in rad) during a 6.00-s interval.

DATA: $\Delta t = 6.00$ s

$$\omega = 3000 \frac{\text{rev}}{\text{min}} \times \frac{2\pi \text{ rad}}{\text{rev}} \times \frac{1 \text{ min}}{60 \text{ s}} = 314 \text{ rad/s}$$

$\Delta\theta = ?$

BASIC EQUATION: $\omega = \dfrac{\Delta\theta}{\Delta t}$

WORKING EQUATION: $\Delta\theta = \omega(\Delta t)$

SUBSTITUTION: $\Delta\theta = (314 \text{ rad/s})(6.00 \text{ s})$

$\qquad\qquad = 1880 \text{ rad}$

example 2   A flywheel of radius 80.0 cm is rotating at 555 rad/s.

(a) Find the linear velocity of a point on the rim.
(b) Find the linear velocity of a point 40.0 cm from the axis.

(a) DATA: $r = 80.0$ cm

$\qquad\qquad\qquad\qquad\qquad \omega = 555 \text{ rad/s}$

$\qquad\qquad\qquad\qquad\qquad v = ?$

BASIC EQUATION: $v = \omega r$

WORKING EQUATION: same

SUBSTITUTION: $v = (555 \text{ rad/s})(80.0 \text{ cm})$

$\qquad\qquad = 44{,}400 \text{ cm/s}$ or 444 m/s

(b) DATA: $r = 40.0$ cm

$\qquad\qquad\qquad\qquad\qquad \omega = 555 \text{ rad/s}$

$\qquad\qquad\qquad\qquad\qquad v = ?$

BASIC EQUATION: $v = \omega r$

WORKING EQUATION: same

SUBSTITUTION: $v = (555 \text{ rad/s})(40.0 \text{ cm})$

$\qquad\qquad = 22{,}200 \text{ cm/s or } 222 \text{ m/s}$

9.3   ANGULAR ACCELERATION

In linear motion, we defined uniform acceleration as the rate of change of velocity; that is,

$$\bar{a} = \frac{\Delta v}{\Delta t}$$

where $\bar{a}$ is the uniform or average acceleration,

Δv is the change in velocity, and

Δt is the time interval.

Again similarly, for rotational motion, we define uniform angular acceleration as the rate of change of angular velocity; that is;

$$\alpha = \frac{\Delta \omega}{\Delta t}$$

where α is the uniform angular acceleration,

$\Delta \omega$ is the change in angular velocity, and

Δt is the time interval.

We shall limit our unit of angular acceleration to rad/s^2.

To find the linear (tangential) acceleration of a point traveling on a body that is rotated about an axis, we substitute the linear velocity equation

$$\Delta \omega = \frac{\Delta v}{r}$$

into the angular acceleration equation

$$\alpha = \frac{\Delta \omega}{\Delta t}$$

This gives us

$$\alpha = \frac{\dfrac{\Delta v}{r}}{\Delta t} = \frac{\dfrac{\Delta v}{\Delta t}}{r} = \frac{a_T}{r}$$

or

$$a_T = \alpha r$$

where a_T is the linear or tangential acceleration,

α is the uniform angular acceleration, in rad/time^2, and

r is the radius or distance from the axis to the point.

Again, let us consider the model airplane traveling in a circular path at a constant speed. Since the plane's speed is constant and the direction is changing uniformly, the plane's velocity is changing uniformly. And, since the plane's velocity is changing, the plane is also accelerating ($a = \Delta v / \Delta t$).

But what is the magnitude and direction of this acceleration? Let us consider the change in velocity of the plane as it rotates counterclockwise at uniform angular velocity ω.

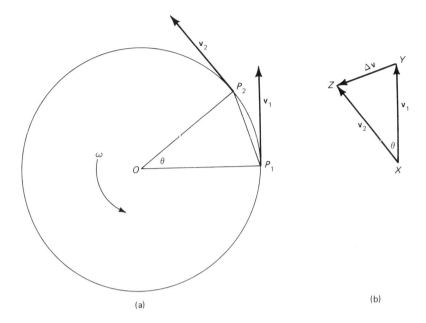

(a)   (b)

The vector diagram in figure (b) indicates the change in velocity, $\Delta \mathbf{v}$, from P_1 to P_2 in figure (a). That is,

$$\mathbf{v}_1 + \Delta \mathbf{v} = \mathbf{v}_2$$

or

$$\Delta \mathbf{v} = \mathbf{v}_2 - \mathbf{v}_1$$

Next, we need to show that angle O in (a) equals angle X in (b). To show this relationship, combine figures (a) and (b) as shown below by drawing $\mathbf{v}_2$ so that its initial point is at point P_1. (See figure below.) Note that $\mathbf{v}_1$ is perpendicular to

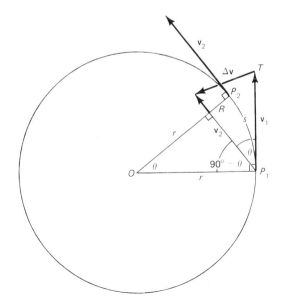

radius $\overline{OP_1}$, and $\mathbf{v}_2$ is perpendicular to radius $\overline{OP_2}$ because each vector is tangent to the circle. Thus, ΔORP_1 is a right triangle with the right angle at R. Then, angle O and angle OP_1R are complementary; or, angle $OP_1R = 90° - \theta$. Angle OP_1R and angle RP_1T are also complementary. Therefore, angle $RP_1T = \theta$ = angle O.

Now, let us refer again to figures (a) and (b). Since $\overline{OP_1} = \overline{OP_2}$ and $|\mathbf{v}_1| = |\mathbf{v}_2|$, ΔOP_1P_2 is similar to ΔXYZ. So the corresponding sides are proportional; that is,

$$\frac{|\Delta \mathbf{v}|}{P_1P_2} = \frac{|\mathbf{v}_1|}{OP_1}$$

For small values of θ, the length of arc P_1P_2 is approximately equal to the length of chord P_1P_2. Thus,

$$\frac{|\Delta \mathbf{v}|}{s} = \frac{|\mathbf{v}_1|}{r}$$

Let $v = |\mathbf{v}_1| = |\mathbf{v}_2|$. Since $s = vt$, we have

$$\frac{|\Delta \mathbf{v}|}{vt} = \frac{v}{r}$$

or

$$\frac{|\Delta \mathbf{v}|}{t} = \frac{v^2}{r}$$

But $|\Delta \mathbf{v}|/t$ is the acceleration. So

$$\boxed{a = \frac{v^2}{r}}$$

where a is the centripetal acceleration,
  v is the speed, and
  r is the radius of the circular path.
And since $v = \omega r$,

$$a = \frac{(\omega r)^2}{r}$$

or

$$\boxed{a = \omega^2 r}$$

where a is the centripetal acceleration,
  ω is the change in angular velocity, and
  r is the radius of the circular path.
  What is the direction of this centripetal acceleration? Again, let us consider the model airplane. Assuming that the speed of the plane remains constant, there are only two alternatives:

1. The wire breaks or is cut and the plane travels at the same constant speed in a straight line that is tangent to the circular path. In this case, since the speed is constant and the motion is in a straight line, the plane is not accelerating.

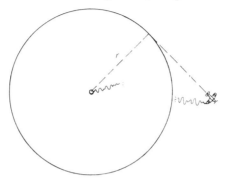

2. The wire is not severed and the plane continues in its circular path at the constant speed.

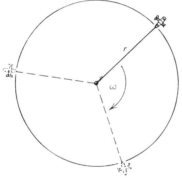

The direction of the velocity is constantly changing. This change in velocity must be directed toward the center of the circular path, because the plane is actually constantly falling toward this center. If not, the plane would not continue in its circular path.

As we know, acceleration is the change in velocity per change in time ($a = \Delta v / \Delta t$). Thus, the centripetal acceleration of the plane in a circular path must be in the same direction as its change in velocity—toward the center.

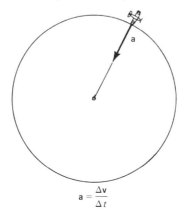

$$a = \frac{\Delta v}{\Delta t}$$

That is, centripetal acceleration causes an object to move and remain in a circular path. The centripetal acceleration is always directed toward the center of the circular path. By itself centripetal acceleration cannot cause the object to change speeds; it only changes the direction of the motion.

The equations for uniformly accelerated rotational motion basically follow from the equations for uniformly accelerated linear motion by substituting θ for s, ω for v, and α for a, as shown below.

Equations for uniformly accelerated motion

Linear motion	*Rotational motion*
$v_f = v_i + a(\Delta t)$	$\omega_f = \omega_i + \alpha(\Delta t)$
$\Delta s = v_i(\Delta t) + \frac{1}{2}a(\Delta t)^2$	$\Delta\theta = \omega_i(\Delta t) + \frac{1}{2}\alpha(\Delta t)^2$
$v_f^2 = v_i^2 + 2a(\Delta s)$	$\omega_f^2 = \omega_i^2 + 2\alpha(\Delta\theta)$

example 1   A flywheel accelerates from rest to 225 rad/s in 75.0 s. Find its angular displacement.

We first need to find the angular acceleration:

DATA:

$\Delta\omega = 225 \text{ rad/s} - 0 \text{ rad/s} = 225 \text{ rad/s}$

$\Delta t = 75.0 \text{ s}$

$\alpha = ?$

BASIC EQUATION:

$\alpha = \dfrac{\Delta\omega}{\Delta t}$

WORKING EQUATION: same

SUBSTITUTION:

$\alpha = \dfrac{225 \text{ rad/s}}{75.0 \text{ s}}$

$= 3.00 \text{ rad/s}^2$

Then,

DATA:

$\omega_i = 0$

$\Delta t = 75.0 \text{ s}$

$\alpha = 3.00 \text{ rad/s}^2$

$\Delta\theta = ?$

BASIC EQUATION:

$\Delta\theta = \omega_i(\Delta t) + \frac{1}{2}\alpha(\Delta t)^2$

WORKING EQUATION: same

SUBSTITUTION:

$\Delta\theta = 0(75.0 \text{ s}) + \frac{1}{2}(3.00 \text{ rad/s}^2)(75.0 \text{ s})^2$

$= 8440 \text{ rad}$ or 1340 rev

example 2   A wind-driven propeller for a power generator is turning at 30.0 rpm. A gust of wind causes the propeller to rotate at 36.0 rpm in 40.0 revolutions. Find its angular acceleration in rad/s^2.

DATA:

$$\omega_i = 30.0 \frac{\text{rev}}{\text{min}} \times \frac{2\pi \text{ rad}}{1 \text{ rev}} \times \frac{1 \text{ min}}{60 \text{ s}} = 3.14 \text{ rad}/\text{s}$$

$$\omega_f = 36.0 \frac{\text{rev}}{\text{min}} \times \frac{2\pi \text{ rad}}{1 \text{ rev}} \times \frac{1 \text{ min}}{60 \text{ s}} = 3.77 \text{ rad}/\text{s}$$

$$\Delta\theta = 40.0 \text{ rev} \times \frac{2\pi \text{ rad}}{1 \text{ rev}} = 251 \text{ rad}$$

$$\alpha = ?$$

BASIC EQUATION:

$$\omega_f^2 = \omega_i^2 + 2\alpha(\Delta\theta)$$

WORKING EQUATION:

$$\alpha = \frac{\omega_f^2 - \omega_i^2}{2(\Delta\theta)}$$

SUBSTITUTION:

$$\alpha = \frac{(3.77 \text{ rad}/\text{s})^2 - (3.14 \text{ rad}/\text{s})^2}{2(251 \text{ rad})}$$

$$= 8.67 \times 10^{-3} \text{ rad}/\text{s}^2$$

9.4   CENTRIPETAL FORCE

Newton's second law of motion states that

$$F = ma$$

By substituting the centripetal acceleration $a = v^2/r$ into Newton's second law, we have

$$\boxed{F = \frac{mv^2}{r}}$$

which is called the centripetal force of an object in circular motion. Since the force and acceleration are in the same direction, the centripetal force is also directed toward the center of the circular path. So then the centripetal force is the force required to produce the centripetal acceleration, which changes the direction of a moving object so that it stays moving in a circular path at a constant speed.

example 1   Find the centripetal force of a $15\overline{0}0$-kg automobile making a circular turn of radius 75.0 m at a speed of 45.0 km/h.

DATA:

$$m = 15\overline{0}0 \text{ kg}$$

$$v = 45.0 \frac{\text{km}}{\text{h}} \times \frac{1000 \text{ m}}{1 \text{ km}} \times \frac{1 \text{ h}}{3600 \text{ s}} = 12.5 \text{ m}/\text{s}$$

$$r = 75.0 \text{ m}$$

$$F = ?$$

BASIC EQUATION: $F = \dfrac{mv^2}{r}$

WORKING EQUATION: same

SUBSTITUTION: $F = \dfrac{(15\overline{0}0 \text{ kg})(12.5 \text{ m/s})^2}{75.0 \text{ m}}$

$= 3130 \text{ kg m/s}^2$

$= 3130 \text{ N}$

example 2   What minimal coefficient of friction must be needed to keep the automobile in Example 1 from skidding?

DATA: $F_f = F = 3130 \text{ N}$ (the frictional force corresponds to the centripetal force)

$F_N = F_w = mg = (15\overline{0}0 \text{ kg})(9.80 \text{ m/s}^2) = 14{,}700 \text{ N}$

$\mu = ?$

BASIC EQUATION: $\mu = \dfrac{F_f}{F_N}$

WORKING EQUATION: same

SUBSTITUTION: $\mu = \dfrac{3130 \text{ N}}{14{,}700 \text{ N}}$

$= 0.21$

The automobile can easily maneuver this turn.

As you are probably aware, many sharp turns on highways are banked to minimize the need for tire traction (frictional force) and minimize skids.

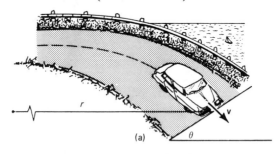

(a)

In figure (a), we have an automobile traveling at velocity v rounding a circular curve of radius r which is banked at an angle of θ. Figure (b) shows the relationships among the various forces. $F_w = mg$ is the weight of the automobile and acts vertically down. F_N is the normal force—the force of the automobile on the roadbed. **R** is the reaction force of the roadbed on the automobile. Forces $\mathbf{F}_N$

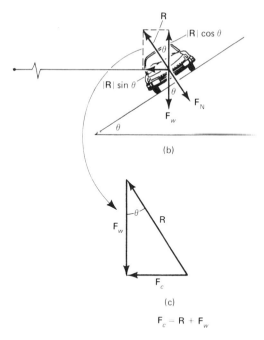

(b)

(c)

$$F_c = R + F_w$$

and **R** are equal and opposite. Because the roadbed is banked at an angle of θ, the vertical component of **R** is $|\mathbf{R}|\cos\theta$ and the horizontal component is $|\mathbf{R}|\sin\theta$. Since there is no vertical motion, the vertical components must be equal and opposite; that is,

$$|\mathbf{F}_w| = mg = |\mathbf{R}|\cos\theta \tag{1}$$

Since the car is in a circular path, the horizontal component of **R** must equal the centripetal force (assuming no friction); that is,

$$|\mathbf{F}_c| = \frac{mv^2}{r} = |\mathbf{R}|\sin\theta \tag{2}$$

Figure (c) shows vectorially that $\mathbf{F}_c = \mathbf{R} + \mathbf{F}_w$. Dividing equation (2) by equation (1), we have

$$\frac{\dfrac{mv^2}{r}}{mg} = \frac{|\mathbf{R}|\sin\theta}{|\mathbf{R}|\cos\theta}$$

or

$$\boxed{\dfrac{v^2}{rg} = \tan\theta}$$

where v is the linear speed of the automobile,
r is the radius of the circular curve,
g is the gravitational acceleration (9.80 m/s²), and
θ is the banking angle.

Note: The mass of the automobile does not affect the value of θ, whereas the velocity of the automobile does.

Civil engineers estimate and set the legal speed limit for a curve and then bank it at an angle of θ for the legal speed limit. They usually assume that no frictional forces are involved in determining θ. Thus, an extra safety factor is automatically included in their calculations. Theoretically, you should be able to maneuver the curve at the legal speed limit even though the curve is a "sheet of ice"!

example 3   A curve of radius 150 m in a roadway is planned. The expected legal speed limit is $7\bar{0}$ km/h. At what angle should the roadbed be banked (assuming no frictional forces)?

DATA:

$$r = 150 \text{ m}$$

$$v = 7\bar{0}\frac{\text{km}}{\text{h}} \times \frac{1000 \text{ m}}{1 \text{ km}} \times \frac{1 \text{ h}}{3600 \text{ s}} = 19 \text{ m/s}$$

$$g = 9.80 \text{ m/s}^2$$

$$\theta = ?$$

BASIC EQUATION:

$$\tan \theta = \frac{v^2}{rg}$$

WORKING EQUATION: same

SUBSTITUTION:

$$\tan \theta = \frac{(19 \text{ m/s})^2}{(150 \text{ m})(9.80 \text{ m/s}^2)}$$

$$= 0.246$$

$$\theta = 14°$$

How do we determine the minimum linear speed that a satellite must attain to stay in a stable circular orbit around the earth? (In this discussion, we will neglect air resistance. At great heights above the earth's surface, the air resistance is very small but enough to eventually slow down and thus bring down any satellite.)

The centripetal force of the satellite is

$$F = \frac{mv^2}{r}$$

where r is the distance from the center of the earth to the satellite.

The gravitational force of the satellite due to its weight is

$$F_w = mg$$

where *g* is the acceleration due to gravity (at the height of the satellite).
    Since these two forces are equal, we have

$$\frac{mv^2}{r} = mg$$

or

$$v = \sqrt{rg}$$

Note: The velocity is independent of the satellite's mass.

example 4 What speed is required for a satellite to be in a stable circular orbit
5̄00 km above the earth's surface? The radius of the earth is 6.4×10^6 m or 6400 km.
(g = 8.3 m/s² at this height.)

DATA: $r = 6400$ km $+ 5\bar{0}0$ km $= 6900$ km $= 6.9 \times 10^6$ m

                                     $g = 8.3$ m/s²

                                     $v = ?$

BASIC EQUATION: $v = \sqrt{rg}$

WORKING EQUATION: same

SUBSTITUTION: $v = \sqrt{(6.9 \times 10^6 \text{ m})(8.3 \text{ m/s}^2)}$

                                $= 7600$ m/s or $27{,}000$ km/h

Note: If this speed is increased, the orbit becomes elliptical. If this speed is increased
to approximately 40,000 km/h, the satellite leaves its earth orbit and goes into orbit
around the sun. This velocity, 40,000 km/h, is called the escape velocity.

9.5 TORQUE

What causes rotational motion of a flywheel that is initially at rest? A force applied
to the rim of the wheel would make it rotate in the same direction as the direction
of the applied force.

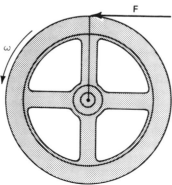

Obviously, the magnitude of the force has a major effect on the resulting angular velocity. But what is the effect of the *direction* of the applied force? If the force is applied directly toward the center of the wheel, no rotational motion results.

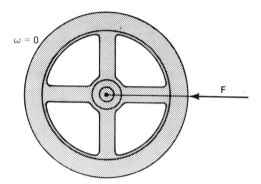

Experimentally, physicists have found that the effect on the rotational motion produced by a given force is directly proportional to the given force and the perpendicular distance between the center of rotation and the given force.

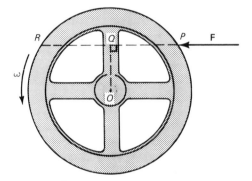

In the figure above, OQ is the perpendicular distance between the center of rotation and the line of force. This perpendicular distance, OQ, is also called the *moment arm* of the force.

The product of the applied force and the length of the moment arm is called *torque*; that is,

$$\tau = Fr$$

where τ is the torque (τ is the Greek letter "tau"),
    F is the applied force, and
    r is the length of the moment arm.
Recall the discussion above where the force was applied directly toward the center of the wheel. The length of the moment arm was zero. Thus, the torque was zero and there was no resulting motion.

Therefore, *torque is the tendency to cause rotational motion*.

A more common experience dealing with torque is that of pedaling a bicycle. If you push "straight down" on the pedal of a bicycle, nothing happens.

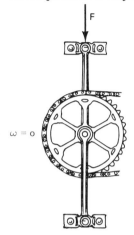

The greatest rotational motion results when the force is tangent to the circular motion of the pedals.

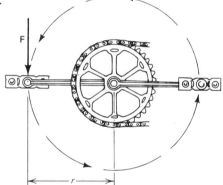

If the force is not exerted tangent to the circular motion of the pedals, the length of the torque arm is *not* the length of the pedal arm. As we know, *r* is measured as the perpendicular distance between the center of rotation and the force.

Since *r* is therefore shorter, the product *Fr* is smaller and the turning effect, the torque, is less.

example   A force of $4\overline{0}$ N is applied to a bicycle pedal as shown. If the length of the pedal arm is 17 cm, what torque is applied to the shaft?

SKETCH:

40 N

17 cm

DATA:

$$F = 4\overline{0} \text{ N}$$

$$r = 17 \text{ cm} = 0.17 \text{ m}$$

$$\tau = ?$$

BASIC EQUATION: $\tau = Fr$

WORKING EQUATION: same

SUBSTITUTION: $\tau = (4\overline{0} \text{ N})(0.17 \text{ m})$

$$= 6.8 \text{ N m}$$

Note that the unit of torque $(\tau = Fr)$ is the same as the unit of work $(W = Fs)$. But remember the difference between the distance, *s*, for work and the length of the moment arm, *r*, for torque.

The torque wrench is a common use of torque in assembly or quality control operations.

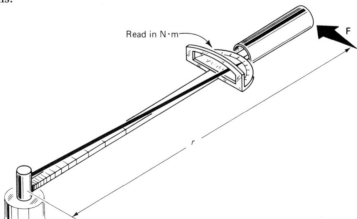

Read in N·m

F

r

Some torque wrenches have dial indicators which read the torque directly in Nm or in whatever units are desired. Other torque wrenches (compressed-air or motorized wrenches) are designed so that a specially designed ratchet begins slipping when the torque reaches a certain predetermined value.

9.6 MOMENT OF INERTIA

The term *rotational dynamics* refers to the study of physical concepts involving the angular acceleration, torque, and mass of rotating objects. The first concept we will study is moment of inertia.

Any rigid body can be considered as a sum of separate particles of masses m_1, m_2, m_3, m_4, and so on. The moment of inertia of a particle is defined as the

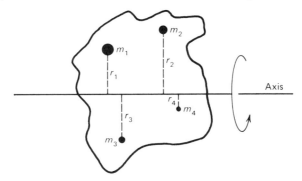

product of its mass and the square of its distance from the axis of rotation. That is, the *moment of inertia of particle* m_1 is given by

$$I_1 = m_1 r_1^2$$

where I_1 is the moment of inertia of the particle,
  m_1 is its mass, and
  r_1 is its distance from the axis of rotation.

The *moment of inertia of the rigid body* may be found by summing the moments of inertia of all the particles:

$$I = m_1 r_1^2 + m_2 r_2^2 + m_3 r_3^2 + m_4 r_4^2 + \cdots = \Sigma m_i r_i^2 = MR^2$$

where I is the moment of inertia of the rigid body,
  M is its total mass, and
  R is some value which best describes the radius of rotation of the rigid body. (*Note*: The Greek letter Σ (sigma) is commonly used in physics and mathematics to mean "sum.") The SI unit of moment of inertia is kg m^2.

example 1 Find the moment of inertia of a thin ring of mass m and radius r about an axis of rotation that passes through its center and is perpendicular to the plane containing the thin ring.
First, divide the thin ring into n particles.

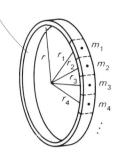

The moment of inertia of the thin ring is then the sum of the moments of inertia of the particles.

$$I = m_1 r_1^2 + m_2 r_2^2 + m_3 r_3^2 + m_4 r_4^2 + \cdots + m_n r_n^2$$
$$= m_1 R^2 + m_2 R^2 + m_3 R^2 + m_4 R^2 + \cdots + m_n R^2$$

Since this is a thin ring,

$$r_1 = r_2 = r_3 = r_4 = \cdots = r_n = R$$
$$I = (m_1 + m_2 + m_3 + m_4 + \cdots + m_n) R^2$$
$$= MR^2, \quad \text{where } M = m_1 + m_2 + m_3 + m_4 + \cdots + m_n$$

We have shown the derivation of one of the simplest moment of inertia equations. Most are quite complicated and require integral calculus to derive. As a result, we merely list the moment of inertia equations for each of the following rigid bodies about the indicated axes of rotation.

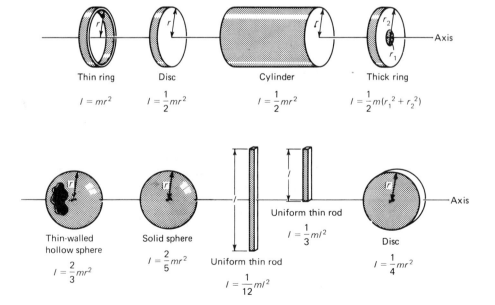

Thin ring
$I = mr^2$

Disc
$I = \dfrac{1}{2} mr^2$

Cylinder
$I = \dfrac{1}{2} mr^2$

Thick ring
$I = \dfrac{1}{2} m(r_1{}^2 + r_2{}^2)$

Thin-walled hollow sphere
$I = \dfrac{2}{3} mr^2$

Solid sphere
$I = \dfrac{2}{5} mr^2$

Uniform thin rod
$I = \dfrac{1}{12} ml^2$

Uniform thin rod
$I = \dfrac{1}{3} ml^2$

Disc
$I = \dfrac{1}{4} mr^2$

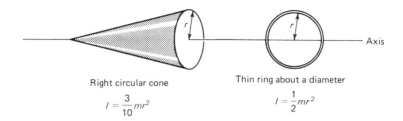

Since inertia must be overcome to start rotational motion from a rest position, to increase rotational velocity from one value to another (angular acceleration), or to decrease rotational velocity (angular deceleration), these equations apply to all such situations.

example 2 Compare the moments of inertia of a 15-kg flywheel

(a) in the shape of a solid disk of radius 40.0 cm, and
(b) in the shape of a thick ring whose outer radius is 40.0 cm and whose inner radius is 20.0 cm.

(a) DATA: $\qquad\qquad\qquad\qquad m = 15$ kg

$\qquad\qquad\qquad\qquad\qquad\qquad r = 40.0$ cm $= 0.400$ m

$\qquad\qquad\qquad\qquad\qquad\qquad I = ?$

BASIC EQUATION: $\qquad\qquad\quad I = \frac{1}{2}mr^2$

WORKING EQUATION: $\qquad\quad$ same

SUBSTITUTION: $\qquad\qquad\quad I = \frac{1}{2}(15 \text{ kg})(0.400 \text{ m})^2$

$\qquad\qquad\qquad\qquad\qquad\qquad = 1.2$ kg m^2

(b) DATA: $\qquad\qquad\qquad\qquad m = 15$ kg

$\qquad\qquad\qquad\qquad\qquad\qquad r_1 = 20.0$ cm $= 0.200$ m

$\qquad\qquad\qquad\qquad\qquad\qquad r_2 = 40.0$ cm $= 0.400$ m

$\qquad\qquad\qquad\qquad\qquad\qquad I = ?$

BASIC EQUATION: $\qquad\qquad\quad I = \frac{1}{2}m(r_1^2 + r_2^2)$

WORKING EQUATION: $\qquad\quad$ same

SUBSTITUTION: $\qquad\qquad\quad I = \frac{1}{2}(15 \text{ kg})[(0.200 \text{ m})^2 + (0.400 \text{ m})^2]$

$\qquad\qquad\qquad\qquad\qquad\qquad = 1.5$ kg m^2

How are the moment of inertia and the torque of a rotating object related? From Newton's second law of motion, we have

$$F = ma_T$$

where a_T is the linear or tangential acceleration. Multiplying both sides of this equation by r, the radius of rotation, we have

$$Fr = ma_T r$$

But Fr is the torque and $a_T = \alpha r$. Thus,

$$\tau = mr^2 \alpha$$

and $I = mr^2$. Therefore,

$$\boxed{\tau = I\alpha}$$

where τ is the torque,
   I is the moment of inertia, and
   α is the angular acceleration.
Note the comparison with linear dynamics:

$$\text{torque} = (\text{moment of inertia}) \times (\text{angular acceleration})$$

$$\text{force} = (\text{mass—a measure of inertia}) \times (\text{linear acceleration})$$

9.7   KINETIC ENERGY OF ROTATING RIGID BODIES

A rotating rigid body possesses kinetic energy because each of its particles that makes up its mass is in motion. The speed of the first particle is

$$v_1 = \omega_1 r_1$$

Its kinetic energy is

$$\text{KE} = \tfrac{1}{2} m_1 v_1^2$$

$$= \tfrac{1}{2} m_1 (\omega_1 r_1)^2$$

$$= \tfrac{1}{2} m_1 \omega_1^2 r_1^2$$

The total kinetic energy of the rigid body is the sum of the kinetic energy of the particles:

$$\text{KE} = \tfrac{1}{2} m_1 v_1^2 + \tfrac{1}{2} m_2 v_2^2 + \tfrac{1}{2} m_3 v_3^2 + \cdots + \tfrac{1}{2} m_n v_n^2$$

$$= \Sigma \tfrac{1}{2} m v^2$$

$$= \Sigma \tfrac{1}{2} m (\omega r)^2 \qquad (v = \omega r)$$

$$= \Sigma \tfrac{1}{2} m \omega^2 r^2$$

$$= \tfrac{1}{2} \omega^2 \Sigma m r^2 \qquad \text{(assuming that the object is rotating at a constant angular velocity)}$$

$$\boxed{KE = \tfrac{1}{2}I\omega^2} \qquad (I = \Sigma mr^2)$$

example 1 Compare the kinetic energies of a 15-kg flywheel rotating at 25 rad/s

(a) in the shape of a solid disk of radius 40.0 cm, and
(b) in the shape of a thick ring whose outer radius is 40.0 cm and whose inner radius is 20.0 cm.

(a) DATA: $\qquad\qquad\qquad \omega = 25$ rad/s

$\qquad\qquad\qquad\qquad\qquad I = 1.2$ kg m^2 $\qquad$ (from Example 2, Section 9.6)

$\qquad\qquad\qquad\qquad\qquad KE = ?$

BASIC EQUATION: $\qquad\quad KE = \tfrac{1}{2}I\omega^2$

WORKING EQUATION: $\qquad$ same

SUBSTITUTION: $\qquad\qquad KE = \tfrac{1}{2}(1.2$ kg m$^2)(25$ rad/s$)^2$

$\qquad\qquad\qquad\qquad\quad = 380$ J $\qquad \boxed{\text{kg m}^2/\text{s}^2 = (\text{kg m}/\text{s}^2)\text{m} = \text{Nm} = \text{J}}$

(b) DATA: $\qquad\qquad\qquad \omega = 25$ rad/s

$\qquad\qquad\qquad\qquad\qquad I = 1.5$ kg m^2 $\qquad$ (from Example 2, Section 9.6)

$\qquad\qquad\qquad\qquad\qquad KE = ?$

BASIC EQUATION: $\qquad\quad KE = \tfrac{1}{2}I\omega^2$

WORKING EQUATION: $\qquad$ same

SUBSTITUTION: $\qquad\qquad KE = \tfrac{1}{2}(1.5$ kg m$^2)(25$ rad/s$)^2$

$\qquad\qquad\qquad\qquad\quad = 470$ J

An object that is both rotating and moving linearly, such as an object rolling along a surface, possesses both rotational and linear kinetic energy. Its total kinetic energy is

$$\boxed{KE_{total} = \tfrac{1}{2}I\omega^2 + \tfrac{1}{2}mv^2}$$

example 2 Consider the earth as a uniform sphere of radius 6.37×10^6 m and mass 5.98×10^{24} kg. Find its total kinetic energy. (Assume the earth's orbit is circular of radius 1.50×10^{11} m.) First, find I:

DATA: $\qquad\qquad\qquad\qquad\qquad m = 5.98 \times 10^{24}$ kg

$\qquad\qquad\qquad\qquad\qquad\qquad r = 6.37 \times 10^6$ m

$\qquad\qquad\qquad\qquad\qquad\qquad I = ?$

BASIC EQUATION: $\qquad\qquad\qquad\qquad I = \frac{2}{5} m r^2$

WORKING EQUATION: $\qquad\qquad\quad$ same

SUBSTITUTION: $\qquad\qquad\qquad I = \frac{2}{5}(5.98 \times 10^{24} \text{ kg})(6.37 \times 10^6 \text{ m})^2$

$\qquad\qquad\qquad\qquad\qquad\qquad = 9.71 \times 10^{37} \text{ kg m}^2$

Then, find ω in rad/s: (ω is the angular velocity of the earth about its own axis.)

$$\omega = \frac{\Delta\theta}{\Delta t} = \frac{2\pi \text{ rad}}{1 \text{ day} \times \dfrac{24 \text{ h}}{1 \text{ day}} \times \dfrac{60 \text{ min}}{1 \text{ h}} \times \dfrac{60 \text{ s}}{1 \text{ min}}} = 7.27 \times 10^{-5} \text{ rad/s}$$

And, find ω' in rad/s: (ω' is the angular velocity of the earth about the sun's axis.)

$$\omega' = \frac{\Delta\theta}{\Delta t} = \frac{2\pi \text{ rad}}{365 \text{ days} \times \dfrac{24 \text{ h}}{1 \text{ day}} \times \dfrac{60 \text{ min}}{1 \text{ h}} \times \dfrac{60 \text{ s}}{1 \text{ min}}} = 1.99 \times 10^{-7} \text{rad/s}$$

Next, find v:

DATA: $\qquad\qquad\quad \omega' = 1.99 \times 10^{-7} \text{ rad/s}$

$\qquad\qquad\qquad\quad r = 1.50 \times 10^{11} \text{ m}$

$\qquad\qquad\qquad\quad v = ?$

BASIC EQUATION: $\qquad\qquad\qquad v = \omega' r$

WORKING EQUATION: $\qquad\qquad\quad$ same

SUBSTITUTION: $\qquad\qquad\qquad v = (1.99 \times 10^{-7} \text{ rad/s})(1.50 \times 10^{11} \text{ m})$

$\qquad\qquad\qquad\qquad\qquad\quad = 2.99 \times 10^4 \text{ m/s}$

Finally, to find its total kinetic energy:

DATA: $\qquad\qquad\qquad\qquad\qquad I = 9.71 \times 10^{37} \text{ kg m}^2$

$\qquad\qquad\qquad\qquad\qquad\quad \omega = 7.27 \times 10^{-5} \text{ rad/s}$

$\qquad\qquad\qquad\qquad\qquad\quad m = 5.98 \times 10^{24} \text{ kg}$

$\qquad\qquad\qquad\qquad\qquad\quad v = 2.99 \times 10^4 \text{ m/s}$

BASIC EQUATION: $\qquad\qquad KE_{\text{total}} = \frac{1}{2} I \omega^2 + \frac{1}{2} m v^2$

WORKING EQUATION: $\qquad\qquad\quad$ same

SUBSTITUTION: $\qquad\qquad KE_{\text{total}} = \frac{1}{2}(9.71 \times 10^{37} \text{ kg m}^2)(7.27 \times 10^{-5} \text{ rad/s})^2$

$\qquad\qquad\qquad\qquad\qquad\quad + \frac{1}{2}(5.98 \times 10^{24} \text{ kg})(2.99 \times 10^4 \text{ m/s})^2$

$$= 2.57 \times 10^{29} \text{ J} + 2.67 \times 10^{33} \text{ J}$$

$$\boxed{\text{kg m}^2/\text{s}^2 = (\text{kg m}/\text{s}^2)\text{m} = \text{Nm} = \text{J}}$$

$$= 2.67 \times 10^{33} \text{ J}$$

As you can see, while the earth's rotational kinetic energy is huge, it is insignificant when compared with its linear kinetic energy. As a comparison, it has been estimated that the United States used 1.0×10^{20} J of total energy in 1980.

9.8 WORK AND POWER

One of the most significant applications of rotational motion is the transmission of mechanical energy. This transmission normally takes place via rotating shafts, rotors, gears, and so on. What is the relationship among power, torque, and angular velocity? We start with the definitions of work and power.

$$W = Fs \quad \text{and} \quad P = \frac{W}{t}$$

Then, apply these definitions to a rotating shaft of radius r and angular velocity ω.

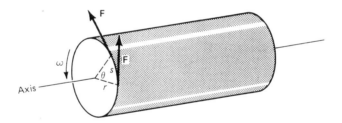

A tangential force **F** is applied to a point on the rim of the shaft for time t. After time t, the shaft has rotated through an angle of θ and the point has moved a distance s. Since

$$s = r\theta$$

the amount of work done is

$$W = Fs = Fr\theta$$

Power is the rate of doing work, so

$$P = \frac{W}{t} = \frac{Fr\theta}{t} = Fr\frac{\theta}{t}$$

But Fr is the torque applied to the shaft and θ/t is the angular velocity of the shaft. Thus,

$$P = \tau\omega$$

where P is the power transmitted,

τ is the applied torque, and

ω is the angular velocity.

Note that the torque and angular velocity are inversely proportional. That is, to transmit a given required power, one may increase the torque and decrease the angular velocity or decrease the torque and increase the angular velocity. These are important in the design of engines, pumps, gear and pulley systems, motors, and so on.

example An automobile engine must deliver 115 kW of power at 3600 rpm. What torque must be applied to develop this power?

DATA: $\qquad\qquad\qquad P = 115\ \text{kW} = 115{,}000\ \text{W}$

$$\omega = 3600\ \frac{\text{rev}}{\text{min}} \times \frac{2\pi\ \text{rad}}{1\ \text{rev}} \times \frac{1\ \text{min}}{60\ \text{s}} = 380\ \text{rad/s}$$

$$\tau = ?$$

BASIC EQUATION: $\qquad\quad P = \tau\omega$

WORKING EQUATION: $\qquad \tau = \dfrac{P}{\omega}$

SUBSTITUTION: $\qquad\qquad \tau = \dfrac{115{,}000\ \text{W}}{380\ \text{rad/s}}$

$$= 3\bar{0}0\ \text{Nm}$$

$$\boxed{\begin{array}{c} \text{Since } 1\ \text{W} = 1\ \text{Nm/s,} \\[4pt] \dfrac{\text{W}}{\text{rad/s}} = \dfrac{\text{Nm/s}}{\text{rad/s}} = \text{Nm} \end{array}}$$

9.9 ANGULAR MOMENTUM

The rotational comparison to linear momentum is angular momentum. The angular momentum of a rotating object is given by the equation

$$L = I\omega$$

where L is the angular momentum,

I is the moment of inertia, and

ω is the angular velocity.

Note the comparison with linear dynamics:

angular momentum = (moment of inertia) × (angular velocity)

linear momentum = (mass—a measure of inertia) × (linear velocity)

When there is no net external torque acting on a rotating system, the angular momentum of the system (that is, the product of its moment of inertia and its angular velocity) is constant.

This conservation of angular momentum can be vividly illustrated by watching an ice skater doing a spin. By keeping her arms and legs near the center of rotation, her moment of inertia decreases and her angular velocity increases significantly, as in figure (a) below. Then by stretching out her arms, she increases her moment of inertia and her angular velocity decreases as in figure (b) below.

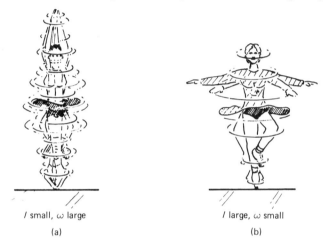

/ small, ω large	/ large, ω small
(a)	(b)

EXERCISES

Sketch	Data	Basic Equation	Working Equation	Substitution
12 cm² $h = ?$ $b = 6.0$ cm	$A = 12$ cm² $b = 6.0$ cm $h = ?$	$A = bh$	$h = \dfrac{A}{b}$	$h = \dfrac{12 \text{ cm}^2}{6.0 \text{ cm}}$ $= 2.0$ cm

1. Convert $6\frac{1}{2}$ revolutions
   (a) to radians.
   (b) to degrees.

2. Convert 2880°
   (a) to revolutions.
   (b) to radians.

3. Convert 10π rad
   (a) to revolutions.
   (b) to degrees.

4. Convert $60\bar{0}$ rpm to rad/s.

5. Convert $60\bar{0}$ rad/s to rpm.

6. A flywheel is rotating at 1050 rpm.

   (a) How long does it take to complete one revolution?
   (b) How many revolutions does it complete in 5.00 s?

7. A rotating wheel completes one revolution in 0.15 s. Find its angular velocity

   (a) in rev/s,
   (b) in rpm, and
   (c) in rad/s.

8. A pendulum of length 1.50 m swings through an arc of 5°. Find the length of the arc through which the pendulum swings.

9. A flywheel of radius 25.0 cm is rotating at 655 rpm.

   (a) Express its angular speed in rad/s.
   (b) Find its angular displacement (in rad) in 3.00 min.
   (c) Find the linear distance traveled by a point on the rim in one complete revolution.
   (d) Find the linear distance traveled by a point on the rim in 3.00 min.
   (e) Find the linear velocity of a point on the rim.

10. An airplane propeller whose blades are 2.00 m long is rotating at $220\bar{0}$ rpm.

    (a) Express the angular speed in rad/s.
    (b) Find the angular displacement in 4.00 s.
    (c) Find the linear velocity of a point on the end of a blade.

11. An automobile is traveling at 60.0 km/h. Its tires have a radius of 33.0 cm.

    (a) Find the angular speed of the tires (in rad/s).
    (b) Find the angular displacement in 30.0 s.
    (c) Find the linear distance traveled by a point on the tread in 30.0 s.

12. Find the angular speed in rad/s of the following hands on a clock:

    (a) second,
    (b) minute, and
    (c) hour.

13. A belt is wrapped around a pulley that is 30.0 cm in diameter. If the pulley rotates at $25\bar{0}$ rpm, what is the linear speed (in m/s) of the belt? (Assume no belt slippage on the pulley.)

14. The earth rotates on its axis at an angular velocity of 1 rev/24 h (page 187).

    (a) Find the linear velocity (in km/h) of a point on the equator where the radius is 6400 km.
    (b) Find the radius of rotation of a point that is at $3\bar{0}°$ north latitude.
    (c) Find the linear velocity of Houston, which is approximately $3\bar{0}°$ north latitude.
    (d) Find the linear velocity of Minneapolis, which is approximately 45° north latitude.
    (e) Find the linear velocity of Anchorage, Alaska, which is approximately $6\bar{0}°$ north latitude.

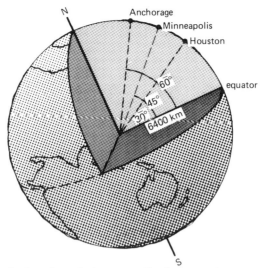

15. The angular velocity of an airplane propeller increases from 1800 rpm to 2200 rpm in 12 s.

    (a) Find its angular acceleration.
    (b) Find its angular displacement during these 12 s.

16. The angular velocity of a flywheel increases from rest to 3500 rpm in 15 s.

    (a) Find its angular acceleration.
    (b) Find its angular displacement during these 15 s.

17. An automobile accelerates at 0.300 m/s² from rest for 6.00 s. The wheels are 33.0 cm in radius.

    (a) Find the angular acceleration.
    (b) Find the angular velocity of the wheels after 6.00 s.
    (c) Find the linear velocity (in m/s) of the wheels after 6.00 s.
    (d) Find the speed of the automobile after 6.00 s.
    (e) Find the angular displacement of the wheels during these 6.00 s.

18. A drum rotating at $40\overline{0}0$ rpm is slowed to $25\overline{0}0$ rpm in 30.0 s.

    (a) Find the angular deceleration.
    (b) Find the angular displacement during these 30.0 s.
    (c) At this same rate of angular deceleration, how long will it take for the drum to come to rest?
    (d) Find the centripetal deceleration during 30.0 s of a point on the rim of the drum, which has a radius of 45.0 cm.

19. A rotating flywheel of diameter 40.0 cm accelerates from rest to $25\overline{0}$ rad/s in 15.0 s.

    (a) Find the angular acceleration of the wheel.
    (b) Find the linear or tangential acceleration of a point on the rim of the wheel.
    (c) Find the linear velocity of a point on the rim of the wheel after 15.0 s.
    (d) Find the centripetal acceleration of a point on the rim of the wheel after 15.0 s.

20. From Exercise 19,

(a) Find the angular displacement in radians of the flywheel during the first 5.0 s.

(b) Find the angular displacement in radians of the flywheel during the last 5.0 s (from t = 10.0 s to t = 15.0 s).

(c) Find the angular velocity of the flywheel after 5.0 s.

21. The earth is revolving about the sun in 1 rev/365 days. Assume an average circular orbit of radius 1.50×10^8 km.

(a) Find the linear speed in km/h of the earth about the sun.

(b) Find the earth's centripetal acceleration.

(c) Find the earth's angular acceleration (in rad/s^2).

(d) Find the earth's linear or tangential acceleration (in m/s^2).

22. A ball of mass $5\overline{0}0$ g is swung at the end of a string in a horizontal circular path at 6.0 m/s. If the length of the string is 1.5 m, what centripetal force does the string exert on the ball?

23. An automobile whose mass is 1650 kg is driven around a circular curve of radius $15\overline{0}$ m at 80.0 km/h. What is the centripetal force of the road on the automobile?

24. (a) What minimal coefficient of friction must be needed to keep the automobile in Exercise 23 from skidding?

(b) The coefficient of friction is 0.65 between the tires and the road. What is the maximum speed (in km/h) for the automobile not to skid?

25. A curve of radius 225 m in a roadway is planned. The expected legal speed limit is $9\overline{0}$ km/h.

(a) At what angle should the roadbed be banked (assuming no frictional forces)?

(b) What is the centripetal force of the road on the automobile of mass 1500 kg?

26. A bobsled run contains a hairpin curve of radius 18 m. The curve is banked at 77°. What is the maximum safe speed (in km/h) for this curve (assuming no frictional forces)?

27. Centrifuges are used to separate substances of different densities. The more dense substances "settle" toward the outer end of the rotating tube and the less dense substances "float" to the top. They are used to separate blood cells from blood plasma and virus cells or bacteria from a liquid. If the centrifuge is 25.0 cm in diameter, how many rpm are required to set up an "artificial gravity" of $20\overline{0}$ g ($g = 9.80$ m/s^2)?

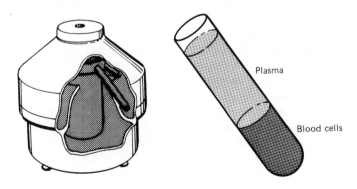

28. One type of space station is designed in the shape of a large toroid, $30\bar{0}$ m in diameter, which rotates about its central axis to produce an artificial gravity for its inhabitants. If the space dwellers "weigh" three-fourths of their earth weight, what is the angular speed of the space station about its axis?

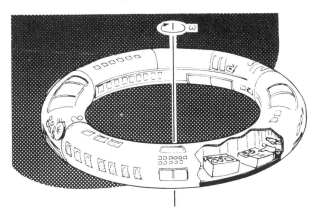

29. What minimal speed must the stunt motorcyclist maintain at the top of the circular loop of diameter 10.0 m below in order to complete it successfully (assuming no frictional forces)? The cyclist and cycle have a combined mass of 235 kg.

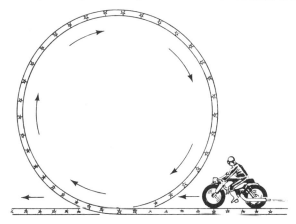

30. If the cyclist in Exercise 29 has a mass of 75 kg, what is his or her "apparent" weight at the top of the loop?

31. What minimum linear speed (in km/h) is necessary for a satellite to attain a stable circular orbit $60\bar{0}$ km above the earth's surface? ($g = 8.1$ m/s² at this height.)

32. An earth satellite at an altitude of 3.58×10^7 m above the earth's surface is called a synchronous satellite because it and the earth have the same period. The satellite stays directly over the same line of the earth's longitude at all times (page 190).

    (a) What is the satellite's angular velocity (in rad/s)?
    (b) What is the satellite's linear velocity (in km/h)?
    (c) If the value of g is inversely proportional to the square of the distance from the center of the earth, what is the value of g at the satellite's altitude?

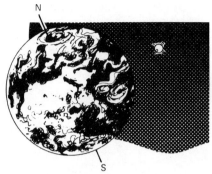

33. A torque wrench reads 14.5 Nm. If its length is 25.0 cm, what force is being applied to the handle?

34. Compare the moments of inertia of a solid circular disk whose radius is 15.0 cm and whose mass is 5.75 kg
    (a) if it is rotated about an axis which passes through its center and which is perpendicular to its flat face
    (b) if it is rotated about an axis that passes through its center and is parallel to its flat face.

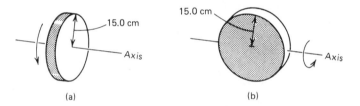

(a)   (b)

35. Find the moment of inertia of a huge cylindrical roller in a milling plant if its diameter is 2.5 m, its length is 2.0 m, and its mass is 76,700 kg.

36. A torque of 125 Nm is applied to a rotor that transmits an angular acceleration of 12 rad/s^2 to the rotor. Find the moment of inertia of the rotor.

37. A torque of 275 Nm is applied to a flywheel for 6.00 s. The flywheel starts from rest and completes 25 revolutions during this 6.00-s period.
    (a) Find the angular acceleration (in rad/s^2).
    (b) Find the angular velocity (in rpm) at the end of 6.00 s.
    (c) Find the moment of inertia of the flywheel.

38. What torque is required to accelerate the cylinder about the axis as shown from rest to 5̄00 rpm in 12 s? Its mass is 2.4 kg.

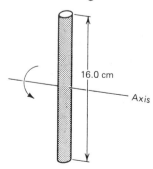

39. What is the kinetic energy of the rotating cylinder in Exercise 38 at the end of 12 s?

40. A 7.1 kg bowling ball of radius 15 cm is rolling down the alley at 6.0 m/s. What is its total kinetic energy? (Assume no slippage.)

41. What is the angular velocity of a motor developing 650 W of power with a torque of 130 Nm?

42. A high-speed drill develops 0.50 kW of power at 1800 rpm. What torque is applied to the drill bit?

43. What power is developed by an engine whose torque of 750 Nm is applied at 4500 rpm?

44. A tangential force of 150 N is applied to a flywheel whose diameter is 0.45 m to maintain a constant angular velocity of 175 rpm. How much work is done per minute?

45. Show that angular momentum has kg m^2/s as its units.

NONCON-
CURRENT
FORCES

10.1 EQUILIBRIUM OF RIGID BODIES

In Section 6.4, we stated that a body is in equilibrium when there is no change in its motion. A body is in equilibrium either when it is at rest or when it is moving along a straight line at constant speed. That is, no outside forces are acting on the object to cause a linear acceleration. The *first condition for equilibrium* stated mathematically is

$$\Sigma \mathbf{F} = 0$$

In other words, the sum of the forces acting on a body in equilibrium must be equal to zero. This statement may also be written in terms of the three perpendicular components of this vector sum as follows:

$$\Sigma F_x = 0$$
$$\Sigma F_y = 0$$
$$\Sigma F_z = 0$$

That is, the sum of each of any three perpendicular components of the forces acting on a body in equilibrium must be equal to zero.

The *second condition for equilibrium* states that when a body is in equilibrium, there may be no rotational motion of the body about any axis. The second condition for equilibrium stated mathematically is

$$\Sigma \tau = 0$$

As we saw in Chapter 6, concurrent forces are those forces which act simultaneously at the same point. When the forces do not act at the same point, these forces are called *nonconcurrent forces*. Nonconcurrent forces may be categorized as parallel forces and nonparallel forces.

Two painters standing on the scaffold below is an example of parallel nonconcurrent forces.

The force diagram is then: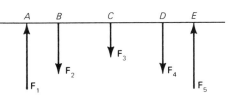

where F_1 is the upward supporting force of the support at point A,
F_2 is the downward force due to the weight of the man at point B,
F_3 is the downward force due to the weight of the plank which acts at point C (more on this in the next section),
F_4 is the downward force due to the weight of the person at point D, and
F_5 is the upward supporting force of the support at point E.
These are parallel nonconcurrent forces because all the forces are parallel and are not acting at the same point.

The boom and cable shown below is an example of nonparallel nonconcurrent forces.

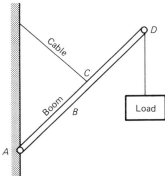

Its force diagram is

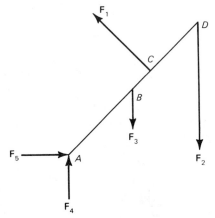

where F_1 is the supporting force or tension in the cable at point C,
F_2 is the downward force due to the weight of the load at point D,
F_3 is the downward force due to the weight of the boom at point B,
F_4 is the upward or vertical supporting force at point A, and
F_5 is the horizontal supporting force at point A.

These are nonparallel nonconcurrent forces because not all the forces are parallel and not all the forces are acting at the same point.

10.2   EQUILIBRIUM WITH PARALLEL FORCES

> The *first condition for equilibrium with parallel forces* may be stated as follows: The sum of the forces in one direction equals the sum of the forces in the opposite direction. If we assign one direction as positive (+) and the opposite direction as negative (−), we have the first condition for equilibrium as stated before; namely,
>
> $$\Sigma\ \mathbf{F} = 0$$

Recall from Section 9.5 that an applied torque causes rotational motion of a body. Further recall that

$$\tau = Fr$$

where τ is the torque,
F is the applied force, and
r is the length of the moment arm.

For example, if a force of 225 N is applied to a wrench that is 35 cm long, a torque of

$$\tau = Fr$$
$$= (225\ \text{N})(0.35\ \text{m})$$
$$= 79\ \text{Nm}$$

is applied to the wrench, which produces the rotational motion shown below:

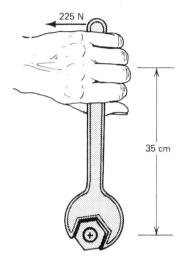

225 N

35 cm

The *second condition for equilibrium with parallel forces* may be stated as follows: The sum of the clockwise torques equals the sum of the counter-clockwise torques about any point or axis of rotation. In essence, this assures no change in rotational motion about any axis. That is,

$$\Sigma \, \tau_{cw} = \Sigma \, \tau_{ccw}$$

The *center of gravity*, or *center of mass*, of any object is that point at which its entire weight can be considered to be concentrated. For a uniform linear object, the center of gravity is at its center. The center of gravity is the point at which the object can be supported. For example, a metre stick is supported on one's finger at the 50-cm mark.

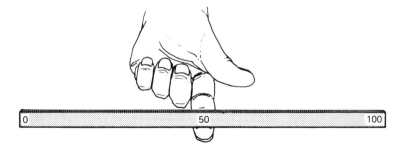

When a long board is carried on your shoulder, you place its center on your shoulder. Its entire weight is concentrated at that point (page 196).

For a nonuniform object, the center of gravity is usually not at its geometric center. Its center of gravity is at the point at which it can be supported in equilibrium by a single force. The center of gravity is also the point about which it spins if allowed to spin freely in space.

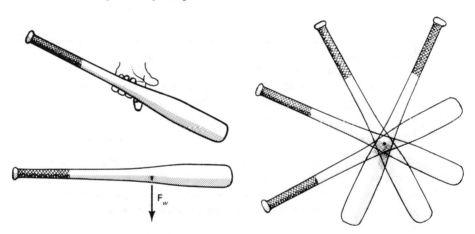

We shall represent the weight of an object by a vector through its center of gravity. We use a vector to show the weight (force due to gravity) of the object. It is placed through the center of gravity to show that all the weight of the object may be considered concentrated at that point. If the center of gravity is not at the middle of the object, its location will be given.

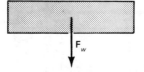

example 1   A $2\bar{0}$-kg child sits on one end of a seesaw 1.5 m from the pivot. How far on the other side from the pivot must a $3\bar{0}$-kg child sit so that the seesaw is in equilibrium?

SKETCH:

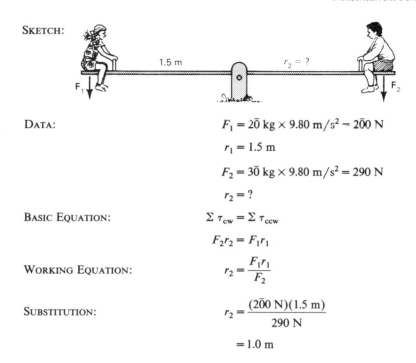

DATA:

$$F_1 = 2\bar{0} \text{ kg} \times 9.80 \text{ m/s}^2 = 2\bar{0}0 \text{ N}$$

$$r_1 = 1.5 \text{ m}$$

$$F_2 = 3\bar{0} \text{ kg} \times 9.80 \text{ m/s}^2 = 290 \text{ N}$$

$$r_2 = ?$$

BASIC EQUATION:

$$\Sigma \tau_{cw} = \Sigma \tau_{ccw}$$

$$F_2 r_2 = F_1 r_1$$

WORKING EQUATION:

$$r_2 = \frac{F_1 r_1}{F_2}$$

SUBSTITUTION:

$$r_2 = \frac{(2\bar{0}0 \text{ N})(1.5 \text{ m})}{290 \text{ N}}$$

$$= 1.0 \text{ m}$$

Sometimes there is a natural axis of rotation, as in Example 1 above. If the body is in equilibrium, the sum of the clockwise torques equals the sum of the counterclockwise torques about *any* point. That is, *any* point or axis of rotation may be chosen from which to calculate the torques. So choose the most convenient point.

example 2 The span of a certain bridge is 32.0 m long and has a mass of 5.38×10^4 kg. A truck of mass 11,700 kg is stopped 11.4 m from one end. Find the upward force that must be exerted by each pier.

SKETCH:

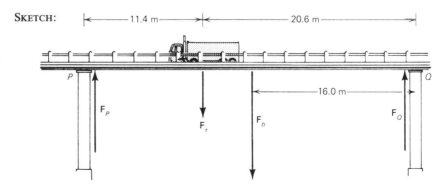

Assume that the span is uniform so that its center of gravity is at its center. Choose point *P* as the point of rotation.

DATA:
$$F_t = (11{,}700 \text{ kg})(9.80 \text{ m/s}^2) = 1.15 \times 10^5 \text{ N}$$

$$r_t = 11.4 \text{ m}$$

$$F_b = (5.38 \times 10^4 \text{ kg})(9.80 \text{ m/s}^2) = 5.27 \times 10^5 \text{ N}$$

$$r_b = 16.0 \text{ m}$$

$$F_Q = ?$$

$$r_Q = 32.0 \text{ m}$$

BASIC EQUATION:
$$\Sigma \tau_{\text{cw}} = \Sigma \tau_{\text{ccw}} \quad \text{(by the second condition for equilibrium)}$$

$$F_t r_t + F_b r_b = F_Q r_Q \quad \text{(note that the torque } F_P r_P = 0 \text{ because } r_P = 0)$$

WORKING EQUATION:
$$F_Q = \frac{F_t r_t + F_b r_b}{r_Q}$$

SUBSTITUTION:
$$F_Q = \frac{(1.15 \times 10^5 \text{ N})(11.4 \text{ m}) + (5.27 \times 10^5 \text{ N})(16.0 \text{ m})}{32.0 \text{ m}}$$

$$= 3.04 \times 10^5 \text{ N}$$

Now to find F_P:

DATA:
$$F_t = 1.15 \times 10^5 \text{ N}$$

$$F_b = 5.27 \times 10^5 \text{ N}$$

$$F_Q = 3.04 \times 10^5 \text{ N}$$

BASIC EQUATION:
$$\Sigma F_{\text{up}} = \Sigma F_{\text{down}} \quad \text{(by the first condition for equilibrium)}$$

$$F_P + F_Q = F_t + F_b$$

WORKING EQUATION:
$$F_P = F_t + F_b - F_Q$$

SUBSTITUTION:
$$F_P = 1.15 \times 10^5 \text{ N} + 5.27 \times 10^5 \text{ N} - 3.04 \times 10^5 \text{ N}$$

$$= 3.38 \times 10^5 \text{ N}$$

That is, pier P must support a weight of 3.38×10^5 N while pier Q supports a weight of 3.04×10^5 N.

example 3   The sign below is 4.50 m long, weighs 1350 N, and is made of uniform material. A weight of 225 N hangs 1.00 m from the end. Find the tension in each support cable.

SKETCH:

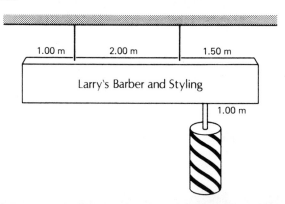

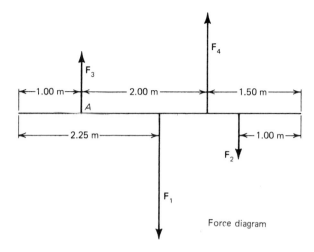

Force diagram

Choose point A as the point of rotation to find F_4:

DATA:

$$F_1 = 1350 \text{ N}$$

$$r_1 = 1.25 \text{ m}$$

$$F_2 = 225 \text{ N}$$

$$r_2 = 2.50 \text{ m}$$

$$F_4 = ?$$

$$r_4 = 2.00 \text{ m}$$

BASIC EQUATION: $\Sigma \tau_{cw} = \Sigma \tau_{ccw}$ (by the second condition for equilibrium)

$$F_1 r_1 + F_2 r_2 = F_4 r_4 \qquad \text{(note that } F_3 r_3 = 0\text{)}$$

WORKING EQUATION: $F_4 = \dfrac{F_1 r_1 + F_2 r_2}{r_4}$

SUBSTITUTION: $F_4 = \dfrac{(1350 \text{ N})(1.25 \text{ m}) + (225 \text{ N})(2.50 \text{ m})}{2.00 \text{ m}}$

$$= 1130 \text{ N}$$

Now to find F_3:

DATA:

$$F_1 = 1350 \text{ N}$$

$$F_2 = 225 \text{ N}$$

$$F_4 = 1130 \text{ N}$$

$$F_3 = ?$$

BASIC EQUATION: $\Sigma F_{up} = \Sigma F_{down}$ (by the first condition for equilibrium)

$$F_3 + F_4 = F_1 + F_2$$

WORKING EQUATION: $F_3 = F_1 + F_2 - F_4$

SUBSTITUTION:
$$F_3 = 1350 \text{ N} + 225 \text{ N} - 1130 \text{ N}$$
$$= 450 \text{ N}$$

example 4   A uniform bar 5.00 m long weighs 775 N. A weight of 225 N is attached to one end and a weight of $34\overline{0}$ N is attached 1.50 from the other end.

(a) What is the tension in the cable?
(b) Where should a cable be tied to lift the bar and its weight so that the bar hangs in a horizontal equilibrium position?

SKETCH:

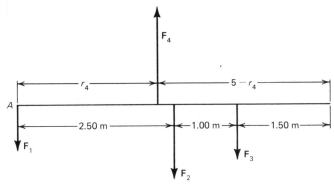

(a) By the first condition for equilibrium, the tension in the cable must equal the sum of the downward forces.

DATA:
$$F_1 = 225 \text{ N}$$
$$F_2 = 775 \text{ N}$$
$$F_3 = 34\overline{0} \text{ N}$$
$$F_4 = ?$$

BASIC EQUATION:
$$\Sigma F_{\text{up}} = \Sigma F_{\text{down}}$$

WORKING EQUATION:
$$F_4 = F_1 + F_2 + F_3$$

SUBSTITUTION:
$$F_4 = 225 \text{ N} + 775 \text{ N} + 34\overline{0} \text{ N}$$
$$= 134\overline{0} \text{ N}$$

(b) Choose point A as the point of rotation:

DATA:
$$F_2 = 775 \text{ N}$$
$$r_2 = 2.50 \text{ m}$$
$$F_3 = 34\overline{0} \text{ N}$$
$$r_3 = 3.50 \text{ m}$$
$$F_4 = 134\overline{0} \text{ N}$$
$$r_4 = ?$$

BASIC EQUATION: $\Sigma \tau_{cw} = \Sigma \tau_{ccw}$ (by the second condition for equilibrium)

$$F_2r_2 + F_3r_3 = F_4r_4$$

WORKING EQUATION: $r_4 = \dfrac{F_2r_2 + F_3r_3}{F_4}$

SUBSTITUTION: $r_4 = \dfrac{(775 \text{ N})(2.50 \text{ m}) + (34\bar{0} \text{ N})(3.50 \text{ m})}{134\bar{0} \text{ N}}$

$$= 2.34 \text{ m} \quad \text{(from point } A\text{)}$$

The center of gravity of a uniform thin plate is also the point at which the object can be supported.

The center of gravity of a uniform thin plate may also be found by suspending it and a plumb bob at two different points on its edge. Use a chalkline plumb bob so that you can mark each line. The point of intersection of any two (or all) such lines is the center of gravity.

(a)

Center of gravity

(b)

If the thin plate is in a regular geometric shape, its center of gravity is its geometric center because of its symmetry. Examples of four common geometric figures with each corresponding center of gravity are given below.

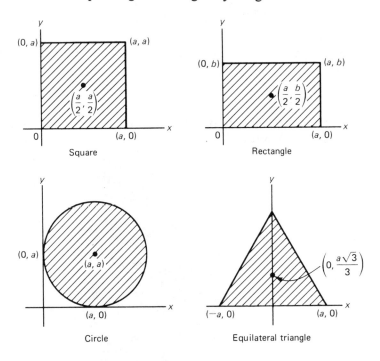

The center of gravity of a more complex but uniform object of weight W that is made up of combinations of the geometric figures above can be found by first subdividing the object into regions of these figures.

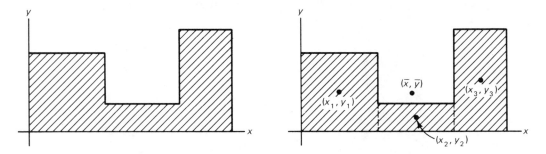

Then, find the center of gravity (geometric center) of each geometric figure. Since the center of gravity is the point at which the total weight of each region is concentrated, consider all the weight of the region to be at its geometric center. For the object to be in equilibrium (supported at its center of gravity), the first condition tells us that the upward force must be W, the weight of the object.

Assume that $(\bar{x}, \bar{y})$ is the center of gravity of the complex object. The torque about the y-axis of the total object is $W\bar{x}$ and must be equal to the sum of the torques of the regions about the y-axis; that is,

$$W\bar{x} = w_1x_1 + w_2x_2 + w_3x_3 + \cdots = \Sigma\, w_n x_n$$

or

$$\bar{x} = \frac{\Sigma\, w_n x_n}{W}$$

Similarly, the torque about the x-axis of the total object is $W\bar{y}$ and must be equal to the sum of the torques of the regions about the x-axis; that is,

$$W\bar{y} = w_1y_1 + w_2y_2 + w_3y_3 + \cdots = \Sigma\, w_n y_n$$

or

$$\bar{y} = \frac{\Sigma\, w_n y_n}{W}$$

example 5 Find the center of gravity of the uniform thin plate below.

SKETCH:

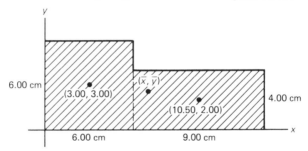

Since the areas of each region are the same, let w be the weight of each.

DATA:
$$W = 2w$$
$$w = w_1 = w_2$$
$$x_1 = 3.00$$
$$x_2 = 10.50$$
$$\bar{x} = ?$$

BASIC EQUATION:
$$W\bar{x} = w_1x_1 + w_2x_2$$

WORKING EQUATION:
$$\bar{x} = \frac{w_1x_1 + w_2x_2}{W}$$

SUBSTITUTION:

$$\bar{x} = \frac{w(3.00) + w(10.50)}{2w}$$

$$= \frac{13.50w}{2w} = 6.75 \text{ cm}$$

DATA:

$$W = 2w$$

$$w = w_1 = w_2$$

$$y_1 = 3.00$$

$$y_2 = 2.00$$

$$\bar{y} = ?$$

BASIC EQUATION:

$$W\bar{y} = w_1 y_1 + w_2 y_2$$

WORKING EQUATION:

$$\bar{y} = \frac{w_1 y_1 + w_2 y_2}{W}$$

SUBSTITUTION:

$$\bar{y} = \frac{w(3.00) + w(2.00)}{2w}$$

$$= \frac{5.00w}{2w} = 2.50 \text{ cm}$$

Thus, the center of gravity is at the point (6.75 cm, 2.50 cm).

example 6   Find the center of gravity of the uniform thin plate below.

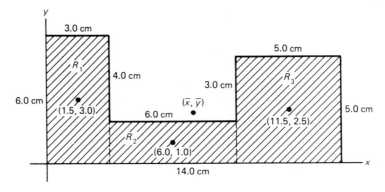

The area of:

$$R_1 = 18 \text{ cm}^2$$

$$R_2 = 12 \text{ cm}^2$$

$$R_3 = \underline{25 \text{ cm}^2}$$

The total area $= 55 \text{ cm}^2$

Let the total weight of the plate be $55w$ and its density be uniform. Then the weight of each segment is:

DATA:
$$R_1 = w_1 = 18w \qquad x_1 = 1.5$$
$$R_2 = w_2 = 12w \qquad x_2 = 6.0$$
$$R_3 = w_3 = 25w \qquad x_3 = 11.5$$
$$W = 55w$$
$$\bar{x} = ?$$

BASIC EQUATION:
$$W\bar{x} = w_1 x_1 + w_2 x_2 + w_3 x_3$$

WORKING EQUATION:
$$\bar{x} = \frac{w_1 x_1 + w_2 x_2 + w_3 x_3}{W}$$

SUBSTITUTION:
$$\bar{x} = \frac{(18w)(1.5) + (12w)(6.0) + (25w)(11.5)}{55w}$$
$$= \frac{387w}{55w} = 7.0 \text{ cm}$$

To find $\bar{y}$:

DATA:
$$W = 55w$$
$$w_1 = 18w \qquad y_1 = 3.0$$
$$w_2 = 12w \qquad y_2 = 1.0$$
$$w_3 = 25w \qquad y_3 = 2.5$$
$$\bar{y} = ?$$

BASIC EQUATION:
$$W\bar{y} = w_1 y_1 + w_2 y_2 + w_3 y_3$$

WORKING EQUATION:
$$\bar{y} = \frac{w_1 y_1 + w_2 y_2 + w_3 y_3}{W}$$

SUBSTITUTION:
$$\bar{y} = \frac{(18w)(3.0) + (12w)(1.0) + (25w)(2.5)}{55w}$$
$$= \frac{129w}{55w} = 2.3 \text{ cm}$$

The center of gravity is at the point (7.0 cm, 2.3 cm).

Note that in this case the center of gravity is located at a point that is not on the surface of the plate.

Also note that as long as the plate is uniform, we do not need to know the actual weights of the plate or its regions. When an object is uniform, the weight is directly proportional to its area. Then, as above, the weights cancel out.

10.3 EQUILIBRIUM WITH NONPARALLEL NONCONCURRENT FORCES

When a force is applied nonperpendicularly to a body, such as a beam, the equivalent perpendicular or normal force must be used to compute the torque.

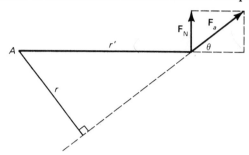

$$F_N = F_a \sin \theta$$

and

$$\tau = F_N r' = F_a r' \sin \theta = F_a r$$

where τ is the torque,

F_a is the applied force,

F_N is the normal force,

r is the length of the moment arm (from point A),

r' is the distance from the pivot point A to the point of application of the force F_a, and

θ is the angle between the applied force and the beam.

Otherwise, the same two conditions for equilibrium are applied as in the previous section.

example   The uniform boom shown below weighs $75\overline{0}$ N and its load weighs $36\overline{0}0$ N.

(a) Find the tension in the cable.

(b) Find the vertical and horizontal components of the supporting force at point A.

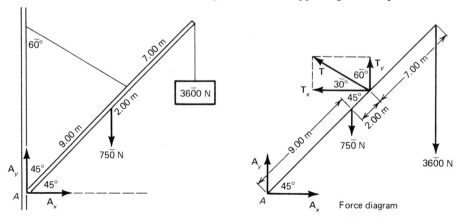

Force diagram

(a) DATA:

$$F_1 = 75\bar{0} \text{ N}$$

$$F_2 = 36\bar{0}0 \text{ N}$$

$$F_3 = T = ?$$

$$\theta_1 = 45.0° = \theta_2$$

$$\theta_3 = 75.0°$$

$$r_1' = 9.00 \text{ m}$$

$$r_2' = 18.0 \text{ m}$$

$$r_3' = 11.0 \text{ m}$$

BASIC EQUATION:

$$\Sigma \tau_{cw} = \Sigma \tau_{ccw} \qquad \text{(by the second condition for equilibrium)}$$

$$F_1 r_1' \sin \theta_1 + F_2 r_2' \sin \theta_2 = F_3 r_3' \sin \theta_3$$

WORKING EQUATION: $F_3 = T = \dfrac{F_1 r_1' \sin \theta_1 + F_2 r_2' \sin \theta_2}{r_3' \sin \theta_3}$

SUBSTITUTION:

$$T = \frac{(75\bar{0} \text{ N})(9.00 \text{ m})(\sin 45.0°) + (36\bar{0}0 \text{ N})(18.0 \text{ m})(\sin 45.0°)}{(11.0 \text{ m})(\sin 75.0°)}$$

$$= \frac{4770 \text{ N m} + 45{,}800 \text{ Nm}}{10.6 \text{ m}}$$

$$= \frac{50{,}600 \text{ Nm}}{10.6 \text{ m}} = 4770\text{N}$$

(b) First, note that

$$T_x = T \cos 30.0°$$

$$T_y = T \cos 60.0°$$

By the first law of equilibrium, we have

BASIC EQUATIONS: $\quad \Sigma F_x = 0 \quad$ and $\quad \Sigma F_y = 0$

or

$$\Sigma F_{\text{right}} = \Sigma F_{\text{left}} \quad \text{and} \quad \Sigma F_{\text{up}} = \Sigma F_{\text{down}}$$

That is,

$$A_x = T_x \quad \text{and} \quad A_y + T_y = 750 \text{ N} + 36\bar{0}0 \text{ N}$$

WORKING EQUATIONS: $\quad A_x = T_x = T \cos 30.0° \qquad A_y = 4350 \text{ N} - T_y$

$$T_x = (4770 \text{ N})(\cos 30.0°) \qquad = 4350 \text{ N} - (4770 \text{ N})(\cos 60.0°)$$

$$= 4130 \text{ N} \qquad\qquad\qquad = 1960 \text{ N}$$

Sketch	Data	Basic Equation	Working Equation	Substitution
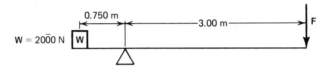 $h = ?$ $b = 6.0$ cm	$A = 12$ cm^2 $b = 6.0$ cm $h = ?$	$A = bh$	$h = \dfrac{A}{b}$	$h = \dfrac{12 \text{ cm}^2}{6.0 \text{ cm}}$ $= 2.0$ cm

1. What vertical force is needed to balance the 20$\overline{0}$0-N weight below?

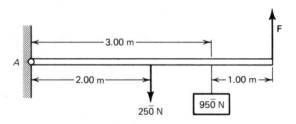

2. What vertical force is needed to support the beam below, which weighs 25$\overline{0}$ N? Its load is 95$\overline{0}$ N.

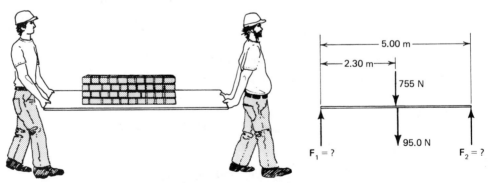

3. Two workers carry a uniform 5.00-m plank, which weighs 95.0 N. On the plank is a load of blocks, which weighs 755 N and is located 2.30 m from the first worker. What force must each worker exert to hold up the plank and the load?

4. A bricklayer weighing 80$\overline{0}$ N stands on a 3.50-m uniform scaffold 1.00 m from one end. A pile of bricks weighing 65$\overline{0}$ N is located 1.20 m from the other end. The scaffold weighs 45$\overline{0}$ N. How much weight must each end support? (See next page.)

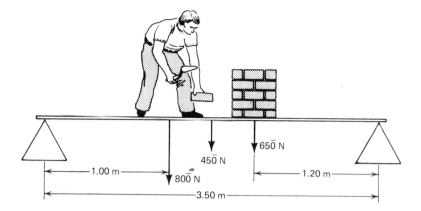

5. A uniform steel beam is 5.00 m long and weighs 3.6×10^5 N. What force is needed to lift one end?

6. A wooden pole is 4.00 m long, weighs 315 N, and has its center of gravity 1.50 from one end. What force is needed to lift each end?

7. A bridge has a mass of 1.60×10^4 kg, is 21.0 m long, and has a 35$\overline{0}$0-kg truck 7.00 m from one end. What force must each end of the bridge support?

8. An auto engine of mass 295 kg is located 1.00 m from one end of a 4.00 m workbench. If the uniform bench has a mass of 45.0 kg, what weight must each end of the bench support?

9. An old covered bridge across a country stream weighs 8.9×10^4 N. A large truck stalls 4.00 m from one end of the 10.5-m bridge. What weight must each of the piers support if the truck weighs 1.0×10^5 N?

10. A window washer's scaffold, 3.50 m long and weighing 325 N, is suspended from each end by a cable. One washer, of mass 70.0 kg, is 1.00 m from one end. The other washer, of mass 85.0 kg, is 1.50 m from the other end. What is the tension of the cable at each end of the scaffold?

11. Find the magnitude, direction, and placement (from point *A*) of a parallel vector F_6 that will produce equilibrium in the force diagram below.

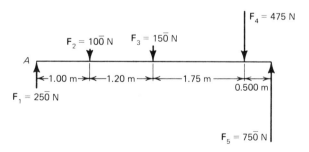

12. Find the magnitude, direction, and placement (from point *A*) of a parallel vector F_6 that will produce equilibrium in the force diagram on page 210.

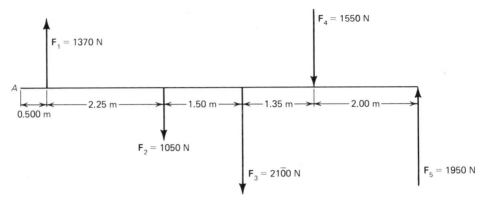

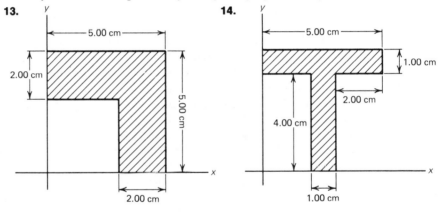

In Exercises 13–18, find the center of gravity of each uniform thin plate. (An x-y coordinate system has been provided for uniformity of answers.)

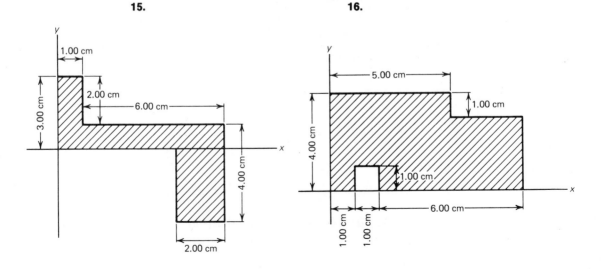

17.

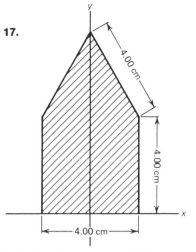

18.

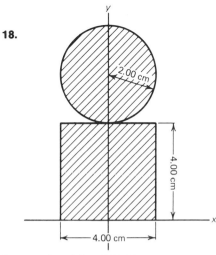

19. The uniform boom below is 6.00 m long and weighs 1250 N. Its load weighs 5750 N.

(a) Find the tension in the cable.

(b) Find the vertical and horizontal components of the supporting force at point A.

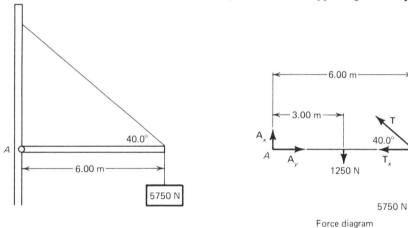

Force diagram

20. The uniform boom below is 8.00 m long and weighs 1850 N. Its load weighs $85\overline{0}0$ N.

(a) Find the tension in the cable.

(b) Find the vertical and horizontal components of the supporting force at point A.

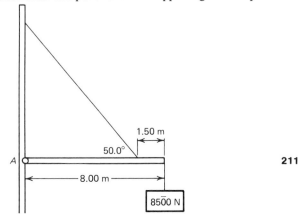

211

21. The uniform boom below is 6.00 m long and weighs 1350 N. Its load weighs 7500 N.

(a) Find the tension in the cable that is attached to the center of the boom.

(b) Find the vertical and horizontal components of the supporting force at point A.

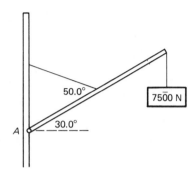

22. The uniform boom above is 6.00 long and weighs 1350 N. Its load weighs 7500 N.

(a) Find the tension in the cable if it is attached 2.50 m from A and makes an angle of 50.0° with the boom.

(b) Find the vertical and horizontal components of the supporting force at point A.

23. The uniform boom above is 6.00 m long and weighs 1350 N. Its load weighs 7500 N.

(a) Find the tension in the cable if it is attached 3.50 m from A and makes an angle of 50.0° with the boom.

(b) Find the vertical and horizontal components of the supporting force at point A.

24. Find the magnitude, direction, and placement (from point A) of F_4 which will produce equilibrium in the force diagram below.

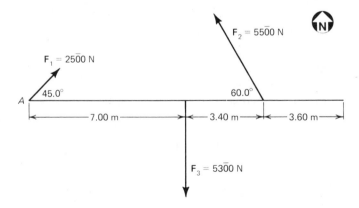

25. Find the magnitude, direction, and placement (from point A) of F_5 which will produce equilibrium in the force diagram below.

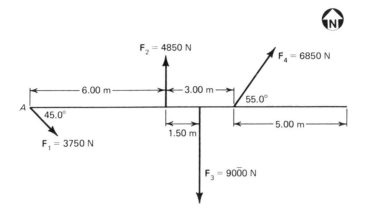

SIMPLE MACHINES

11.1 FUNCTIONS OF MACHINES

Machines are used to transfer energy from one place to another. Energy is transferred from the ignition of the hot gases in the cylinders of an automobile to the rear wheels via the connecting rods, crankshaft, drive shaft, and the rear axle.

Machines are sometimes used to multiply force. For example, diagonal pliers allow a person to cut a wire with the strength in his or her hand. A pulley system allows one person to lift easily an engine from an automobile. A jack allows one person to raise an automobile by exerting a small force. These machines allow us to gain force but at the expense of speed. Note that the engine raises more slowly than the rope being pulled; the jack handle moves through a greater total distance than the automobile; and the handles of the pliers move a greater distance than the wire cutter portion.

Machines are sometimes used to multiply speed. The gears on a bicycle allow a person to gain speed, but at the expense of force. That is, a greater force must be exerted to multiply or gain speed. An aircraft catapult is also an example of a machine that increases speed. Unfortunately, no machine can be designed which gains both a force advantage and a speed advantage at the same time.

Finally, machines may be used to change the direction of a force. For example, a single pulley is used to pull up a weight with neither a force advantage nor a speed advantage.

There are six basic or *simple machines*:

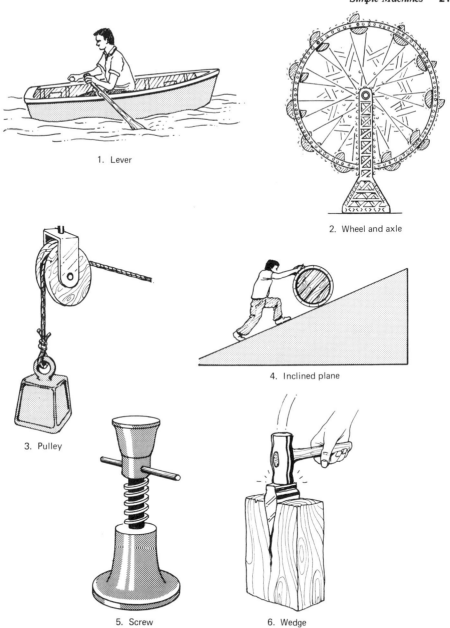

1. Lever

2. Wheel and axle

3. Pulley

4. Inclined plane

5. Screw

6. Wedge

All other machines—no matter how complex—are combinations of two or more of these simple machines.

In every ideal machine we are concerned with two forces—effort and resistance. The *effort* is the force *applied to* the machine. The *resistance* is the force *overcome by* the machine.

The person below applies 150 N on the jack handle to produce a lifting force of 4500 N on the vehicle's bumper. The effort force is 150 N. The resistance force is 4500 N.

In every ideal machine, the effort and resistance forces are related according to the following law:

LAW OF SIMPLE MACHINES

(Resistance force) × (resistance distance) = (effort force) × (effort distance)

$$F_R \cdot s_R = F_E \cdot s_E$$

11.2   MECHANICAL ADVANTAGE

The mechanical advantage of a machine may be defined in two ways:

1. The *ideal mechanical advantage* (IMA) is the ratio of the distance that the effort force moves, s_E, to the distance that the resistance force moves, s_R. That is,

$$\boxed{\text{IMA} = \frac{s_E}{s_R}}$$

2. The *actual mechanical advantage* (AMA) is the ratio of the resistance force, F_R, to the effort force, F_E. That is,

$$\boxed{\text{AMA} = \frac{F_R}{F_E}}$$

Since some of the effort force is used to overcome friction and move intermediate parts of a machine, the AMA is in practice always less than the IMA. The efficiency of a machine is the ratio of the AMA to the IMA expressed as a percent. In other words,

$$\boxed{\text{efficiency} = \frac{\text{AMA}}{\text{IMA}} \times 100 \text{ percent}}$$

Another useful variation of efficiency follows from the definitions of AMA and IMA:

$$\frac{\text{AMA}}{\text{IMA}} = \frac{\dfrac{F_R}{F_E}}{\dfrac{s_E}{s_R}} = \frac{F_R s_R}{F_E s_E} = \frac{W_{\text{output}}}{W_{\text{input}}}$$

Note that the product $F_R s_R$ is the work done *by* the machine, while the product $F_E s_E$ is the work put *into* the machine. So

$$\text{efficiency} = \frac{W_{\text{output}}}{W_{\text{input}}} \times 100 \text{ percent}$$

example   A force of $6\bar{0}$ N on a pulley system moves the rope 15 m, which lifts a weight of 144 N to a height of 5.0 m.

(a) Find the IMA of the pulley system.
(b) Find its AMA.
(c) Find its efficiency.

DATA:
$$F_R = 144 \text{ N}$$

$$s_R = 5.0 \text{ m}$$

$$F_E = 6\bar{0} \text{ N}$$

$$s_E = 15 \text{ m}$$

(a) $\text{IMA} = \dfrac{s_E}{s_R} = \dfrac{15 \text{ m}}{5.0 \text{ m}} = \dfrac{3.0}{1}$ or   3.0

(b) $\text{AMA} = \dfrac{F_R}{F_E} = \dfrac{144 \text{ N}}{60 \text{ N}} = \dfrac{2.4}{1}$ or   2.4

(c) $\text{efficiency} = \dfrac{\text{AMA}}{\text{IMA}} \times 100 \text{ percent} = \dfrac{2.4}{3.0} \times 100 \text{ percent} = 8\bar{0} \text{ percent}$

11.3   LEVER

A *lever* consists of a rigid bar free to turn on a pivot called a fulcrum (page 218).

The ideal mechanical advantage (IMA) is the ratio of the length of the effort arm (s_E) to the length of the resistance arm (s_R). That is,

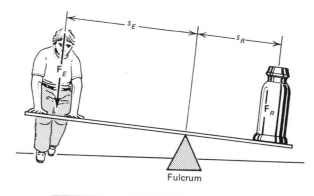

$$\text{IMA}_{\text{lever}} = \frac{\text{effort arm}}{\text{resistance arm}} = \frac{s_E}{s_R}$$

The *effort arm* is the distance from the effort to the fulcrum. The *resistance arm* is the distance from the fulcrum to the resistance.

There are three types of classes of levers:

1. First class: The fulcrum is between the resistance force (F_R) and the effort force (F_E).

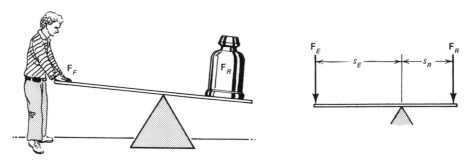

2. Second class: The resistance force (F_R) is between the fulcrum and the effort force (F_E).

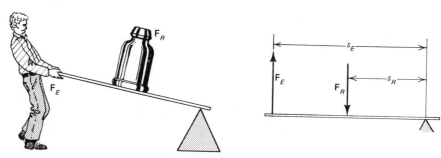

3. Third class: The effort force (F_E) is between the fulcrum and the resistance force (F_R).

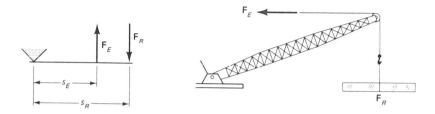

Law of simple machines as applied to levers (basic equation):

$$F_R \cdot s_R = F_E \cdot s_E$$

example 1 A crowbar is used to raise a 2400-N stone. The pivot is placed 24 cm from the stone. The worker pushes 288 cm from the pivot.

(a) What is the ideal mechanical advantage?
(b) What force does the worker exert?

SKETCH:

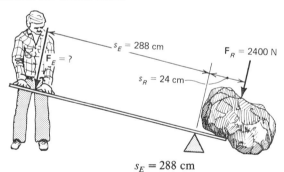

(a) DATA:

$$s_E = 288 \text{ cm}$$

$$s_R = 24 \text{ cm}$$

$$\text{IMA} = ?$$

$$\text{IMA} = \frac{s_E}{s_R} = \frac{288 \text{ cm}}{24 \text{ cm}} = \frac{12}{1}$$

To find the force:
(b) DATA:

$$s_E = 288 \text{ cm}$$

$$s_R = 24 \text{ cm}$$

$$F_R = 2400 \text{ N}$$

$$F_E = ?$$

BASIC EQUATION: $F_R \cdot s_R = F_E \cdot s_E$

WORKING EQUATION: $F_E = \dfrac{F_R \cdot s_R}{s_E}$

SUBSTITUTION: $F_E = \dfrac{(2400 \text{ N})(24 \text{ cm})}{288 \text{ cm}} = 2\bar{0}0 \text{ N}$

example 2   A wheelbarrow 2.00 m long has a $9\bar{0}0$-N load 0.50 m from the axle.

(a) What is the IMA?
(b) What force is needed to lift the wheelbarrow?

SKETCH:

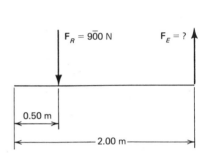

(a) $\text{IMA} = \dfrac{s_E}{s_R} = \dfrac{2.00 \text{ m}}{0.50 \text{ m}} = 4.0$

To find the force:

(b) DATA: $s_E = 2.00 \text{ m}$

                        $s_R = 0.50 \text{ m}$

                        $F_R = 9\bar{0}0 \text{ N}$

                        $F_E = ?$

BASIC EQUATION: $F_R \cdot s_R = F_E \cdot s_E$

WORKING EQUATION: $F_E = \dfrac{F_R \cdot s_R}{s_E}$

SUBSTITUTION: $F_E = \dfrac{(9\bar{0}0 \text{ N})(0.50 \text{ m})}{2.00 \text{ m}} = 230 \text{ N}$

example 3   The AMA of a pair of pliers is 6.0. A force of 25 N is exerted on the handle. What force is exerted on a wire in the pliers?

AMA = 6.0 means that, for each newton of force applied on the handle, 6.0 N is exerted on the wire. Therefore, if a force of 25 N is applied on the handle, a force of (6.0) (25 N), or 150 N, is exerted on the wire.

11.4   WHEEL AND AXLE

This simple machine consists of a large *wheel* attached to an *axle* so that both turn together. Both the wheel and the axle have the same angular velocity.

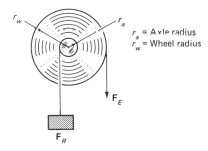

r_a = Axle radius
r_w = Wheel radius

Note that during one complete revolution, the distance that F_E moves is $2\pi r_w$, the circumference of the wheel. And the distance that F_R moves is $2\pi r_a$, the circumference of the axle. Thus, $s_E = C_w = 2\pi r_w$ and $s_R = C_a = 2\pi r_a$. Then, the IMA of the wheel and axle is

$$\text{IMA} = \frac{s_E}{s_R} = \frac{2\pi r_w}{2\pi r_a} = \frac{r_w}{r_a}$$

That is,

$$\boxed{\text{IMA} = \frac{\text{radius of wheel}}{\text{radius of axle}} = \frac{r_w}{r_a} = \frac{D_w}{D_a} = \frac{\text{diameter of wheel}}{\text{diameter of axle}}}$$

and

$$\text{AMA} = \frac{F_R}{F_E}$$

Then, the law of simple machines as applied to the wheel and axle (basic equation) is

$$\boxed{F_R \cdot r_a = F_E \cdot r_w}$$

example   The winch shown on page 222 has a handle that turns in a radius of 36 cm. The radius of the drum or axle is 12 cm.

(a) Find the IMA of the winch.
(b) Find the force required to lift a bucket weighing 420 N.

(a) DATA:
$$\text{IMA} = \frac{r_w}{r_a} = \frac{36 \text{ cm}}{12 \text{ cm}} = \frac{3.0}{1}$$

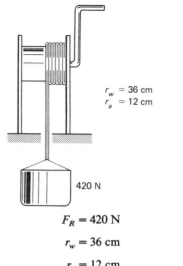

$r_w = 36$ cm
$r_a = 12$ cm

420 N

(b) DATA:

$$F_R = 420 \text{ N}$$
$$r_w = 36 \text{ cm}$$
$$r_a = 12 \text{ cm}$$
$$F_E = ?$$

BASIC EQUATION: $$F_R \cdot r_a = F_E \cdot r_w$$

WORKING EQUATION: $$F_E = \frac{F_R \cdot r_a}{r_w}$$

SUBSTITUTION: $$F_E = \frac{(420 \text{ N})(12 \text{ cm})}{(36 \text{ cm})} = 140 \text{ N}$$

11.5   PULLEY

A *pulley* is a grooved wheel that turns readily on an axle and is supported in a frame. It can be fastened to a fixed object or it can be fastened to the resistance that is to be moved. If the pulley is fastened to a fixed object, it is called a *fixed pulley*. If the pulley is fastened to the resistance to be moved, it is called a *movable pulley*. (See top of page 223.)

The law of simple machines as applied to pulleys (basic equation) is

$$\boxed{F_R \cdot s_R = F_E \cdot s_E}$$

From the last equation,

$$\frac{F_R}{F_E} = \frac{s_E}{s_R} \qquad \text{(See bottom of page 223.)}$$

$$\text{AMA} = \frac{F_R}{F_E}$$

$$\text{IMA} = \frac{s_E}{s_R}$$

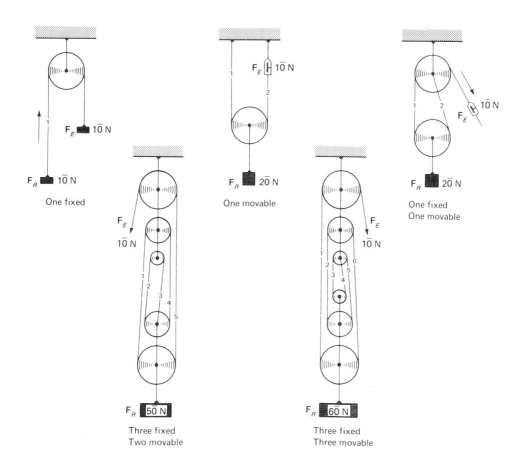

One fixed

One movable

One fixed
One movable

Three fixed
Two movable

Three fixed
Three movable

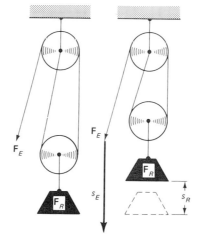

However, when one continuous cord is used, this ratio reduces to the number of strands holding the resistance in the pulley system. Therefore,

> IMA = *the number of strands holding the resistance*

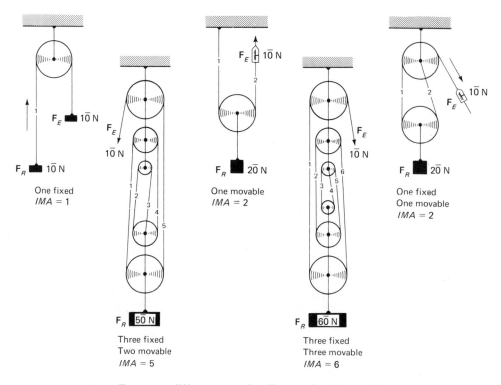

One fixed
IMA = 1

Three fixed
Two movable
IMA = 5

One movable
IMA = 2

Three fixed
Three movable
IMA = 6

One fixed
One movable
IMA = 2

example 1 Draw two different sets of pulleys, each with an IMA of 4.

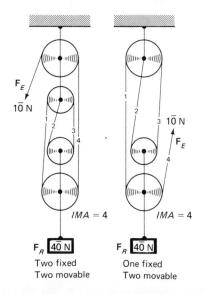

IMA = 4

Two fixed
Two movable

IMA = 4

One fixed
Two movable

example 2   Ideally, what effort will lift a resistance of 240 N in the pulley system in Example 1?

DATA: $\text{IMA} = 4$

$$F_R = 240 \text{ N}$$

$$F_E = ?$$

BASIC EQUATION: $$\text{IMA} = \frac{F_R}{F_E}$$

WORKING EQUATION: $$F_E = \frac{F_R}{\text{IMA}}$$

SUBSTITUTION: $$F_E = \frac{240 \text{ N}}{4} = 6\bar{0} \text{ N}$$

example 3   If the resistance moves 2.00 m, what is the effort distance of the pulley system in Example 1?

DATA: $\text{IMA} = 4$

$$s_R = 2.00 \text{ m}$$

$$s_E = ?$$

BASIC EQUATION: $$\text{IMA} = \frac{s_E}{s_R}$$

WORKING EQUATION: $$s_E = (s_R)(\text{IMA})$$

SUBSTITUTION: $$s_E = (2.00 \text{ m})(4)$$

$$= 8.00 \text{ m}$$

example 4   The pulley system shown is used to raise a 650-N object 15 m.

(a) What is the ideal mechanical advantage?
(b) What ideal force is exerted?
(c) What is the effort distance; that is, how many metres of rope were pulled through the person's hand?

(a) IMA = number of strands holding up the resistance

$$= 5$$

To find the ideal force exerted:

(b) DATA: $$F_R = 650 \text{ N}$$

$$\text{IMA} = 5$$

$$F_E = ?$$

BASIC EQUATION: $$\text{IMA} = \frac{F_R}{F_E}$$

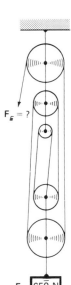

$F_E = ?$

F_R $\boxed{65\bar{0} \text{ N}}$

Three fixed
Two movable
IMA = ?

WORKING EQUATION:

$$F_E = \frac{F_R}{IMA}$$

SUBSTITUTION:

$$F_E = \frac{650 \text{ N}}{5}$$

$$= 130 \text{ N}$$

(c) DATA:

$$IMA = 5$$

$$s_R = 15 \text{ m}$$

$$s_E = ?$$

BASIC EQUATION:

$$IMA = \frac{s_E}{s_R}$$

WORKING EQUATION:

$$s_E = (IMA)(s_R)$$

SUBSTITUTION:

$$s_E = (5)(15 \text{ m})$$

$$= 75 \text{ m}$$

example 5   In Example 4, the actual effort force needed to raise the 650-N object was 180 N.

(a) What is the actual mechanical advantage?
(b) What is the efficiency of this pulley system?

(a)

$$AMA = \frac{F_R}{F_E} = \frac{650 \text{ N}}{180 \text{ N}} = \frac{3.6}{1} \quad \text{or} \quad 3.6$$

(b)

$$efficiency = \frac{AMA}{IMA} \times 100 \text{ percent}$$

$$= \frac{3.6}{5} \times 100 \text{ percent}$$

$$= 72 \text{ percent}$$

11.6   INCLINED PLANE

Gangplanks, chutes, and ramps are all examples of the *inclined plane*. Inclined planes are used to raise heavy objects that are too heavy to lift vertically.

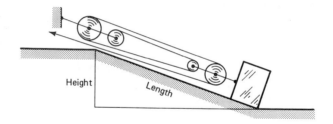

The work done in raising a resistance using the inclined plane equals the resistance times the height. This must also equal the work input (if there is no friction), which can be found by multiplying the effort times the length of the plane.

$$F_R \cdot s_R = F_E \cdot s_E \qquad \text{(law of machines)}$$

$$F_R \cdot (\text{height}) = F_E \cdot (\text{length})$$

From the last equation:

$$\frac{F_R}{F_E} = \frac{\text{length of plane}}{\text{height of plane}}$$

Then

$$\text{AMA} = \frac{F_R}{F_E}$$

and

$$\text{IMA} = \frac{\text{length of plane}}{\text{height of plane}} = \frac{l}{h}$$

example   A worker is pushing a box weighing 1650 N up a ramp 4.50 m long onto a platform 1.50 m above the ground.

(a) What is the ideal mechanical advantage?
(b) Ideally, what effort force is required?
(c) If the coefficient of sliding friction is 0.40, what actual effort force is required?
(d) What is the actual mechanical advantage?

SKETCH:

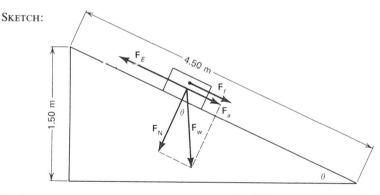

(a) DATA:

$$l = 4.50 \text{ m}$$

$$h = 1.50 \text{ m}$$

$$\text{IMA} = ?$$

BASIC EQUATION: $$\text{IMA} = \frac{l}{h}$$

WORKING EQUATION: same

SUBSTITUTION: $$\text{IMA} = \frac{4.50 \text{ m}}{1.50 \text{ m}}$$

$$= \frac{3.00}{1} \quad \text{or} \quad 3.00$$

(b) DATA: $$F_w = F_R' = 1650 \text{ N}$$

$$\text{IMA} = 3.00$$

$$F_a = F_E' = ?$$

BASIC EQUATION: $$\text{IMA} = \frac{F_R'}{F_E'}$$

WORKING EQUATION: $$F_a = \frac{F_w}{\text{IMA}}$$

SUBSTITUTION: $$F_a = \frac{1650 \text{ N}}{3.00}$$

$$= 55\bar{0} \text{ N}$$

(c) Note: $F_E = F_a + F_f$. [Assuming that the box is sliding at a constant speed (that is, in equilibrium), the sum of the forces up the plane equals the sum of the forces down the plane.]

First, find angle θ:

DATA: $$h = 1.50 \text{ m}$$

$$l = 4.50 \text{ m}$$

$$\theta = ?$$

BASIC EQUATION: $$\sin \theta = \frac{h}{l}$$

WORKING EQUATION: same

SUBSTITUTION: $$\sin \theta = \frac{1.50 \text{ m}}{4.50 \text{ m}} = 0.333$$

$$\theta = 19.5°$$

To find F_f, we first need to find F_N:

DATA: $$F_w = 1650 \text{ N}$$

$$\theta = 19.5°$$

$$F_N = ?$$

BASIC EQUATION: $$F_N = F_w \cos \theta$$

WORKING EQUATION: same

SUBSTITUTION:	$F_N = (1650 \text{ N})(\cos 19.5°)$
	$= 1560 \text{ N}$

Now, to find F_f:

DATA:	$\mu = 0.40$
	$F_N = 1560 \text{ N}$
	$F_f = ?$
BASIC EQUATION:	$\mu = \dfrac{F_f}{F_N}$
WORKING EQUATION:	$F_f = \mu F_N$
SUBSTITUTION:	$F_f = (0.40)(1560 \text{ N})$
	$= 620 \text{ N}$
Then,	$F_E = F_a + F_f$
	$= 55\overline{0} \text{ N} + 620 \text{ N}$
	$= 1170 \text{ N}$
(d) DATA:	$F_w = F_R = 1650 \text{ N}$
	$F_E = 1170 \text{ N}$
	AMA $= ?$
BASIC EQUATION:	AMA $= \dfrac{F_R}{F_E}$
WORKING EQUATION:	same
SUBSTITUTION:	AMA $= \dfrac{1650 \text{ N}}{1170 \text{ N}}$
	$= 1.41$

11.7 SCREW

A *screw* is an inclined plane wrapped around a cylinder. The jack screw and wood screw are examples of this simple machine.

Wood screw

Jack screw

The distance the jack screw rises or the distance the wood screw advances into a piece of wood in one revolution is called the pitch of the screw. Therefore, the pitch of a screw is actually the distance between two successive threads.

From the law of machines:

$$F_R \cdot s_R = F_E \cdot s_E$$

However,

$$s_R = \text{pitch}$$

$$s_E = \text{circumference of the handle of the screwdriver}$$

But $C = 2\pi r$, so

$$s_E = 2\pi r$$

where r is the radius of the handle of the screwdriver. Therefore, the law of simple machines as applied to the screw is

$$\boxed{F_R \cdot (\text{pitch}) = F_E \cdot (2\pi r)}$$

or

$$\frac{F_R}{F_E} = \frac{2\pi r}{\text{pitch}}$$

Then

$$\text{AMA} = \frac{F_R}{F_E}$$

and

$$\boxed{\text{IMA} = \frac{2\pi r}{\text{pitch}}}$$

example 1   A jack screw has a pitch of 3.2 mm and a handle radius of 24 cm.

(a) Find the ideal mechanical advantage.
(b) Ideally, what resistance can be lifted if a force of 120 N is exerted?

(a) DATA: pitch = 3.2 mm = 0.32 cm

$$r = 24 \text{ cm}$$

$$\text{IMA} = ?$$

BASIC EQUATION: $\text{IMA} = \dfrac{2\pi r}{\text{pitch}}$

WORKING EQUATION: same

SUBSTITUTION: $IMA = \dfrac{2\pi(24 \text{ cm})}{0.32 \text{ cm}}$

$= 470$

(b) DATA: $F_E = 120 \text{ N}$

$IMA = 470$

$F_R = ?$

BASIC EQUATION: $IMA = \dfrac{F_R}{F_E}$

WORKING EQUATION: $F_R = (IMA)(F_E)$

SUBSTITUTION: $F_R = (470)(120 \text{ N})$

$= 56{,}000 \text{ N}$

example 2   A truck of mass 2100 kg is raised using a jack screw having a pitch of 7.1 mm and a handle radius of $5\bar{0}$ cm.

(a) If its efficiency is 15 percent, what is the actual mechanical advantage?
(b) What actual effort force is needed?

First, find the IMA:

(a) DATA: $r = 5\bar{0}$ cm

pitch $= 7.1$ mm $= 0.71$ cm

IMA $= ?$

BASIC EQUATION: $IMA = \dfrac{2\pi r}{\text{pitch}}$

WORKING EQUATION: same

SUBSTITUTION: $IMA = \dfrac{2\pi(5\bar{0} \text{ cm})}{0.71 \text{ cm}}$

$= 440$

Now, find the AMA:

DATA: efficiency $= 15$ percent $= 0.15$

IMA $= 440$

AMA $= ?$

BASIC EQUATION: efficiency $= \dfrac{AMA}{IMA}$

WORKING EQUATION: $AMA = (IMA)(\text{efficiency})$

SUBSTITUTION: $AMA = (440)(0.15)$

$= 66$

(b) DATA:

$$F_R = (2100 \text{ kg})(9.80 \text{ m/s}^2) = 2.1 \times 10^4 \text{ N}$$

$$\text{AMA} = 66$$

$$F_E = ?$$

BASIC EQUATION:

$$\text{AMA} = \frac{F_R}{F_E}$$

WORKING EQUATION:

$$F_E = \frac{F_R}{\text{AMA}}$$

SUBSTITUTION:

$$F_E = \frac{2.1 \times 10^4 \text{ N}}{66} = 320 \text{ N}$$

11.8   WEDGE

A *wedge* is a double inclined plane in which the plane is moved instead of the resistance.

Thus,

$$\boxed{\text{IMA} = \frac{L}{t}}$$

However, the ideal mechanical advantage of a wedge is not very practical, because of the large amount of friction involved. And the AMA is considerably less than the IMA. The friction can be lessened by sharpening and smoothing the surfaces. Any quantitative discussion of the wedge is of little significance, at best. Examples of a wedge include a hatchet, a nail, and a wood chisel.

11.9 COMPOUND MACHINES

A *compound machine* is a combination of simple machines. In almost all compound machines, the *total ideal mechanical advantage is the product of the ideal mechanical advantages of each of the simple machines.*

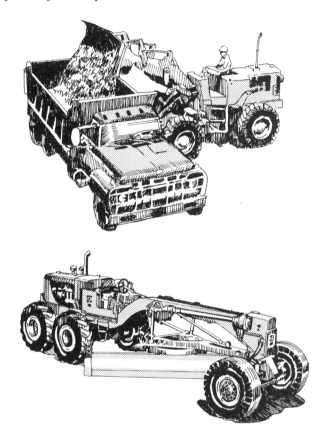

example The 18,000-N box below is pulled up the inclined plane using the pulley system as shown.

(a) Find the IMA of the total system (compound machine).
(b) What effort force must be exerted (assume no frictional forces)?

SKETCH:

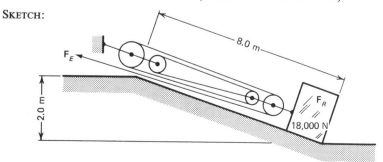

First, find the IMA_{plane}:

(a) DATA:

$$l = 8.0 \text{ m}$$

$$h = 2.0 \text{ m}$$

$$IMA_{plane} = ?$$

BASIC EQUATION:

$$IMA_{plane} = \frac{l}{h}$$

WORKING EQUATION: same

SUBSTITUTION:

$$IMA_{plane} = \frac{8.0 \text{ m}}{2.0 \text{ m}}$$

$$= 4.0$$

Next, $IMA_{pulley} = 5$, and

$$IMA_{system} = (IMA_{plane})(IMA_{pulley})$$

$$= (4.0)(5) = 2\bar{0}$$

(b) DATA:

$$IMA = 2\bar{0}$$

$$F_R = 18,000 \text{ N}$$

$$F_E = ?$$

BASIC EQUATION:

$$IMA = \frac{F_R}{F_E}$$

WORKING EQUATION:

$$F_E = \frac{F_R}{IMA}$$

SUBSTITUTION:

$$F_E = \frac{18,000 \text{ N}}{2\bar{0}}$$

$$= 9\bar{0}0 \text{ N}$$

EXERCISES

Sketch	Data	Basic Equation	Working Equation	Substitution
12 cm² $h = ?$ $b = 6.0$ cm	$A = 12$ cm² $b = 6.0$ cm $h = ?$	$A = bh$	$h = \dfrac{A}{b}$	$h = \dfrac{12 \text{ cm}^2}{6.0 \text{ cm}}$ $= 2.0$ cm

1. A pole is used to lift an automobile that fell off a jack (see page 235). The pivot is 0.75 m from the automobile. Two men exert a force of 1200 N at a distance of 2.0 m from the pivot.

   (a) Find the IMA of the lever.
   (b) What force is applied to the automobile (assuming no friction)?

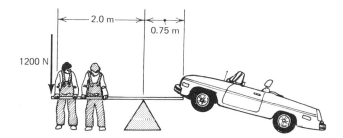

2. A crowbar is used to lift a $10\overline{0}$-kg block of concrete. The pivot is 1.00 m from the block. A person pushes down on the other end of the bar a distance of 2.50 m from the pivot.

   (a) Find the IMA of the lever.

   (b) What force must he apply (assuming no friction)?

3. The handles of a wheelbarrow are 1.50 m from the front wheel. An $8\overline{0}$-kg load is placed 25 cm behind the wheel.

   (a) Find its IMA.

   (b) What lifting force is needed (assuming no friction)?

4. (a) Find the IMA of the lever below.

   (b) Find the force, F_E, pulling up on the beam (the tension in the cable) assuming no friction.

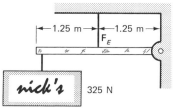

5. The radius of the axle of a winch is 7.5 cm. The length of the handle (radius of wheel) is 45 cm.

   (a) Find the IMA.

   (b) What weight will be lifted by an effort of 325 N (assuming no friction)?

6. A wheel having a radius of 0.70 m is attached to an axle that has a radius of 0.20 m.

   (a) Find the IMA.

   (b) What force must be applied to the rim of the wheel to raise a weight of 1500 N (assuming no friction)?

   If the actual effort force is 620 N,

   (c) Find the AMA.

   (d) Determine its efficiency.

7. The diameter of the wheel of a wheel and axle is 10.0 m. A force of 475 N is to be raised by applying a force of 142 N.

   (a) Find the IMA if frictional forces are ignored.

   (b) Find the diameter of the axle if frictional forces are ignored.

   (c) Find the diameter of the axle if the efficiency is $5\overline{0}$ percent.

What is the ideal mechanical advantage of each of the following systems?

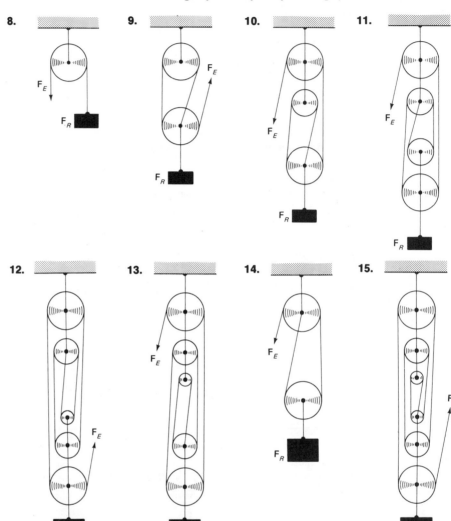

In Exercises 16–21, draw each of the following pulley systems.

16. One fixed and two movable.
17. Two fixed and two movable with an IMA of 5.
18. Three fixed and three movable with an IMA of 6.
19. Four fixed and three movable.
20. Four fixed and four movable with an IMA of 8.
21. Three fixed and three movable with an IMA of 7.

22. What is the IMA of a single movable pulley?

23. (a) What effort force is needed to lift a 1200-N weight by using a single movable pulley (assuming no frictional forces)?
    (b) If the weight is lifted 2.5 m, how much rope must be pulled by the person exerting the force?
    (c) If 675 N are actually required to lift the weight, what is the AMA?
    (d) What is the system's efficiency?

24. (a) Draw a system consisting of two fixed pulleys and two movable pulleys with an ideal mechanical advantage of 4.
    (b) If a force of 115 N is to be exerted, what weight can be raised (assuming no frictional forces)?
    (c) If the weight is raised 20.5 m, what length of rope is pulled?
    (d) If the system's efficiency is $6\bar{0}$ percent, what is the AMA?
    (e) If the system's efficiency is $6\bar{0}$ percent, what weight does the 115-N effort force raise?

25. (a) Draw a system of one fixed and two movable pulleys.
    (b) What is the ideal mechanical advantage of this system?
    (c) A $40\bar{0}$-N weight is to be lifted 30.0 m. Find the effort force and effort distance (assuming no frictional forces).
    (d) An effort force of 125 N is applied through an effort distance of 13.0 m. Find the weight of the load and the distance it is moved if the efficiency is $5\bar{0}$ percent.

26. An inclined plane is 10.0 m long and 2.50 m high.
    (a) What is the ideal mechanical advantage?
    (b) A resistance of 727 N is to be pushed up the plane. What effort is needed (assuming no frictional forces)?
    (c) An effort of $20\bar{0}$ N is applied to push an 815-N resistance up the inclined plane. Is the effort enough (assuming no frictional forces)?
    (d) If the coefficient of sliding friction is 0.30, what actual effort is required?
    (e) What is the AMA?

27. A safe is to be lifted onto a truck whose bed in 1.75 m off the ground. The mass of the safe is 345 kg.
    (a) If the effort to be applied is $60\bar{0}$ N, what length of ramp is needed to raise the safe (assuming no frictional forces)?
    (b) What is the IMA?
    (c) If the coefficient of sliding friction is 0.25, what actual effort is required to push the safe *up* the inclined plane?
    (d) If the coefficient of sliding friction is 0.25, what actual effort is required to push the safe *down* the inclined plane?
    (e) What is the actual AMA?

28. An 1800-kg automobile is to be raised using a jack screw having a pitch of 3.2 mm and a handle $4\bar{0}$ cm long.
    (a) Find the IMA.
    (b) Ideally, what effort force is required to lift the automobile?
    (c) If its efficiency is 12 percent, what is the AMA?
    (d) What actual effort force is required to lift the automobile?

29. (a) The handle of a jack screw is $6\bar{0}$ cm long. If the ideal mechanical advantage is 350, what is the pitch?

   (b) How much weight can be raised by applying a force of 430 N to the jack screw handle (assuming no frictional forces)?

   (c) If the efficiency is 8.0 percent, what actual force is raised by the 430-N effort force?

30. (a) Find the ideal mechanical advantage of the compound machine below. The radius of the wheel is 36.0 cm and the radius of the axle is 6.00 cm.

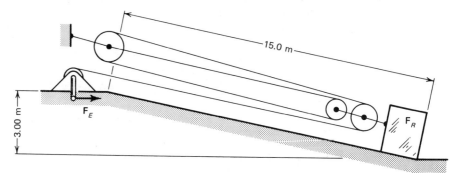

   (b) If an effort force of $30\bar{0}$ N is exerted, what weight can be moved up the inclined plane? (Assume no frictional forces.)

   (c) What effort force is required to move 1.50 metric tons up the inclined plane? (Assume no frictional forces.)

   (d) The coefficient of sliding friction of the inclined plane is 0.45; the pulley system is 75 percent efficient; and the wheel and axle is $6\bar{0}$ percent efficient. What effort force is required to move the 1.50-metric-ton mass up the inclined plane? What is the AMA?

GEARS AND PULLEYS

12.1 INTRODUCTION

Suppose that we have two disks touching each other as shown below. Disk A is driven by a motor and turns disk B by making use of the friction between them.

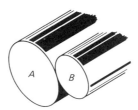

Let r_A be the radius of disk A and ω_A be its angular velocity. Then the linear velocity of any point P on its rim is

$$v = \omega_A r_A$$

And the linear velocity of any point Q on the rim of disk B is

$$v = \omega_B r_B$$

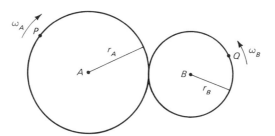

Assuming no slippage, the linear velocity of point P is equal to the linear velocity of point Q. That is,

$$\omega_A r_A = \omega_B r_B$$

or

$$\omega_A \text{dia}_A = \omega_B \text{dia}_B$$

If we divide both sides of this equation by $\omega_A \text{dia}_B$, we have

$$\boxed{\frac{\text{dia}_A}{\text{dia}_B} = \frac{\omega_B}{\omega_A}}$$

Thus, the ratio of the diameters is inversely proportional to the ratio of the angular velocities. That is (assuming that disk A remains a constant size and angular velocity) as the diameter of disk B increases, its angular velocity decreases, and vice versa.

However, using two disks to transfer rotational motion is not very efficient, because of the slippage that usually occurs between disks. The most common ways to prevent disk slippage are the placing of teeth on the edge of the disk and the connecting of disks with a belt. Therefore, instead of using disks, we use gears or belt-driven pulleys to transfer this motion. The teeth on the gears eliminate the slippage, and the belt connecting the pulleys helps reduce the slippage.

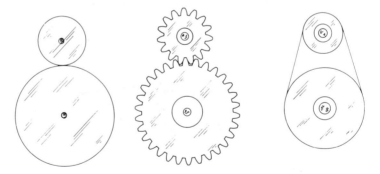

The linear velocity of points on the edges of two meshed gears are equal. Similarly, the linear velocity of points on the rims (and on the belt itself) of two pulleys connected by a belt are also equal. Thus, the basic relationship

$$\boxed{\omega_A \text{dia}_A = \omega_B \text{dia}_B}$$

remains the same for two pulleys connected by a belt and becomes

$$\boxed{\omega_A(\text{number of teeth}_A) = \omega_B(\text{number of teeth}_B)}$$

for gears. This basic relationship, sometimes written using different letters, becomes the basis for the remainder of this chapter.

12.2 GEARS

Gears are used to transfer rotational motion from one gear to another. The gear that causes the motion is called the *driver* gear. The gear to which the motion is transferred is called the *driven* gear.

There are many different sizes, shapes, and types of gears. A few examples are shown below and on the following page.

Some of these gears and more are shown in the transmission below.

Photograph by R. T. Gladin.

Photograph courtesy Illinois Gear,
Wallace-Murray Corporation.

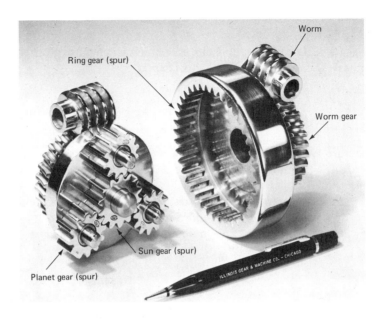

Ring gear (spur)

Worm

Worm gear

Sun gear (spur)

Planet gear (spur)

Photograph courtesy Illinois Gear,
Wallace-Murray Corporation.

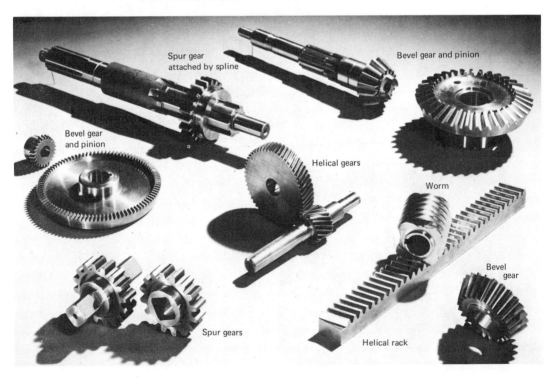

Spur gear
attached by spline

Bevel gear and pinion

Bevel gear
and pinion

Helical gears

Worm

Bevel
gear

Spur gears

Helical rack

Photograph courtesy Illinois Gear, Wallace-Murray Corporation.

For any pair of gears, we use one basic formula:

$$T \cdot \omega_N = t \cdot \omega_n$$

where T is the number of teeth on the driver,

ω_N is the angular velocity of the driver,

t is the number of teeth on the driven, and

ω_n is the angular velocity of the driven.

example 1   A driven gear of 70 teeth makes 63 revolutions per minute (rpm). The driver gear makes 90 rpm. What is the number of teeth required for the driver gear?

DATA: $\omega_N = 90$ rpm

$t = 70$ teeth

$\omega_n = 63$ rpm

$T = ?$

BASIC EQUATION: $T \cdot \omega_N = t \cdot \omega_n$

WORKING EQUATION: $T = \dfrac{t \cdot \omega_n}{\omega_N}$

SUBSTITUTION: $T = \dfrac{(70 \text{ teeth})(63 \text{ rpm})}{90 \text{ rpm}}$

$= 49$ teeth

example 2   A driver gear has 30 teeth. How many revolutions does the driver gear with 20 teeth make while the driver makes 1 revolution?

DATA: $T = 30$ teeth

$\theta_N = 1$ revolution

$t = 20$ teeth

$\theta_n = ?$

BASIC EQUATION: $T \cdot \theta_N = t \cdot \theta_n$

WORKING EQUATION: $\theta_n = \dfrac{T \cdot \theta_N}{t}$

SUBSTITUTION: $\theta_n = \dfrac{(30 \text{ teeth})(1 \text{ rev})}{20 \text{ teeth}}$

$= 1.5$ rev

12.3 GEAR TRAINS

When two gears mesh as shown below,* they turn in opposite directions. If gear A turns clockwise, gear B turns counterclockwise. If gear A turns counterclockwise, gear B turns clockwise.

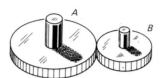

If a third gear is inserted between the two, as below, gears A and B are rotating in the same direction. This third gear is called an *idler*, and such an arrangement of gears is called a *gear train*.

When the number of shafts in a gear train is odd (such as 1, 3, 5, . . .), the first gear and last gear rotate in the same direction. When the number of shafts is even, the gears rotate in opposite directions.

When a very complicated gear train is considered, the relationship between angular velocity and number of teeth is still present. This relationship is: the angular velocity of the first driver times the product of the numbers of teeth of all the driver gears equals the angular velocity of the final driven gear times the product of the numbers of teeth on all the driven gears. That is,

$$\omega_N T_1 T_2 T_3 T_4 \cdots = \omega_n t_1 t_2 t_3 t_4 \cdots$$

where ω_N is the angular velocity of first driver,
   T_1 is the number of teeth on first driver,
   T_2 is the number of teeth on second driver,
   T_3 is the number of teeth on third driver,
   T_4 is the number of teeth on fourth driver,
   ω_n is the angular velocity of last driven,
   t_1 is the number of teeth on first driven,

*Although gears have teeth, in technical work they are usually shown as cylinders.

t_2 is the number of teeth on second driven,

t_3 is the number of teeth on third driven, and

t_4 is the number of teeth on fourth driven.

example 1 Determine the relative rotation of gears A and B below.

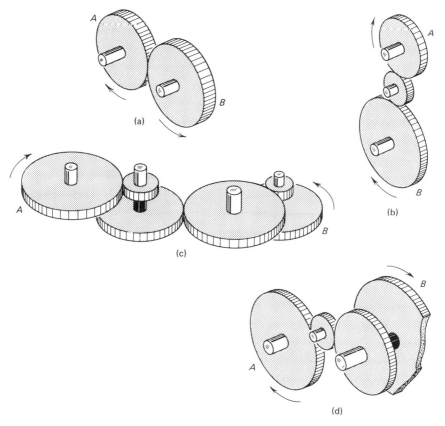

(a)

(b)

(c)

(d)

example 2 Find the angular velocity of gear D below if gear A rotates at 20.0 revolutions per minute.

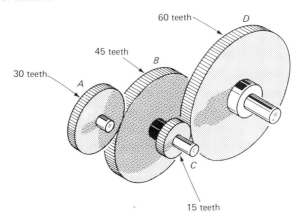

60 teeth — D

45 teeth — B

30 teeth — A

C

15 teeth

Gears A and C are drivers and gears B and D are driven.

DATA:

$$\omega_N = 20.0 \text{ rpm}$$
$$T_1 = 30 \text{ teeth}$$
$$T_2 = 15 \text{ teeth}$$
$$t_1 = 45 \text{ teeth}$$
$$t_2 = 60 \text{ teeth}$$
$$\omega_n = ?$$

BASIC EQUATION:

$$\omega_N T_1 T_2 = \omega_n t_1 t_2$$

WORKING EQUATION:

$$\omega_n = \frac{\omega_N T_1 T_2}{t_1 t_2}$$

SUBSTITUTION:

$$\omega_n = \frac{(20.0 \text{ rpm})(30 \text{ teeth})(15 \text{ teeth})}{(45 \text{ teeth})(60 \text{ teeth})}$$

$$= 3.33 \text{ rpm}$$

example 3 Find the angular velocity of gear D in the train below.

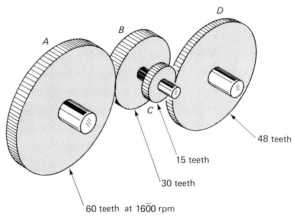

Gears A and C are drivers and gears B and D are driven.

DATA:

$$\omega_N = 16\overline{0}0 \text{ rpm}$$
$$T_1 = 60 \text{ teeth}$$
$$T_2 = 15 \text{ teeth}$$
$$t_1 = 30 \text{ teeth}$$
$$t_2 = 48 \text{ teeth}$$
$$\omega_n = ?$$

BASIC EQUATION:

$$\omega_N T_1 T_2 = \omega_n t_1 t_2$$

Working Equation:

$$\omega_n = \frac{\omega_N T_1 T_2}{t_1 t_2}$$

Substitution:

$$\omega_n = \frac{(16\bar{0}0 \text{ rpm})(60 \text{ teeth})(15 \text{ teeth})}{(30 \text{ teeth})(48 \text{ teeth})}$$

$$= 10\bar{0}0 \text{ rpm}$$

example 4 In the gear train below, find the angular velocity of gear A.

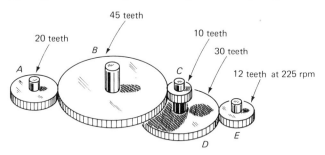

Data:

$$T_1 = 20 \text{ teeth}$$

$$T_2 = 45 \text{ teeth}$$

$$T_3 = 30 \text{ teeth}$$

$$t_1 = 45 \text{ teeth}$$

$$t_2 = 10 \text{ teeth}$$

$$t_3 = 12 \text{ teeth}$$

$$\omega_n = 225 \text{ rpm}$$

$$\omega_N = ?$$

Gear B is both a driver and a driven gear.

Basic Equation:

$$\omega_N T_1 T_2 T_3 = \omega_n t_1 t_2 t_3$$

Working Equation:

$$\omega_N = \frac{\omega_n t_1 t_2 t_3}{T_1 T_2 T_3}$$

Substitution:

$$\omega_N = \frac{(225 \text{ rpm})(45 \text{ teeth})(10 \text{ teeth})(12 \text{ teeth})}{(20 \text{ teeth})(45 \text{ teeth})(30 \text{ teeth})}$$

$$= 45 \text{ rpm}$$

In a gear train, when a gear is both a driver gear and a driven gear, it may be left out of the computation.

example 5 Example 4 could have been worked as follows, since gear B is both a driver and a driven.

Basic Equation:

$$\omega_N T_1 T_3 = \omega_n t_2 t_3$$

WORKING EQUATION:

$$\omega_N = \frac{\omega_n t_2 t_3}{T_1 T_3}$$

SUBSTITUTION:

$$\omega_N = \frac{(225 \text{ rpm})(10 \text{ teeth})(12 \text{ teeth})}{(20 \text{ teeth})(30 \text{ teeth})}$$

$$= 45 \text{ rpm}$$

12.4   PULLEYS CONNECTED WITH A BELT

Pulleys connected with a belt are used to transfer rotational motion from one shaft to another.

Two pulleys connected with a belt have a relationship similar to gears. Assuming no slippage, when two pulleys are connected

$$\boxed{D \cdot \omega_N = d \cdot \omega_n}$$

where D is the diameter of the driver pulley,
  ω_N is the angular velocity of the driver pulley,
  d is the diameter of the driven pulley, and
  ω_n is the angular velocity of the driven pulley.

The preceding equation may be generalized in the same manner as for gear trains to get

$$\boxed{\omega_N D_1 D_2 D_3 \cdots = \omega_n d_1 d_2 d_3 \cdots}$$

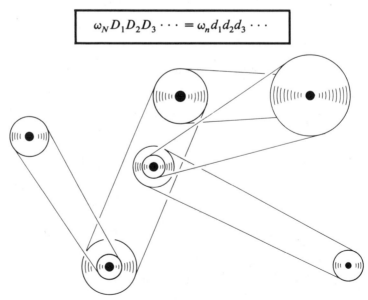

Power Transmission

example   Find the angular velocity of pulley A below.

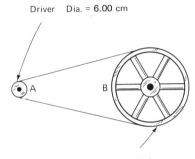

Driver   Dia. = 6.00 cm

A   B

Dia. = 30.0 cm
$35\overline{0}$ rpm

DATA:	$D = 6.00$ cm
	$d = 30.0$ cm
	$\omega_n = 35\overline{0}$ rpm
	$\omega_N = ?$
BASIC EQUATION:	$D \cdot \omega_N = d \cdot \omega_n$
WORKING EQUATION:	$\omega_N = \dfrac{d\omega_n}{D}$
SUBSTITUTION:	$\omega_N = \dfrac{(30.0 \text{ cm})(35\overline{0} \text{ rpm})}{6.00 \text{ cm}}$
	$= 1750$ rpm

When two pulleys are connected with an open type belt, the pulleys turn in the same direction. When two pulleys are connected with a cross type belt, the pulleys turn in opposite directions. This is illustrated below.

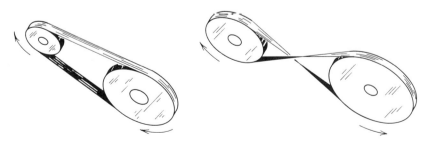

Open type   Cross type

12.5 MECHANICAL ADVANTAGE OF GEARS AND PULLEYS

As we know, torque is the product of the applied force and the length of the moment arm. That is,

$$\tau = Fr \sin \theta$$

if the force is applied at an angle θ with respect to the moment arm

or simply

$$\tau = Fr$$

if the force is applied perpendicular to the moment arm

In gear and pulley systems work is done by transmitting and applying torques.

As we saw in Section 11.2, the ideal mechanical advantage of a machine is defined by

$$\text{IMA} = \frac{s_E}{s_R}$$

For gears (and pulleys), the effort distance for one gear and the resistance distance for the other gear correspond to the radius of the opposite gear (or pulley).

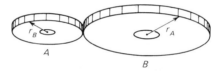

From Section 9.1, we know that the linear distance traveled by a point in circular motion is given by

$$s = r\theta \qquad \text{or} \qquad \frac{s}{\theta} = r$$

For driver gear A (or pulley A), its radius may be expressed as

$$r_B = \frac{s_A}{\theta_A} \qquad \text{(see figure above)}$$

where r_B is the radius of gear A and the moment arm of the resistance of gear B,
θ_A is the angular displacement of gear A, and
s_A is the arc length traveled by a point on the rim of gear A as it rotates θ_A.
Similarly, for driven gear B (or pulley B), its radius may be expressed as

$$r_A = \frac{s_B}{\theta_B}$$

Then

$$\text{IMA} = \frac{s_E}{s_R} = \frac{r_A}{r_B} = \frac{\text{radius of driven}}{\text{radius of driver}} = \frac{\text{diameter of driven}}{\text{diameter of driver}}$$

or, assuming no slippage $(s_A = s_B)$, then

$$\text{IMA} = \frac{r_A}{r_B} = \frac{s_B/\theta_B}{s_A/\theta_A} = \frac{\theta_A}{\theta_B}$$

Or, since $\theta = \omega t$,

$$\boxed{\text{IMA} = \frac{\theta_A}{\theta_B} = \frac{\omega_A t}{\omega_B t} = \frac{\omega_A}{\omega_B} = \frac{\text{angular velocity of driver}}{\text{angular velocity of driven}}}$$

Or, more generally,

$$\boxed{\text{IMA} = \frac{\omega_N}{\omega_n} = \frac{\text{angular velocity of driver}}{\text{angular velocity of driven}}}$$

Since the number of teeth in a gear is directly proportional to its radius (or diameter),

$$\boxed{\text{IMA} = \frac{t}{T} = \frac{\text{number of teeth of driven gear}}{\text{number of teeth of driver gear}}}$$

As we said in Section 11.1, no machine can gain both mechanical advantage and speed at the same time. This idea can be vividly illustrated with gears (and pulleys). In the figure below, gear A is the driver and gear B is driven. Here the IMA > 1 because $\omega_A > \omega_B$. Note that mechanical advantage is gained at the expense of speed delivered to gear B.

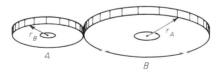

In the figure below, gear A is the driver and gear B is driven. But here the IMA < 1 because $\omega_A < \omega_B$. Note that the speed advantage delivered to gear B is gained at the expense of less mechanical advantage.

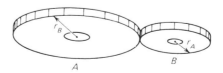

example 1 A gear with 48 teeth revolving at 162 rpm drives another gear with 36 teeth.

(a) What is the angular velocity of the driven gear?

(b) What is the IMA?

(a) DATA:

$$T = 48 \text{ teeth}$$

$$\omega_N = 162 \text{ rpm}$$

$$t = 36 \text{ teeth}$$

$$\omega_n = ?$$

BASIC EQUATION:

$$T \cdot \omega_N = t \cdot \omega_n$$

WORKING EQUATION:

$$\omega_n = \frac{T \cdot \omega_N}{t}$$

SUBSTITUTION:

$$\omega_n = \frac{(48 \text{ teeth})(162 \text{ rpm})}{36 \text{ teeth}}$$

$$= 216 \text{ rpm}$$

(b) Using the angular velocity:

DATA:

$$\omega_N = 162 \text{ rpm}$$

$$\omega_n = 216 \text{ rpm}$$

$$\text{IMA} = ?$$

BASIC EQUATION:

$$\text{IMA} = \frac{\omega_N}{\omega_n}$$

WORKING EQUATION: same

SUBSTITUTION:

$$\text{IMA} = \frac{162 \text{ rpm}}{216 \text{ rpm}}$$

$$= 0.75$$

Using the number of teeth:

DATA:

$$T = 48 \text{ teeth}$$

$$t = 36 \text{ teeth}$$

$$\text{IMA} = ?$$

BASIC EQUATION:

$$\text{IMA} = \frac{t}{T}$$

WORKING EQUATION: same

SUBSTITUTION:

$$\text{IMA} = \frac{36 \text{ teeth}}{48 \text{ teeth}}$$

$$= 0.75$$

Note that angular velocity was gained at the expense of mechanical advantage.

The AMA of two gears (or pulleys) is given by the equation

$$AMA = \frac{\tau_{driven}}{\tau_{driver}}$$

The power of the driver gear (or pulley) may be found by

$$P_{driver} = \tau_{driver}\omega_{driver}$$

And the power of the driven gear (or pulley) may be found by

$$P_{driven} = \tau_{driven}\omega_{driven}$$

The differential pulley or chain hoist is a compound machine commonly used in industry and machine shops to lift heavy objects because of its large mechanical advantage.

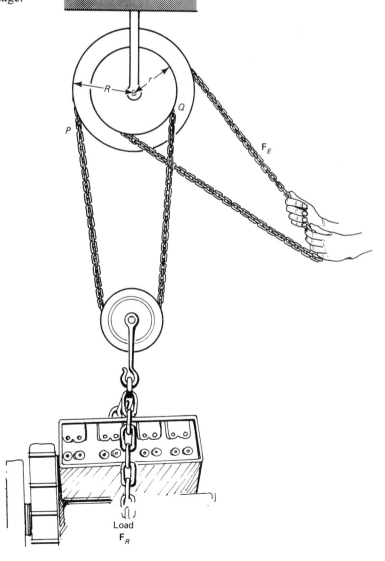

The differential pulley is a combination wheel and axle and block and tackle. It is commonly called a chain hoist because it has a continuous chain loop as shown on page 253 that runs over notched pulleys, which prevents slipping. We let R be the radius of the larger pulley (wheel) and r be the radius of the smaller pulley (axle). The effort distance, which is the circumference of the larger pulley, is $2\pi R$. That is, during one complete rotation of the larger pulley, a length of $2\pi R$ of chain is pulled through it. Through one complete rotation of the smaller pulley, a length of $2\pi r$ of chain is pulled through it. As the chain is pulled a length of $2\pi R$, the loop of chain PQ is shortened by a length equal to the difference of the circumferences of the two pulleys; that is, $2\pi R - 2\pi r = 2\pi(R - r)$. One-half of this difference is taken off each side of loop PQ. Thus, the load is lifted a distance of $\pi(R - r)$, which is the resistance distance. Then the IMA is

$$\text{IMA} = \frac{s_E}{s_R} = \frac{2\pi R}{\pi(R - r)}$$

or

$$\boxed{\text{IMA} = \frac{2R}{R - r}}$$

Obviously, the greatest IMA is obtained when r is only slightly less than R. But then the effort distance becomes very large and much pulling on the chain is necessary to raise the load even a small distance. The AMA is much less than the IMA because of the relatively large amount of friction between the chain and the notched pulleys.

example 2   A 10$\overline{0}$0-kg mass is lifted by a differential pulley that has two upper pulleys of radii 37.0 cm and 35.0 cm.

(a) What is its IMA?
(b) To lift the mass 50.0 cm, what length of chain must be pulled through it?
(c) If the differential pulley is 55 percent efficient, what is its AMA?

(a) DATA: $R = 37.0$ cm

$r = 35.0$ cm

$\text{IMA} = ?$

BASIC EQUATION: $\text{IMA} = \dfrac{2R}{R - r}$

WORKING EQUATION: same

SUBSTITUTION: $\text{IMA} = \dfrac{2(37.0 \text{ cm})}{37.0 \text{ cm} - 35.0 \text{ cm}}$

$= 37.0$

(b) DATA:

$$s_R = 50.0 \text{ cm}$$

$$\text{IMA} = 37.0$$

$$s_E = ?$$

BASIC EQUATION:

$$\text{IMA} = \frac{s_E}{s_R}$$

WORKING EQUATION:

$$s_E - (\text{IMA})(s_R)$$

SUBSTITUTION:

$$s_E = (37.0)(50.0 \text{ cm})$$

$$= 1850 \text{ cm} \quad \text{or} \quad 18.5 \text{ m}$$

(c) DATA:

$$\text{efficiency} = 55 \text{ percent} = 0.55$$

$$\text{IMA} = 37.0$$

$$\text{AMA} = ?$$

BASIC EQUATION:

$$\text{efficiency} = \frac{\text{AMA}}{\text{IMA}}$$

WORKING EQUATION:

$$\text{AMA} = (\text{efficiency})(\text{IMA})$$

SUBSTITUTION:

$$\text{AMA} = (0.55)(37.0)$$

$$= 2\bar{0}$$

EXERCISES

Sketch	Data	Basic Equation	Working Equation	Substitution
$12 \text{ cm}^2 \quad h = ?$ $b = 6.0 \text{ cm}$	$A = 12 \text{ cm}^2$ $b = 6.0 \text{ cm}$ $h = ?$	$A = bh$	$h = \dfrac{A}{b}$	$h = \dfrac{12 \text{ cm}^2}{6.0 \text{ cm}}$ $= 2.0 \text{ cm}$

Fill in the blanks in Exercises 1–6.

	Number of teeth		rpm	
	Driver	Driven	Driver	Driven
1.	58	—	$23\bar{0}$	145
2.	—	150	$24\bar{0}$	$12\bar{0}$
3.	200	240	—	$16\bar{0}$
4.	70	—	$42\bar{0}$	$70\bar{0}$
5.	80	65	$26\bar{0}$	—
6.	—	80	$48\bar{0}$	$78\bar{0}$

7. A driver gear has 72 teeth and makes 45 rpm. What is the angular velocity of the driven gear with 38 teeth?

8. A motor turning at 1250 rpm is fitted with a gear having 56 teeth. Find the angular velocity of the driven gear if it has 66 teeth.

9. A gear running at $25\overline{0}$ rpm drives another revolving at $10\overline{0}$ rpm. If the smaller gear has 30 teeth, how many teeth does the larger gear have?

10. A driven gear with 40 teeth makes 154 rpm. How many teeth must the driver have if it makes $22\overline{0}$ rpm?

11. Two gears have a speed ratio of 4.6 to 1. If the smaller gear has 15 teeth, what must be the number of teeth on the larger gear?

12. What size gear should be mated with a 15-tooth pinion to achieve a speed reduction of 10 to 3?

If gear A turns in a clockwise motion, determine the rotation of gear B in Exercises 13–22.

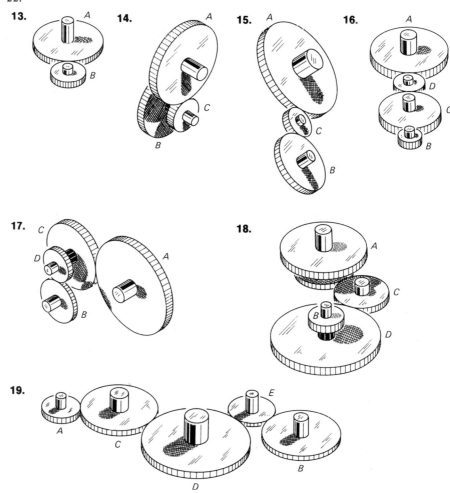

20.

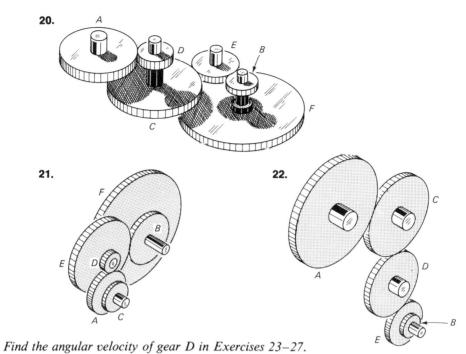

21.   **22.**

Find the angular velocity of gear D in Exercises 23–27.

23.

A = 60 teeth
at 1850 rpm

B = 30 teeth

D = 48 teeth
at ? rpm

C = 15 teeth

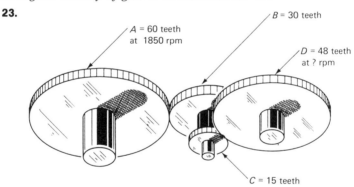

24.

D = 48 teeth
at ? rpm

C = 20 teeth

A = 30 teeth
at 740 rpm B = 45 teeth

25.

R = 30 teeth

C = 48 teeth

A = 45 teeth
at 160 rpm

D = 20 teeth
at ? rpm

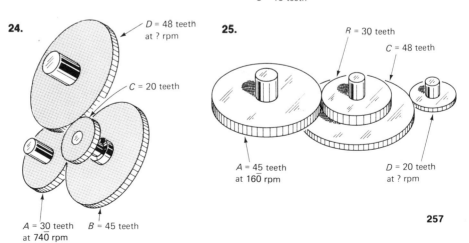

257

26.   A = 20 teeth
at 250 rpm

D = 12 teeth
at ? rpm

C = 10 teeth

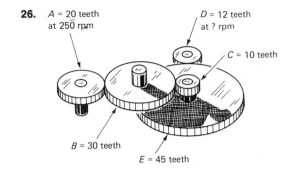

B = 30 teeth

E = 45 teeth

27.

C = 45 teeth

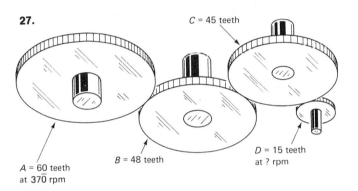

A = 60 teeth
at 370 rpm

B = 48 teeth

D = 15 teeth
at ? rpm

Find the number of teeth for gear D in Exercises 28–32.

28.   D = ? teeth
at 1500 rpm

29.

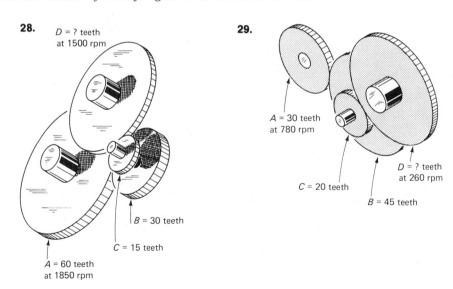

A = 30 teeth
at 780 rpm

C = 20 teeth

D = ? teeth
at 260 rpm

B = 45 teeth

B = 30 teeth

C = 15 teeth

A = 60 teeth
at 1850 rpm

30.

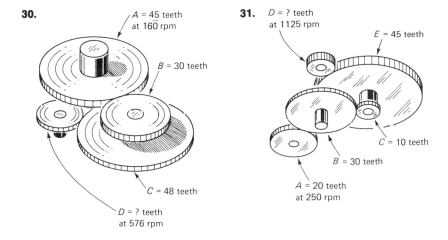

A = 45 teeth at 160 rpm

B = 30 teeth

C = 48 teeth

D = ? teeth at 576 rpm

31. *D* = ? teeth at 1125 rpm

E = 45 teeth

C = 10 teeth

B = 30 teeth

A = 20 teeth at 250 rpm

32.

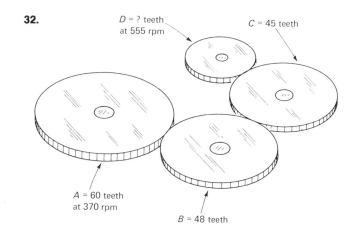

D = ? teeth at 555 rpm

C = 45 teeth

A = 60 teeth at 370 rpm

B = 48 teeth

33. The diameter of a driving pulley is 18.0 cm and revolves at 1650 rpm. At what speed will the driven pulley revolve if it is 26.0 cm in diameter?

34. The diameter of a driving pulley is 25 cm and makes 120 rpm. At what speed will the driven pulley turn if it is 42 cm in diameter?

35. The diameter of a driving pulley is 18 cm and makes 600 rpm. What is the diameter of the driven pulley if it rotates at 360 rpm?

36. A driving pulley rotates at 440 rpm. The diameter of the driven pulley is 15 cm and makes 680 rpm. What is the diameter of the driving pulley?

37. The radius of a driving pulley is 40 cm and rotates at 120 rpm. The radius of the driven pulley is 30 cm. Find the rpm of the driven pulley.

Determine the rotation of pulley B in Exercises 38–42.

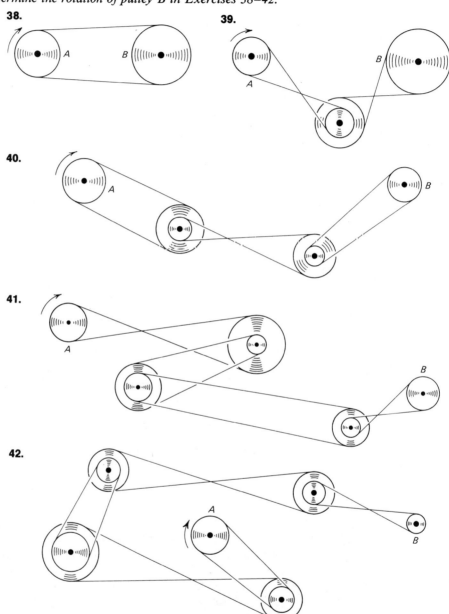

43. What size pulley should be placed on a countershaft turning 150 rpm to drive a grinder with a $4\bar{0}$ cm pulley that is to turn at 1200 rpm?

44. Find the IMA of the gear system in Exercise 8.

45. Find the IMA of the gear system in Exercise 7.

46. Find the IMA of the gear system in Exercise 10.

47. Find the IMA of the gear system in Exercise 9.

48. Find the IMA of the pulley system in Exercise 34.

49. Find the IMA of the pulley system in Exercise 36.

50. A 75$\overline{0}$-kg mass is lifted by a chain hoist that has two upper pulleys of radii 24.0 cm and 22.5 cm.

    (a) What is its IMA?

    (b) Ideally, what effort force is required to lift the mass?

    (c) To lift the mass 50.0 cm, what length of chain must be pulled through it?

51. (a) If the chain hoist in Exercise 50 is 45.0 percent efficient, what is its AMA?

    (b) What effort is actually needed to raise the mass?

    (c) If the mass was lifted 50.0 cm, how much work was done by the machine?

    (d) If the mass was lifted 50.0 cm, how much work was done by the effort force?

    (e) If the mass was lifted 50.0 cm in 30.0 s, how much power was developed by the machine?

    (f) If the mass was lifted 50.0 cm in 30.0 s, how much power was delivered to the machine by the effort force?

13

PROPERTIES OF MATTER

13.1 NATURE OF MATTER

What are the building blocks of matter? First, matter is anything that occupies space and has mass. Suppose that we take a cube of sugar and divide it into two pieces. Then, divide a resulting piece into two pieces. Can we continue this process indefinitely and get smaller and smaller particles of sugar each time? No, at some point in time the subdivision will result in something different from sugar.

A *molecule* is the smallest particle of a substance that exists in a stable and independent state. Most simple molecules are about 3×10^{-10} m in diameter, which is indeed small.

What do we get if we divide the sugar molecule into pieces? We would find that these resulting particles are three simpler kinds of matter—carbon, hydrogen, and oxygen—which are called chemical elements or atoms. An *atom* is the smallest particle of an element that can exist either alone or in combination with other atoms of the same or different elements. Thus, a sugar molecule is made up of carbon, hydrogen, and oxygen atoms.

Models of water and sugar molecules are shown below and on page 263.

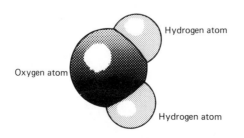

Oxygen atom

Hydrogen atom

Hydrogen atom

(a) The water molecule is composed of two hydrogen atoms and one oxygen atom (H_2O).

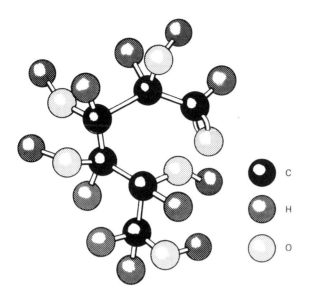

(b) The sugar (glucose) molecule shown above is composed of six carbon atoms, twelve hydrogen atoms, and six oxygen atoms.

C

H

O

The size of atoms varies. The hydrogen atom is the smallest. Its diameter is 6×10^{-11} m and its mass is 1.67×10^{-27} kg. Uranium is one of the heaviest atoms, 3.96×10^{-25} kg.

What happens if an atom is subdivided? Several particles of the atom have been discovered. Of these, the three most important particles of the atom are the proton, the electron, and the neutron. We will limit our discussion here to these three. The table below provides some of the basic information known about these three particles.

Particle	Mass	Diameter	Charge
Proton	1.672×10^{-27} kg	2–3×10^{-15} m	1
Electron	9.108×10^{-31} kg	10^{-14} (approx.)	-1
Neutron	1.675×10^{-27} kg	2–3×10^{-15} m	0

Note that the size of an electron is not accurately known.

Models of the hydrogen atom and the carbon atom are shown below.

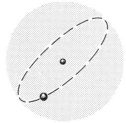

(a) The hydrogen atom is composed of a nucleus which contains one proton. Its one electron moves about or orbits the nucleus.

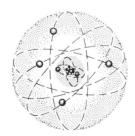

(b) The carbon atom is composed of a nucleus which contains six protons and six neutrons. Its six electrons move about or orbit the nucleus as shown above.

The nucleus of an atom is made up of protons and neutrons while the electrons orbit the nucleus. The atoms are held together by strong nuclear forces. The molecules are held together by electrical forces.

Some physicists say that physics basically deals with only four forces:

1. Gravitational—holds universe together.
2. Electrical—holds molecules together.
3. Strong nuclear—holds nuclei together.
4. Weak nuclear—only present when one nucleus is scattered by another.

In summary, the most basic building blocks of matter are protons, electrons, and neutrons. These particles, formed in various combinations, give us the 105 known atoms or chemical elements. The atoms, formed in various combinations, give us the very long list of known molecules.

Matter exists in three states: solids, liquids, and gases. A solid is a substance that has a definite shape and a definite volume. A liquid is a substance that takes the shape of its container and has a definite volume. A gas is a substance that takes the shape of its container and has the same volume as its container.

The molecules of a solid are fixed in relation to each other. They vibrate in a back-and-forth motion. They are so close that a solid can be compressed only slightly. Solids are usually crystalline substances, meaning that their molecules are arranged in a definite pattern. This is why a solid tends to hold its shape and has a definite volume.

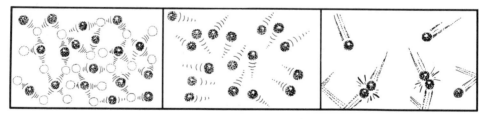

Solid molecules vibrate in fixed positions

Liquid molecules flow over each other

Gas molecules move rapidly in all directions and collide

The molecules of a liquid are not fixed in relation to each other. They normally move in a flowing type of motion but yet are so close together that they are practically incompressible, thus having a definite volume. Because the molecules move in a smooth flowing motion and not in any fixed manner, a liquid takes the shape of its container.

The molecules of a gas are not fixed in relation to each other and move rapidly in all directions, colliding with each other. They are much farther apart than molecules in a liquid, and they are extremely far apart when compared to the distance between molecules in solids. The movement of the molecules is limited only by the container. Therefore, a gas takes the shape of its container. Because the molecules are far apart, a gas can easily be compressed and it has the same volume as its container.

13.2 PROPERTIES OF SOLIDS

Solids have a definite shape and a definite volume. Solids have molecules which are usually arranged in a definite pattern. Let us now study the following properties which most solids have in common.

Cohesion and adhesion The molecules of a solid are held together by rather large internal molecular forces (try pulling a solid apart!). *Cohesion* is the force of attraction between like molecules. The cohesive forces hold the closely packed molecules of a solid together, which keep its shape and volume from being easily changed.

Cohesion, this force of attraction between like molecules, can also be shown by grinding and polishing the surfaces of two like solids and then sliding their surfaces together. For example, take two pieces of polished plate glass and slide them together. Try to pull them apart. It is only the force of attraction of the like molecules of the two pieces of glass that makes it difficult to pull them apart.

Adhesion is the force of attraction between different or unlike molecules. Common examples include glue and wood, adhesive tape and skin, and tar and road (and tires and car and shoes and . . .).

Tensile strength The *tensile strength* of a solid is a measure of its resistance to being pulled apart. That is, the tensile strength of a solid is a measure of its cohesive forces between adjacent molecules. The tensile strength of a rod or wire is found by putting it in a machine which pulls the rod or wire until it breaks (see the figure below). The tensile strength is the ratio:

$$\frac{\text{force required to break the rod or wire}}{\text{cross-sectional area of the rod or wire}}$$

Hardness   The *hardness* of a solid is a measure of the internal resistance of its molecules being forced farther apart or closer together. More commonly, we talk of the hardness of a solid in terms of its difficulty in being scratched. As a matter of classifying the relative hardness of a material, a "scratch test" is used. The given material is scratched in a certain way. Its scratch is then compared with a series of standard scratches of materials which form an arbitrary hardness scale from very soft solids to the hardest known substance, diamond.

The *Brinell method* is a common industrial method used to measure the hardness of a metal. A machine is used to press a 10-mm hardened chrome-steel ball with an equivalent mass of 3000 kg into the metal being tested. The diameter of the resulting impression is used as a measure of the metal's hardness. The Brinell value or number is the ratio:

$$\frac{\text{the load (in kg)}}{\text{the surface area of the impression (in mm}^2)}$$

This value can also be compared with a scale of the accepted hardnesses of given metals. The larger the Brinell number, the harder the metal.

Steel can be hardened by heating it to a very high temperature, then suddenly cooling it by putting it in water. However, it then becomes brittle. This cooled steel can then be tempered (toughened) by reheating it and allowing it to cool slowly. As the steel cools, it loses hardness and gains toughness. If the steel cools down slowly and completely, we say that it is annealed. Annealed steel is soft and tough but not brittle.

Ductility   A metal rod that can be drawn through a die to produce a wire is said to have a property called *ductility*. As the rod is pulled through the die, its diameter is decreased and its length is increased as it becomes a wire.

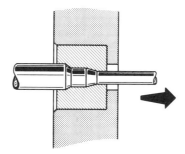

Ductility: A metal being drawn into a wire

Malleability   A metal that can be hammered and rolled into sheets is said to have a property called *malleability*. As the metal is hammered or rolled, its shape or thickness is changed. During this process, the atoms slide over each other and change positions. The cohesive forces are relatively strong; thus, the atoms do not become widely separated during their rearrangement and the resulting shape remains relatively stable.

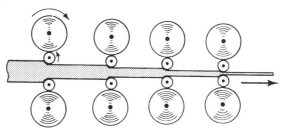

Elasticity   An object becomes deformed when outside forces change its shape or size. The object's ability to return to its original size and shape—when the outside forces are removed—is called its *elasticity*. When the solid is being deformed, these molecules sometimes attract each other and sometimes repel each other. For instance, take a rubber ball and try to pull it apart. You notice that the ball stretches out of shape.

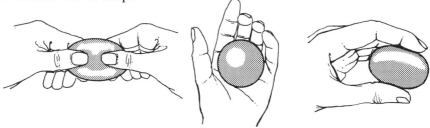

However, when you release the pulling force, the ball returns to its original shape because the molecules, being farther apart than normal, attract each other. If you squeeze the ball together, the ball will again become out of shape. Now release the pressure and the ball will again return to its original shape because the molecules, being too close together, repel each other. Therefore, we can see that when molecules are slightly pulled out of position, they attract each other. When they are pressed too close together, they repel each other. Most solids have the property of elasticity; however, some are only slightly elastic. For example, wood and styrofoam are two solids whose elasticity is small.

Sometimes an elastic object does not return to its original shape. If too large a deforming force is applied, it will become permanently deformed. Take a small spring and pull it apart as far as you can. When you let go, it will probably not return to its same original shape. When a solid is deformed as such, it is said to have been deformed past its elastic limit. Its molecules have been pulled far enough apart that they have slid past one another beyond which the original molecular forces could return the spring to its original shape. If the deforming force is enough greater, the body breaks apart.

Spring before stretching.

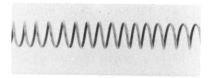

Spring stretched near its elastic limit.

Spring stretched beyond its elastic limit.

Spring stretched much beyond its elastic limit . . . break occurs!

Elasticity properties can be measured in a variety of ways. Whenever an object is deformed by an outside force, the internal molecular forces tend to resist the change in shape and/or volume. Stress is a measure of the tendency of an object to return to its original shape and size after a deforming force has been applied. There are five basic stresses: tension, compression, shear, twisting, and bending.

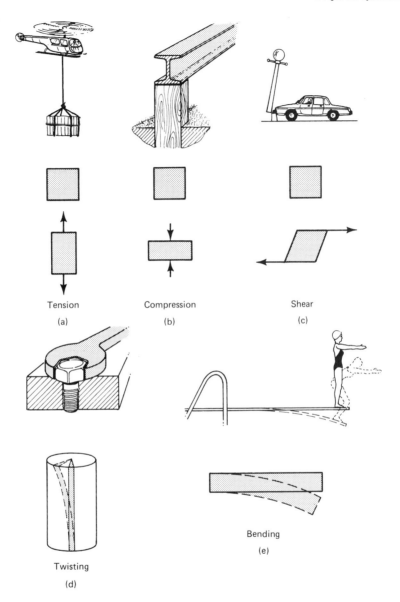

Tension
(a)

Compression
(b)

Shear
(c)

Twisting
(d)

Bending
(e)

More specifically, *stress* may be defined as the ratio of the outside applied force which tends to cause a distortion to the area over which the force acts. In other words,

$$\text{stress} = \frac{\text{applied force}}{\text{area over which the force acts}}$$

or

$$S = \frac{F}{A}$$

The metric stress unit is usually the pascal (Pa).

$$1 \text{ Pa} = 1 \text{ N/m}^2$$

Whenever a stress is applied to an object, the object will be changed minutely, at least. If you stand on a large steel beam, it will bend—at least slightly. The resulting deformation is called *strain*. That is, strain is the relative amount of deformation of a body that is under stress. Or strain is change in length per unit of length, change in volume per unit of volume, and so on. Strain is a direct and necessary consequence of stress.

Before continuing this discussion of elasticity, we need to discuss pressure and the pascal unit.

13.3   PRESSURE

When a force acts perpendicular to a surface, the pressure exerted on the surface is given by

$$P = \frac{F}{A}$$

where P is the pressure,
    F is the force, and
    A is the area of the surface.

Since the SI metric unit for force is the newton (N) and for area is the square metre (m^2), the corresponding pressure unit is N/m^2. This unit is given the special name pascal (Pa), named after Blaise Pascal, a French physicist (1623–1662) who made some important discoveries in the studies of pressure. That is,

$$1 \text{ N/m}^2 = 1 \text{ Pa}$$

Imagine a brick weighing 12.0 N first lying on its side on a table and then standing on one end (page 271). The weight of the brick is the same no matter what its position. So the total force (the weight of the brick) on the table is the same in both cases.

However, the position of the brick does make a difference on the pressure exerted on the table. In which case is the pressure greater? When standing on end, the brick exerts a greater pressure on the table. The reason is that the area of contact on the end is *smaller* than on the side. Using $P = F/A$, let us find the pressure exerted in each case.

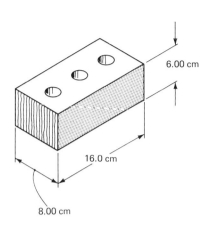

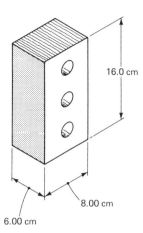

Case 1:

$$F = 12.0 \text{ N}$$
$$A = 8.00 \text{ cm} \times 16.0 \text{ cm} = 128 \text{ cm}^2$$
$$P = \frac{F}{A} = \frac{12.0 \text{ N}}{128 \text{ cm}^2} \times \frac{10^4 \text{ cm}^2}{1 \text{ m}^2}$$
$$= 938 \text{ N/m}^2 = 938 \text{ Pa}$$

Case 2:

$$F = 12.0 \text{ N}$$
$$A = 6.00 \text{ cm} \times 8.00 \text{ cm} = 48.0 \text{ cm}^2$$
$$P = \frac{F}{A} = \frac{12.0 \text{ N}}{48.0 \text{ cm}^2} \times \frac{10^4 \text{ cm}^2}{1 \text{ m}^2}$$
$$= 25\overline{0}0 \text{ N/m}^2 = 25\overline{0}0 \text{ Pa}$$

This shows that when the same force is applied to a smaller area, the pressure is greater.

From the discussion so far, would you rather a woman step on your foot with a pointed-heel shoe or with a flat-heel shoe? Before you snicker, thinking this question is silly, you should be aware that the aircraft industry does not think it is. They must design and construct floors light in weight but strong enough to stand the pressure of women wearing pointed-heel shoes. This was a serious problem for them.

For example, if a $7\overline{0}$-kg woman rests her weight on a 25-cm² heel, the pressure is

$$P = \frac{F}{A}$$

$$P = \frac{(7\overline{0}\ \text{kg})(9.80\ \text{m/s}^2)}{25\ \text{cm}^2 \times \dfrac{1\ \text{m}^2}{10^4\ \text{cm}^2}} = 2.7 \times 10^5\ \text{N/m}^2$$

$$= 2.7 \times 10^5\ \text{Pa} = 270\ \text{kPa}$$

But if she rests her weight on a pointed heel of 1.0 cm², the pressure is

$$P = \frac{F}{A}$$

$$P = \frac{(7\overline{0}\ \text{kg})(9.80\ \text{m/s}^2)}{1.0\ \text{cm}^2 \times \dfrac{1\ \text{m}^2}{10^4\ \text{cm}^2}} = 6.9 \times 10^6\ \text{N/m}^2$$

$$= 6.9 \times 10^6\ \text{Pa}$$

$$= 6900\ \text{kPa}$$

Since the pascal is a relatively small unit, the kilopascal (kPa) is a commonly used unit of pressure.

Note that pressure is a scalar and the force, $F = PA$, has a direction perpendicular to the area.

13.4   HOOKE'S LAW

One of the most basic principles related to the elasticity of solids is called Hooke's law: *Stress is directly proportional to strain as long as the elastic limit has not been exceeded.*

This law is named after Robert Hooke, who first discovered the principle in the seventeenth century while inventing the balance spring for spring-driven clocks.

To illustrate this law, consider a simple hanging spring. Let us hang three different weights from the spring and measure each resulting stretch of the spring.

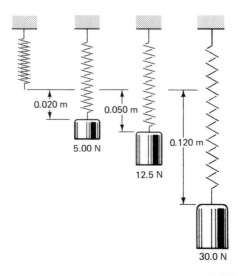

Weight	Stretch	$\dfrac{Weight}{Stretch}$
5.00 N	0.020 m	250 N/m
12.5 N	0.050 m	250 N/m
30.0 N	0.120 m	250 N/m

Note that in each case the stress is directly proportional to the strain; that is,

$$F = ks$$

where F is the applied force (weight),

  s is the deformation of the spring (stretch), and

  k is the proportionality constant or spring constant (250 N/m in this case).
If you were to hang a 40.0-N weight on the spring, how much would the spring
stretch?

example   A spring stretches 6.0 cm with a 150-N weight attached.

(a) Find the spring constant.
(b) How much will the spring stretch if a $4\bar{0}0$-N weight is attached (assuming that the
  elastic limit is not reached)?

(a) DATA: $s = 6.0$ cm $= 0.060$ m

                                         $F = 150$ N

                                         $k = ?$

BASIC EQUATION: $F = ks$

WORKING EQUATION: $k = \dfrac{F}{s}$

SUBSTITUTION: $k = \dfrac{150 \text{ N}}{0.060 \text{ m}}$

                                         $= 2500$ N/m

(b) DATA: $k = 2500$ N/m

                                         $F = 4\bar{0}0$ N

                                         $s = ?$

BASIC EQUATION: $F = ks$

WORKING EQUATION: $s = \dfrac{F}{k}$

SUBSTITUTION: $s = \dfrac{4\bar{0}0 \text{ N}}{2500 \text{ N/m}}$

                                         $= 0.16$ m or 16 cm

The proportionality constant in Hooke's law may be thought of as the constant of elasticity. This proportionality constant between stress and strain is most often called the elastic modulus (modulus means a constant). In general, Hooke's law states:

$$\text{stress} = (\text{elastic modulus}) \times (\text{strain})$$

Hooke's law may be expressed in terms of each specific kind of stress. (We shall study these in the next three sections.)

13.5   YOUNG'S MODULUS: LENGTH ELASTICITY

The elastic modulus used for finding length elasticity is called *Young's modulus*. That is, from

$$\text{stress} = (\text{elastic modulus}) \times (\text{strain})$$

we get

$$\frac{F}{A} = Y \times \frac{\Delta l}{l}$$

or

$$Y = \frac{F/A}{\Delta l/l} = \frac{\text{stress}}{\text{strain}}$$

where Y is Young's modulus for length elasticity.

example 1   A 12,000-N load is hanging from a steel cable that is 10.000 m long and 16 mm in diameter. This load stretches the cable 3 mm.

(a) Find the stress.
(b) Find the strain.
(c) Find Young's modulus for this steel.

(a) DATA:
$$F = 12,000 \text{ N}$$
$$A = \pi r^2 \quad \text{where } r = 8 \text{ mm} = 0.008 \text{ m}$$
$$\text{stress} = ?$$

BASIC EQUATION:
$$\text{stress} = \frac{F}{A}$$

WORKING EQUATION: same

SUBSTITUTION:
$$\text{stress} = \frac{12,000 \text{ N}}{\pi(0.008 \text{ m})^2}$$
$$= 6 \times 10^7 \text{ N/m}^2$$
$$= 6 \times 10^7 \text{ Pa} \quad \text{or} \quad 60 \text{ MPa}$$

(b) DATA:

$$l = 10.000 \text{ m}$$
$$\Delta l = 3 \text{ mm} = 0.003 \text{ m}$$
$$\text{strain} = ?$$

BASIC EQUATION:

$$\text{strain} = \frac{\Delta l}{l}$$

WORKING EQUATION: same

SUBSTITUTION:

$$\text{strain} = \frac{0.003 \text{ m}}{10.000 \text{ m}}$$
$$= 3 \times 10^{-4}$$

(c) DATA:

$$\text{stress} = 6 \times 10^7 \text{ Pa}$$
$$\text{strain} = 3 \times 10^{-4}$$
$$Y = ?$$

BASIC EQUATION:

$$Y = \frac{\text{stress}}{\text{strain}}$$

WORKING EQUATION: same

SUBSTITUTION:

$$Y = \frac{6 \times 10^7 \text{ Pa}}{3 \times 10^{-4}}$$
$$= 2 \times 10^{11} \text{ Pa} \quad \text{or} \quad 200 \text{ GPa}$$

The table below shows some common materials and their corresponding values of Young's modulus.

Material	Young's modulus (Pa)
Aluminum, rolled	7.0×10^{10}
Brass	9.0×10^{10}
Copper, wire	1.2×10^{11}
Gold	7.9×10^{10}
Iron, cast	9.1×10^{10}
Iron, wrought	1.9×10^{11}
Lead, rolled	1.6×10^{10}
Steel, annealed	2.0×10^{11}
Rubber	1.4×10^5

example 2 How much will a cast-iron bar 8.00 cm by 15.0 cm by 3.50 m long be compressed by a load of 15,500 N?

DATA:

$$F = 15,500 \text{ N}$$
$$A = 8.00 \text{ cm} \times 15.0 \text{ cm} = 0.0800 \text{ m} \times 0.150 \text{ m}$$
$$l = 3.50 \text{ m}$$
$$Y = 9.1 \times 10^{10} \text{ Pa}$$
$$\Delta l = ?$$

BASIC EQUATION:
$$Y = \frac{F/A}{\Delta l / l}$$

WORKING EQUATION:
$$\Delta l = \frac{Fl}{AY}$$

SUBSTITUTION:

$$\Delta l = \frac{(15,500 \text{ N})(3.50 \text{ m})}{(0.0800 \text{ m} \times 0.150 \text{ m})(9.1 \times 10^{10} \text{ Pa})}$$

$$\frac{\text{N m}}{\text{m}^2 \text{ Pa}} = \frac{\text{N m}}{\text{m}^2(\text{N/m}^2)} = \text{m}$$

$$= 5.0 \times 10^{-5} \text{ m} \quad \text{or} \quad 0.050 \text{ mm}$$

Up to this point in this discussion, we have been assuming that the elastic limit has not been exceeded because Hooke's law is valid only for values of stresses less than the elastic limit. What happens when the elastic limit is exceeded? (We already know that the object becomes permanently deformed and may even break if the applied force is great enough.) After the elastic limit is reached, the deformation continues with only slight increases of applied forces. Then eventually, a point is reached when the deformation continues with no increases or even decreases in the applied forces. Just before the object breaks, it reaches the *ultimate strength* of which it is capable, which corresponds to the highest point on the curve. This is shown graphically below. The curve varies with the material used and its conditioning.

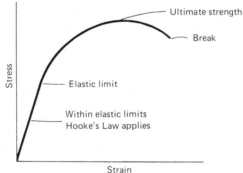

Material	Elastic limit (Pa)	Ultimate strength (Pa)
Aluminum, rolled	1.3×10^8	1.4×10^8
Brass	3.8×10^8	4.6×10^8
Copper	1.5×10^8	3.4×10^8
Iron, wrought	1.6×10^8	3.2×10^8
Steel, annealed	2.5×10^8	5.0×10^8
Steel, spring-tempered	1.2×10^9	1.4×10^9

example 3 A 20$\bar{0}$-kg mass is supported by a copper wire 12.000 m long.

(a) What is the minimum diameter of the wire needed to support the mass if the elastic limit is not to be exceeded?
(b) By how much does this wire stretch when supporting the 20$\bar{0}$ kg?
(c) What is the maximum mass that this wire can support without breaking?

(a) First, find the cross-sectional area of the wire:

DATA: stress $= 1.5 \times 10^8$ Pa   (at elastic limit)

$$F = mg = (20\bar{0} \text{ kg})(9.80 \text{ m/s}^2) = 1960 \text{ kg m/s}^2 = 1960 \text{ N}$$

$$A = ?$$

BASIC EQUATION: $\text{stress} = \dfrac{F}{A}$

WORKING EQUATION: $A = \dfrac{F}{\text{stress}}$

SUBSTITUTION: $A = \dfrac{1960 \text{ N}}{1.5 \times 10^8 \text{ Pa}}$ $\boxed{\dfrac{\text{N}}{\text{Pa}} = \dfrac{\text{N}}{\text{N/m}^2} = \text{m}^2}$

$$= 1.3 \times 10^{-5} \text{ m}^2$$

and

$$A = \pi r^2$$

or

$$r = \sqrt{\dfrac{A}{\pi}} = \sqrt{\dfrac{1.3 \times 10^{-5} \text{ m}^2}{\pi}}$$

$$= 2.0 \times 10^{-3} \text{ m} = 2.0 \text{ mm}$$

Thus,

$$\text{diameter} = 2r = 2(2.0 \text{ mm}) = 4.0 \text{ mm}$$

(b) DATA: $F = 1960$ N

$$A = 1.3 \times 10^{-5} \text{ m}^2$$

$$l = 12.000 \text{ m}$$

$$Y = 1.2 \times 10^{11} \text{ Pa}$$

$$\Delta l = ?$$

BASIC EQUATION: $Y = \dfrac{F/A}{\Delta l / l}$

WORKING EQUATION: $\Delta l = \dfrac{Fl}{AY}$

SUBSTITUTION:
$$\Delta l = \frac{(1960\ \text{N})(12.000\ \text{m})}{(1.3 \times 10^{-5}\ \text{m}^2)(1.2 \times 10^{11}\ \text{Pa})}$$

$$= 0.015\ \text{m} = 15\ \text{mm}$$

(c) First, find the maximum force:

DATA:
$$\text{stress} = 3.4 \times 10^8\ \text{Pa} \qquad \text{(at ultimate strength)}$$

$$A = 1.3 \times 10^{-5}\ \text{m}^2$$

$$F = ?$$

BASIC EQUATION:
$$\text{stress} = \frac{F}{A}$$

WORKING EQUATION:
$$F = (\text{stress})(A)$$

SUBSTITUTION:
$$F = (3.4 \times 10^8\ \text{Pa})(1.3 \times 10^{-5}\ \text{m}^2)$$

$$\boxed{(\text{Pa})(\text{m}^2) = (\text{N/m}^2)(\text{m}^2) = \text{N}}$$

$$= 4400\ \text{N}$$

And, since
$$F = mg$$

$$m = \frac{F}{g}$$

$$= \frac{4400\ \text{N}}{9.80\ \text{m/s}^2} = 450\ \text{kg} \quad \text{(the maximum mass)}$$

13.6 BULK MODULUS: VOLUME ELASTICITY

The volume of any substance can be decreased when subjected to pressure on all sides. As a bathysphere is lowered deeper into the ocean for deep sea observation, the tremendous pressure of the water causes it to decrease in volume (see page 279).

As the bathysphere is lowered, the change in volume is directly proportional to the change in pressure. And, we have another variation of Hooke's law:

$$\text{stress} = (\text{elastic modulus}) \times (\text{strain})$$

where here the elastic modulus is called the *bulk modulus* and is represented by B. The stress is the change in force per unit area (or change in pressure). The strain is the change in volume per unit of volume or $\Delta V / V$. Thus, the bulk modulus of volume elasticity may be expressed:

$$\boxed{B = \frac{\text{stress}}{\text{strain}} = \frac{\Delta F / A}{-\Delta V / V} = \frac{\Delta P}{-\Delta V / V}}$$

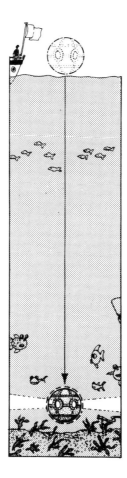

If the elastic limit is not exceeded, the volume returns to the original volume after the pressure is released.

The table below shows some common materials and their corresponding values of bulk modulus.

Material	Bulk modulus (Pa)
Solids	
Aluminum	7.7×10^{10}
Brass	6.1×10^{10}
Copper	1.4×10^{11}
Glass	3.7×10^{10}
Iron, cast	9.6×10^{10}
Lead	7.7×10^{9}
Steel	1.6×10^{11}

Material	Bulk modulus (Pa)
Liquids	
Ethyl alcohol	1.1×10^9
Kerosene	1.3×10^9
Lubricating oil	1.7×10^9
Mercury	2.8×10^{10}
Water	2.1×10^9

At times in technical work, the compressibility, c, may be used. The compressibility of a material is the reciprocal of the bulk modulus. That is,

$$c = \frac{1}{B}$$

Compressibility expresses the fractional decrease in volume per unit increase in pressure.

example A lead ball of radius 1.00 m is lowered to the bottom of a deep sea where the pressure is $7\bar{0}$ MPa. Find its decrease in volume in cm^3.

DATA:
$$\Delta P = 7\bar{0} \text{ MPa} = 7.0 \times 10^7 \text{ Pa}$$

$$r = 1.00 \text{ m}$$

$$V = \tfrac{4}{3}\pi r^3 = \tfrac{4}{3}\pi(1.00 \text{ m})^3$$

$$B = 7.7 \times 10^9 \text{ Pa}$$

$$\Delta V = ?$$

BASIC EQUATION:
$$B = \frac{\Delta P}{-\Delta V / V}$$

WORKING EQUATION:
$$\Delta V = \frac{-(\Delta P)V}{B}$$

SUBSTITUTION:
$$\Delta V = \frac{-(7.0 \times 10^7 \text{ Pa})\left[\tfrac{4}{3}\pi(1.00 \text{ m})^3\right]}{7.7 \times 10^9 \text{ Pa}}$$

$$= -0.038 \text{ m}^3 \times \left(\frac{100 \text{ cm}}{1 \text{ m}}\right)^3 = -38,000 \text{ cm}^3$$

The minus sign indicates a *decrease* in volume.

13.7 SHEAR MODULUS

The shape of an object can be changed by applying two equal but opposite forces on the object along parallel lines. This is called a *shear stress*. To illustrate this concept, lay a rather large book on a flat surface. Then push the top and bottom covers in opposite directions as in the following figure.

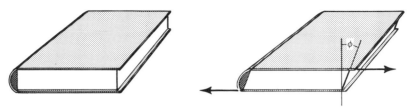

The layers of the atoms in an object correspond to the pages in the book. As the shearing forces are applied, the layers of atoms are displaced but no volume changes occur. Angle ϕ is called the angle of shear.

More generally, let us consider the rectangular solid below. As a result of applying the two equal but opposite shearing forces, **F** and **F'**, the shape of the rectangular solid changes such that the rectangular side *MNOP* becomes the parallelogram *MNO'P'*. Actually, edges *PM* and *ON* rotate through angle ϕ. Note that the area of the top and bottom layers (actually all layers parallel to the top and bottom layers) remains A (as the area of the pages remained the same above).

The shearing stress is defined as the change in force per unit area ($\Delta F/A$). The shearing strain is defined as the amount of lateral displacement per unit height (x/h). Again, we have a variation of Hooke's law:

$$\text{stress} = (\text{elastic modulus}) \times (\text{strain})$$

where here the elastic modulus is called the *shear modulus*.

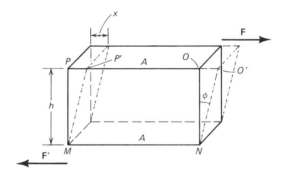

The shear modulus may be expressed

$$\boxed{S = \frac{\text{stress}}{\text{strain}} = \frac{\Delta F/A}{x/h}}$$

where S is the shear modulus. Note that $\tan \phi = x/h$ and for small angles $\tan \phi$ is approximately equal to ϕ, when ϕ is expressed in radians. Therefore,

$$S = \frac{\Delta F/A}{\phi} = \frac{\Delta F}{A\phi}$$

If the elastic limit is not exceeded, the solid returns to its original shape after the forces are withdrawn.

The table below shows some common materials and their corresponding values of shear modulus.

Material	Shear modulus (Pa)
Aluminum	2.4×10^{10}
Brass	3.6×10^{10}
Copper	4.2×10^{10}
Glass	2.3×10^{10}
Iron	7.0×10^{10}
Lead	5.6×10^{9}
Steel	8.4×10^{10}

example 1   The shearing force of 125,000 N is applied to the steel rivet as shown below. The rivet has a diameter of 20.0 mm.

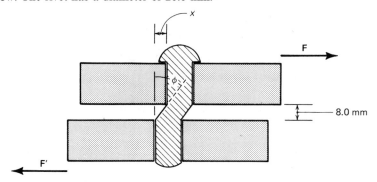

(a) Find the shearing stress.
(b) Find the lateral displacement, x.
(c) Find the angle of shear, ϕ.

(a) DATA:
$$\Delta F = 125,000 \text{ N}$$
$$r = 10.0 \text{ mm} = 0.0100 \text{ m}$$
$$A = \pi r^2$$
$$\text{stress} = ?$$

BASIC EQUATION:
$$\text{stress} = \frac{\Delta F}{A}$$

WORKING EQUATION: same

SUBSTITUTION:
$$\text{stress} = \frac{125,000 \text{ N}}{\pi (0.0100 \text{ m})^2}$$
$$= 3.98 \times 10^8 \text{ N/m}^2$$
$$= 3.98 \times 10^8 \text{ Pa} \quad \text{or} \quad 398 \text{ MPa}$$

(b) DATA:

$$\text{stress} = 3.98 \times 10^8 \text{ Pa}$$

$$h = 8.0 \text{ mm}$$

$$S = 8.4 \times 10^{10} \text{ Pa}$$

$$x = ?$$

BASIC EQUATION:

$$S = \frac{\text{stress}}{x/h}$$

WORKING EQUATION:

$$x = \frac{(\text{stress})(h)}{S}$$

SUBSTITUTION:

$$x = \frac{(3.98 \times 10^8 \text{ Pa})(8.0 \text{ mm})}{8.4 \times 10^{10} \text{ Pa}}$$

$$= 0.038 \text{ mm}$$

(c) DATA:

$$x = 0.038 \text{ mm}$$

$$h = 8.00 \text{ mm}$$

$$\phi = ?$$

BASIC EQUATION:

$$\tan \phi = \frac{x}{h}$$

WORKING EQUATION:

same

SUBSTITUTION:

$$\tan \phi = \frac{0.038 \text{ mm}}{8.00 \text{ mm}} = 4.75 \times 10^{-3}$$

$$\phi = 0.27°$$

Or, an alternative solution for part (c):

DATA:

$$\text{stress} = 3.98 \times 10^8 \text{ Pa}$$

$$S = 8.4 \times 10^{10} \text{ Pa}$$

$$\phi = ?$$

BASIC EQUATION:

$$S = \frac{\Delta F/A}{\phi} = \frac{\text{stress}}{\phi}$$

WORKING EQUATION:

$$\phi = \frac{\text{stress}}{S}$$

SUBSTITUTION:

$$\phi = \frac{3.98 \times 10^8 \text{ Pa}}{8.4 \times 10^{10} \text{ Pa}}$$

$$= 4.7 \times 10^{-3} \text{ rad} \times \frac{180°}{\pi \text{ rad}}$$

$$= 0.27°$$

Shear modulus is also used to calculate the twisting or torsional stress of a straight rod or shaft.

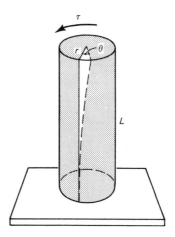

A torque τ is applied to the top of the cylinder, which results in a twist of θ as shown. (The bottom of the cylinder is fixed.) To express angle θ in terms of the other quantities, calculus is required. So we will only give the result:

$$\theta = \frac{2\tau L}{\pi S r^4}$$

where θ is the twist or torsion, in radians,
   τ is the torque,
   L is the length of the cylinder,
   S is the shear modulus, and
   r is the radius of the cylinder.

For a hollow cylinder,

$$\theta = \frac{\tau L}{2\pi S r^3 t}$$

where t is the thickness of the wall of the cylinder.

example 2   In drilling an oil well, a steel pipe whose diameter is 10.0 cm and whose thickness is 6.5 mm is used to turn the drill bit. The pipe is 30$\overline{0}$0 m long, rotates at 210 rpm, and is driven by a 4$\overline{0}$0-kW motor. Find the angle θ through which the pipe twists. First, we need to find the torque, τ, transmitted by the motor.

DATA: $P = 4\overline{0}0 \text{ kW} = 4.0 \times 10^5 \text{ W}$

$\omega = 210 \text{ rev/min}$

$\tau = ?$

BASIC EQUATION: $P = \tau\omega$

WORKING EQUATION: $\tau = \dfrac{P}{\omega}$

SUBSTITUTION: $\tau = \dfrac{4.0 \times 10^5 \text{ W}}{210 \text{ rev/min}}$

$$= \frac{4.0 \times 10^5 \text{ Nm/s}}{\dfrac{210 \text{ rev}}{\text{min}} \times \dfrac{2\pi \text{ rad}}{\text{rev}} \times \dfrac{1 \text{ min}}{60 \text{ s}}}$$

$$= 1.8 \times 10^4 \text{ Nm}$$

Now, to find θ:

DATA: $\tau = 1.8 \times 10^4 \text{ Nm}$

$L = 30\overline{0}0 \text{ m}$

$S = 8.4 \times 10^{10} \text{ Pa}$

$r = 5.0 \text{ cm} = 0.050 \text{ m}$

$t = 6.5 \text{ mm} = 0.0065 \text{ m}$

$\theta = ?$

BASIC EQUATION: $\theta = \dfrac{\tau L}{2\pi S r^3 t}$

WORKING EQUATION: same

SUBSTITUTION: $\theta = \dfrac{(1.8 \times 10^4 \text{ Nm})(30\overline{0}0 \text{ m})}{2\pi (8.4 \times 10^{10} \text{ Pa})(0.050 \text{ m})^3 (0.0065 \text{ m})}$

$$\boxed{\frac{(\text{Nm})(\text{m})}{\text{Pa}(\text{m}^3)(\text{m})} = \frac{(\text{Nm})(\text{m})}{(\text{N/m}^2)(\text{m}^3)(\text{m})} = 1 = \text{rad}}$$

$$= 130 \text{ rad} \times \frac{180°}{\pi \text{ rad}}$$

$$= 7400°$$

That is, there is one full turn of twist for each 150 m of pipe.

13.8   PROPERTIES OF LIQUIDS

As we noted in Section 13.1, a liquid is a substance that has a definite volume and takes the shape of its container. The molecules move in a flowing motion, yet are so close together that it is very difficult to compress a liquid. (The distances between molecules in a liquid are roughly the same as the distances between molecules in a solid.) In addition, most liquids share the following common properties.

Cohesion and adhesion   Cohesion, the force of attraction between like molecules, causes a liquid such as molasses to be sticky. Adhesion, the force of attraction between unlike molecules, also causes the molasses to stick to your finger. In the case of water, its adhesive forces are greater than its cohesive forces. Put a plate of glass in a pan of water and pull it out. Some water remains on the glass. In the case of mercury, the opposite is true. Mercury's cohesive forces are greater than its adhesive forces. If the glass is submerged in a vat of mercury and then pulled out, virtually no mercury remains on the glass.

We say that a liquid whose adhesive forces are greater than its cohesive forces tends to wet any surface that comes in contact with it. And a liquid whose cohesive forces are greater than its adhesive forces tends to leave dry, or not wet, any surface that comes in contact with it.

Surface tension   The ability of the surface of water to support a needle is an example of surface tension. The water's surface acts like a thin, flexible surface film. The surface tension of water can be reduced by adding soap to the water. Soaps are added to laundry water to decrease the surface tension of the water so that the water more easily penetrates the fibers of the clothes being washed.

Water   Soap added

Surface tension causes a raindrop to hold together. Surface tension causes a small drop of mercury to keep an almost spherical shape. A liquid drop suspended in space would be spherical. A falling raindrop's shape is due to the friction with the air.

Mercury drop on
a surface

(a)

Raindrop in space

(b)

Falling raindrop

(c)

Viscosity One characteristic of liquids is called flow. In liquids there is friction, which is called *viscosity*. The greater the molecular attraction, the more the friction and the greater the viscosity. For example, it takes more force to move a block of wood through oil than through water. This is because oil is more viscous than water.

If a liquid's temperature is increased, its viscosity decreases. For example, the viscosity of oil in a car engine before it is started on a winter morning at $-10°C$ is greater than after the engine has been running for an hour.

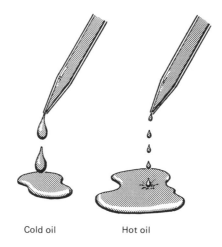

Cold oil   Hot oil

Capillary action Liquids keep the same level in tubes whose ends are submerged in a liquid if the tubes have a large enough diameter.

In tubes of varying small diameters, water does not stand at the same level. The smaller the diameter, the higher the water rises. If mercury is used, it does not stand at the same level either. But instead of rising up the tube, the mercury level falls or its level is depressed. The smaller the diameter, the lower its level is depressed. This behavior of liquids in small tubes is called *capillary action*.

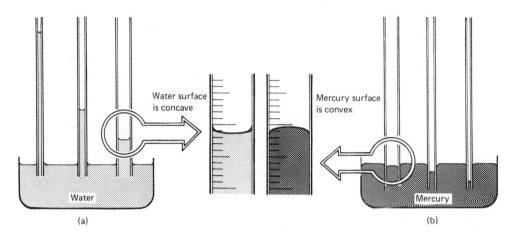

Capillary action is due to both adhesion of the liquid molecules with the tube and the surface tension of the liquid. In the case of water, the adhesive forces are greater than the cohesive molecular forces. Thus, water creeps up the sides of the tube and produces a concave water surface. The surface tension of the water tends to flatten the concave surface. Together these two forces raise the water up the tubes until it is counterbalanced by the weight of the water column itself.

In the case of mercury, the cohesive molecular forces are greater than the adhesive forces, producing a convex mercury surface. The surface tension of the mercury tends to further hold down the mercury level.

Experimentally, scientists have found that:

1. Liquids rise in capillary tubes they tend to wet and are depressed in tubes they tend not to wet.
2. Elevation or depression in the tube is inversely proportional to the diameter of the tube.
3. The elevation or depression decreases as the temperature increases.

13.9 PROPERTIES OF GASES

Because of the rapid random movement of its molecules, a gas spreads to completely occupy the volume of its container. This property is called *expansion*.

Diffusion of a gas is the process by which molecules of the gas mix with the molecules of a solid, liquid, or another gas. If you remove the cap from a bottle of rubbing alcohol, you soon smell the fumes of the alcohol. The air molecules and the alcohol molecules mix throughout the room because of diffusion.

A balloon inflates because of the pressure of the air molecules on the inside surface of the balloon. This pressure is caused by the bombardment on the walls by the moving molecules. The pressure may be increased by increasing the number of molecules by blowing more air in the balloon. Pressure may also be increased by heating the air molecules already in the balloon. Heat increases the velocity of the molecules.

The behavior of liquids and gases is very similar in many cases. The term *fluid* is used when discussing principles and behaviors common to both liquids and gases.

13.10 ATOMIC OSCILLATORS

The atoms of a solid have an orderly geometric arrangement which repeats itself throughout the solid. There are many different geometric arrangements possible; however, each solid material has a specific arrangement of atoms with well-determined atomic positions or sites. Atoms remain on sites for long periods of time and provide a stableness and rigidity to solids. The thermal energy in the solid demands that the atoms have kinetic energy; hence, the atoms vibrate back and forth, acting as mechanical oscillators, with their sites as their respective equilibrium positions.

Molecules in fluids have kinetic energy because of their mobility and freedom to move. Molecules that consist of two or more atoms have freedom to vibrate, and indeed do as they receive thermal energy.

13.11 DENSITY

Density is a property that all matter has. Density is defined as the mass per unit volume. That is,

$$D = \frac{m}{V}$$

where D is the density,
   m is the mass, and
   V is the volume.

To say that the density of water is 1000 kg/m^3 means that 1 m^3 of water has a mass of 1000 kg. Often, when we speak of metals as being "heavy" or "light," we speak of their relative densities. Densities are usually expressed in kg/m^3.

In all forms of matter, the density usually decreases as the temperature increases and increases as the temperature decreases. The densities of various substances are listed in the table on page 290.

Substance	Mass density (kg/m³)
Solids	
Aluminum	2,700
Concrete	2,300
Copper	8,960
Cork	240
Gold	19,300
Ice	917
Iron	7,870
Lead	11,300
Silver	10,500
Steel	7,800
Wood, maple	690
Wood, white pine	420
Zinc	7,130
Liquids	
Alcohol	790
Gasoline	680
Kerosene	820
Mercury	13,600
Oil	870
Seawater	1,025
Water	1,00$\overline{0}$
Gases[a] (at 0°C and 1 atm pressure)	
Air	1.29
Ammonia	0.76
Carbon dioxide	1.96
Carbon monoxide	1.25
Helium	0.177
Hydrogen	0.0899
Nitrogen	1.25
Oxygen	1.43
Propane	2.02

[a]The density of a gas is found by pumping the gas into a container, by measuring its volume and mass, and then by using the appropriate density formula.

example 1   What is the density of a block of wood having dimensions 45 cm × 3$\overline{0}$ cm × 25 cm with a mass of 20.3 kg?

DATA:

$$m = 20.3 \text{ kg}$$

$$V = lwh = 45 \text{ cm} \times 3\overline{0} \text{ cm} \times 25 \text{ cm}$$

$$= 0.45 \text{ m} \times 0.30 \text{ m} \times 0.25 \text{ m}$$

$$D = ?$$

BASIC EQUATION: $$D = \frac{m}{V}$$

WORKING EQUATION: same

SUBSTITUTION:
$$D = \frac{20.3 \text{ kg}}{0.45 \text{ m} \times 0.30 \text{ m} \times 0.25 \text{ m}}$$

$$= 6\bar{0}0 \text{ kg/m}^3$$

example 2 Find the density of a ball bearing with mass 22.0 g and radius 0.875 cm.

DATA:
$$r = 0.875 \text{ cm} = 0.00875 \text{ m}$$

$$V = \tfrac{4}{3}\pi r^3 = \tfrac{4}{3}\pi(0.00875 \text{ m})^3$$

$$m = 22.0 \text{ g} = 0.0220 \text{ kg}$$

$$D = ?$$

BASIC EQUATION: $$D = \frac{m}{V}$$

WORKING EQUATION: same

SUBSTITUTION:
$$D = \frac{0.0220 \text{ kg}}{\tfrac{4}{3}\pi(0.00875 \text{ m})^3}$$

$$= 7840 \text{ kg/m}^3$$

example 3 What is the mass of 1.00 kL of gasoline?

DATA:
$$D = 680 \text{ kg/m}^3$$

$$V = 1.00 \text{ kL} \times \frac{1000 \text{ L}}{1 \text{ kL}} \times \frac{1000 \text{ cm}^3}{1 \text{ L}} \times \frac{1 \text{ m}^3}{10^6 \text{ cm}^3} = 1.00 \text{ m}^3$$

$$m = ?$$

BASIC EQUATION: $$D = \frac{m}{V}$$

WORKING EQUATION: $$m = DV$$

SUBSTITUTION: $$m = (680 \text{ kg/m}^3)(1.00 \text{ m}^3)$$

$$= 680 \text{ kg}$$

The density of an irregular solid (rock) cannot be found directly because of the difficulty of finding its volume. However, we could find the amount of water the solid displaced, which is the same as the volume of the irregular solid. Volume of water in beaker = volume of rock (see page 292).

example 4 A rock of mass 5.8 kg displaces 1870 cm³ of water. What is the density of the rock?

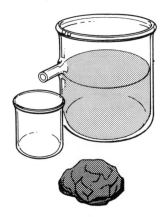

DATA:	$m = 5.8$ kg
	$V = 1870$ cm$^3 = 1.87 \times 10^{-3}$ m^3
	$D = ?$
BASIC EQUATION:	$D = \dfrac{m}{V}$
WORKING EQUATION:	same
SUBSTITUTION:	$D = \dfrac{5.8 \text{ kg}}{1.87 \times 10^{-3} \text{ m}^3}$
	$= 3100$ kg/m^3

To compare the densities of two substances, it would be easier if we had a standard substance with which to compare. Scientists have chosen water as the standard. The ratio of the density of any solid or liquid to the density of water is called its *specific gravity*. That is,

$$\boxed{\text{specific gravity (sp gr)} = \dfrac{D_{\text{substance}}}{D_{\text{water}}}}$$

example 5   The density of steel is 7800 kg/m^3. Find its specific gravity.

DATA:	$D_{\text{substance}} = 7800$ kg/m^3
	$D_{\text{water}} = 100\bar{0}$ kg/m^3
	sp gr $= ?$
BASIC EQUATION:	sp gr $= \dfrac{D_{\text{substance}}}{D_{\text{water}}}$
WORKING EQUATION:	same

SUBSTITUTION: $$\text{sp gr} = \frac{7800 \text{ kg/m}^3}{100\bar{0} \text{ kg/m}^3}$$

$$= 7.8$$

That is, steel is 7.8 times as dense as water and thus it sinks in water.

example 6   The density of oil is 870 kg/m^3. Find its specific gravity.

DATA: $$D_{\text{substance}} = 870 \text{ kg/m}^3$$

$$D_{\text{water}} = 100\bar{0} \text{ kg/m}^3$$

$$\text{sp gr} = ?$$

BASIC EQUATION: $$\text{sp gr} = \frac{D_{\text{substance}}}{D_{\text{water}}}$$

WORKING EQUATION: same

SUBSTITUTION: $$\text{sp gr} = \frac{870 \text{ kg/m}^3}{100\bar{0} \text{ kg/m}^3}$$

$$= 0.87$$

That is, oil is 0.87 times as dense as water, and thus it floats on water.

In general, the specific gravity of

water $= 1$,

a substance denser than water > 1, and

a substance less dense than water < 1.

Methods for finding the specific gravity of a given substance are found in Section 14.4.

EXERCISES

Sketch	Data	Basic Equation	Working Equation	Substitution
	$A = 12 \text{ cm}^2$ $b = 6.0 \text{ cm}$ $h = ?$	$A = bh$	$h = \dfrac{A}{b}$	$h = \dfrac{12 \text{ cm}^2}{6.0 \text{ cm}}$ $= 2.0 \text{ cm}$

1. A packing crate is 1.50 m × 0.80 m × 0.45 m and weighs 4.1 × 10^4 N. Find the pressure (in kPa) exerted by the crate on the floor in each of its three possible positions.

2. A packing crate is 2.50 m $\times$ $2\bar{0}$ cm $\times$ $3\bar{0}$ cm and has a mass of 950 kg. Find the pressure (in kPa) exerted by the crate on the floor in each of its three possible positions.

3. A force of 36.0 N stretches a spring 18.0 cm. Find the spring constant (in N/m).

4. A force of 5.00 N is applied to a spring whose spring constant is 0.250 N/cm. Find its change in length (in cm).

5. If a 17.0-N force stretches a wire 0.650 cm, what force will stretch a similar piece of wire 1.87 cm?

6. If a force of 21.3 N is applied on a similar piece of wire as in Exercise 5, how far will it stretch?

7. A coiled spring is stretched 30.0 cm by a 5.00-N weight. How far will it be stretched by a 15.0-N weight?

8. In Exercise 7, what weight will stretch the spring 40.0 cm?

9. The vertical steel columns of an office building each support a weight of 1.3×10^5 N at the second floor, with each being compressed 5.9×10^{-3} cm.
   (a) What will be the compression of each column if a weight of 5.5×10^5 N is supported?
   (b) If the compression for each steel column is 0.071 cm, what weight is supported by each column?

10. In a Hooke's law experiment the following weights were attached to a spring, which resulted in the following elongations.

Weight (N)	Elongation (cm)
$2\bar{0}$	1.4
25	1.7
35	2.3
$5\bar{0}$	3.4
$8\bar{0}$	5.3
95	6.2
$12\bar{0}$	7.9

   (a) Plot the graph of weight versus elongation.
   (b) Draw the best straight line through the data to determine the spring constant (in N/cm).
   (c) From the graph, what weight would result from an elongation of 6.9 cm?
   (d) From the graph, what elongation would result from a weight of $7\bar{0}$ N?

11. A rectangular cast-iron column 25 cm $\times$ 25 cm $\times$ 5.0000 m supports a weight of 6.8×10^6 N.
   (a) Find the stress in pascals on the top portion of the column.
   (b) This weight compresses the column 6.0 mm. Find the strain.
   (c) Find Young's modulus for this cast iron.

12. A wire 3.50 mm in diameter and 2.000 m long stretches 0.45 mm when a force of 275 N is applied.

(a) Find the stress in pascals on the wire.

(b) Find the strain on the wire.

(c) Find Young's modulus for the wire.

13. How much stress in a copper wire will increase its length by 0.10 percent?

14. The maximum stress allowed in a given steel cable is 105 MPa.

(a) Find the largest weight that a cable 15.0 mm in diameter can support.

(b) Find the smallest diameter cable with the same maximum stress that can support 3.35×10^4 N.

15. An aluminum wire 3.00 mm in diameter breaks under a load of 995 N. Find its breaking stress.

16. A nylon cord $1\bar{0}$ mm in diameter breaks under a load of 1.4×10^4 N. Find its breaking stress.

17. One metric ton ($100\bar{0}$ kg) of mass is supported by an annealed steel wire 10.0 m long.

(a) What is the minimum diameter of the wire to support the mass if the elastic limit is not to be exceeded?

(b) By how much does this wire stretch when supporting the mass?

(c) What is the maximum mass this wire can support without breaking?

18. A 13,500-kg mass is supported by a square wrought-iron rod 8.0 m long.

(a) What is the minimum size of square rod needed to support the mass if the elastic limit is not to be exceeded?

(b) By how much does this rod stretch when supporting the mass?

(c) What is the maximum mass that this rod can support without breaking?

19. A hydraulic system contains $5\bar{0}$ L of lubricating oil.

(a) What is the decrease in the volume of the oil under a pressure of 43 MPa?

(b) What is the compressibility of the oil?

20. A steel ball of radius 50.0 cm is lowered to the bottom of a deep sea, where the pressure is 65 MPa. Find its decrease in volume.

21. How much pressure does it take to decrease a given volume of water 0.10 percent?

22. How much pressure does it take to decrease a given volume of oil 0.20 percent?

23. Six steel rivets, each of diameter 6.5 mm, hold together two steel plates which are 12.0 mm apart. A total shearing force of 8.5×10^5 N is applied.

(a) Find the shearing stress on each rivet.

(b) Find the displacement on each rivet.

(c) Find the angle of shear. See Example 1, Section 13.7.

24. A solid steel shaft is 4.00 m long and 6.00 cm in radius. If a torque of 1.6×10^5 N m is applied to the shaft, find the angle through which the shaft twists.

25. A 120-kW engine drives a solid aluminum driveshaft which is 5.00 m long and 8.00 cm in diameter at 850 rpm. Find the angle through which the shaft is twisted.

26. In drilling an oil well, a steel pipe whose diameter is 10.0 cm and thickness is 6.5 mm is used to turn the drill bit. The pipe is $100\bar{0}$ m long, rotates at 250 rpm, and is driven by a 600-kW motor. Find the angle through which the pipe twists.

27. What is the density (in kg/m^3) of a metal block having dimensions 18 cm $\times$ 24 cm $\times$ 8.0 cm with a mass of 9.76 kg?

28. What is the mass density (in kg/m^3) of a chunk of rock of mass 210 g which displaces a volume of 75 cm^3 of water?

29. What volume (in cm^3) does 1300 g of mercury occupy?

30. What volume (in cm^3) does 1300 g of cork occupy?

31. What volume does 1300 g of nitrogen occupy at 0°C and 1 atm pressure?

32. Find the mass of a rectangular bar of gold 4.00 cm × 6.00 cm × 20.00 cm.

33. Find the volume of 5.0 kg of propane at 0°C and 1 atm pressure.

34. Find the mass (in kg) of 1.00 m^3 of
   (a) water
   (b) gasoline
   (c) copper
   (d) mercury
   (e) air at 0°C and 1 atm pressure.

35. What size tank (in L) is needed for $10\overline{0}0$ kg of
   (a) water?
   (b) gasoline?
   (c) mercury?

36. Granite has a density of 2650 kg/m^3. Find its density in g/cm^3.

From the table on page 290, find the specific gravity of the substances in Exercises 37–42.

37. aluminum   **38.** cork   **39.** gasoline

40. air   **41.** ice   **42.** mercury

43. The specific gravity of substance X is 1.8. Does it float or sink in water?

44. The specific gravity of substance Y is 0.76. Does it float or sink in water?

45. The specific gravity of substance Z is 12.6. Does it float or sink in a vat of mercury?

46. The specific gravity of substance W is 0.5. Does it float or sink in a tank of gasoline?

47. A proton has a mass of 1.67×10^{-27} kg and a diameter of 3.0×10^{-15} m. Find its specific gravity.

FLUIDS

14.1 HYDROSTATIC PRESSURE

Hydrostatic pressure deals with the characteristics of liquids at rest and the pressure the liquid exerts on an immersed object. As you probably know, the pressure increases as you go deeper in water. Liquids are different in this respect from solids in that, where solids exert only a downward force due to gravity, the force exerted by liquids is in all directions.

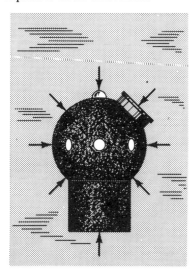

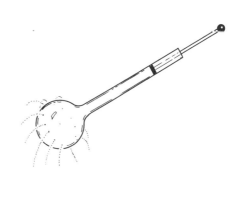

The pressure at any given depth in a liquid is due to its weight. The mass of a column of liquid is given by

$$m = DV$$

or

$$m = DAh$$

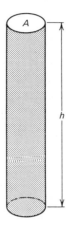

where m is the mass,

      D is the density of the liquid,

      V is the volume of the liquid,

      A is the cross-sectional area of the column, and

      h is the height of the column.

The weight of the column or force that the column exerts downward is then

$$W = F = DAhg$$

Substituting this into $P = F/A$, we find that the pressure exerted by this column of liquid is

$$P = Dhg$$

where P is the pressure,

      D is the density,

      h is the height of the column of liquid (or depth of the liquid), and

      g is the gravitational acceleration (9.80 m/s^2).

That is, the pressure depends only on the depth and the density of the liquid. Because the pressure exerted by water increases with depth, dams are built much thicker at the base than at the top.

An apparatus called Pascal's vases was devised by Pascal to show liquids of different volumes in different shaped containers *but of equal depths* yield the same pressure.

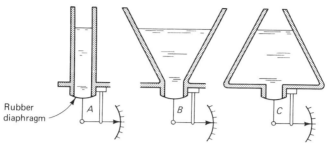

Pascal's vases: The pressure at each point is the same.

example 1   Find the pressure at a depth of 3.00 m in a swimming pool filled with water.

DATA:

$$D = 100\bar{0} \text{ kg/m}^3$$

$$h = 3.00 \text{ m}$$

$$g = 9.80 \text{ m/s}^2$$

$$P = ?$$

BASIC EQUATION: $P = Dhg$

WORKING EQUATION: same

SUBSTITUTION:

$$P = (100\bar{0} \text{ kg/m}^3)(3.00 \text{ m})(9.80 \text{ m/s}^2)$$

$$= 29{,}400 \text{ (kg m/s}^2)/\text{m}^2$$

$$= 29{,}400 \text{ N/m}^2$$

$$= 29{,}400 \text{ Pa}$$

$$= 29.4 \text{ kPa}$$

In the discussions here, we will ignore the additional pressure due to the weight of the atmosphere. This topic is addressed in Section 17.2.

example 2   Find the depth in a sea at which the pressure is ten times normal atmospheric pressure, which is 101.3 kPa.

DATA:

$$P = 10(101.3 \text{ kPa})$$

$$D = 1025 \text{ kg/m}^3$$

$$g = 9.80 \text{ m/s}^2$$

$$h = ?$$

BASIC EQUATION: $P = Dhg$

WORKING EQUATION: $h = \dfrac{P}{Dg}$

SUBSTITUTION: $h = \dfrac{10(101.3 \text{ kPa})}{(1025 \text{ kg/m}^3)(9.80 \text{ m/s}^2)}$

$$= \dfrac{1.013 \times 10^6 \text{Pa}}{(1025 \text{ kg/m}^3)(9.80 \text{ m/s}^2)}$$

$$\dfrac{\text{Pa}}{\dfrac{\text{kg}}{\text{m}^3}\dfrac{\text{m}}{\text{s}^2}} = \dfrac{\dfrac{\text{N}}{\text{m}^2}}{\dfrac{\text{kg}}{\text{m}^3}\dfrac{\text{m}}{\text{s}^2}} = \dfrac{\dfrac{\text{kg m/s}^2}{\text{m}^2}}{\dfrac{\text{kg}}{\text{m}^3}\dfrac{\text{m}}{\text{s}^2}} = \text{m}$$

$$= 101 \text{ m}$$

To find the total force exerted by a liquid on a horizontal surface (such as the bottom of a tank), substitute $P = F/A$ into $P = Dhg$. Thus,

$$\frac{F}{A} = Dhg$$

or

$$\boxed{F_t = ADhg}$$

where F_t is the total force exerted on and perpendicular to the surface and
 A is the area of the horizontal surface.

To find the total force exerted by a liquid on the vertical sides of a container, you have to take into account that the pressure varies linearly with the depth. In this case, we use the average height, $h/2$, to compute the total force on the vertical sides; that is,

$$F_s = \frac{ADhg}{2}$$

or

$$\boxed{F_s = \tfrac{1}{2}ADhg}$$

where F_s is the total force on the vertical side and
 A is the area of the vertical surface.

example 3 A water tank of radius 8.00 m and height 12.0 m and 40.0 m above the ground is filled with water.

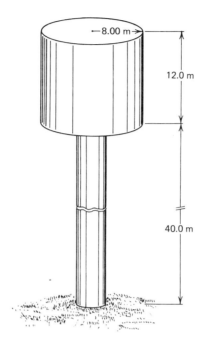

(a) What is the total force on the bottom of the tank?
(b) What is the total force on the sides of the tank?
(c) What is the pressure on the bottom of the tank?
(d) What is the pressure at ground level?

(a) DATA:

$$D = 100\bar{0} \text{ kg/m}^3$$

$$h = 12.0 \text{ m}$$

$$r = 8.00 \text{ m}$$

$$A = \pi r^2 = \pi(8.00 \text{ m})^2$$

$$g = 9.80 \text{ m/s}^2$$

$$F_t = ?$$

BASIC EQUATION:

$$F_t = ADhg$$

WORKING EQUATION: same

SUBSTITUTION:

$$F_t = \pi(8.00 \text{ m})^2(100\bar{0} \text{ kg/m}^3)(12.0 \text{ m})(9.80 \text{ m/s}^2)$$

$$= 2.36 \times 10^7 \text{ kg m/s}^2$$

$$= 2.36 \times 10^7 \text{ N}$$

(b) DATA:
$$D = 100\bar{0} \text{ kg/m}^3$$
$$h = 12.0 \text{ m}$$
$$r = 8.00 \text{ m}$$
$$A = 2\pi rh \quad \text{(lateral surface area of a cylinder)}$$
$$g = 9.80 \text{ m/s}^2$$
$$F_s = ?$$

BASIC EQUATION:
$$F_s = \tfrac{1}{2}ADhg$$
$$= \tfrac{1}{2}(2\pi rh)Dhg$$

WORKING EQUATION:
$$F_s = \pi rh^2 Dg$$

SUBSTITUTION:
$$F_s = \pi(8.00 \text{ m})(12.0 \text{ m})^2(100\bar{0} \text{ kg/m}^3)(9.80 \text{ m/s}^2)$$
$$= 3.55 \times 10^7 \text{ N}$$

(c) DATA:
$$F = F_t = 2.36 \times 10^7 \text{ N}$$
$$r = 8.00 \text{ m}$$
$$A = \pi r^2$$
$$P = ?$$

BASIC EQUATION:
$$P = \frac{F}{A}$$

WORKING EQUATION: same

SUBSTITUTION:
$$P = \frac{2.36 \times 10^7 \text{ N}}{\pi(8.00 \text{ m})^2}$$
$$= 1.17 \times 10^5 \text{ N/m}^2$$
$$= 1.17 \times 10^5 \text{ Pa}$$
$$= 117 \text{ kPa}$$

(d) DATA:
$$h = 12.0 \text{ m} + 40.0 \text{ m} = 52.0 \text{ m}$$
$$D = 100\bar{0} \text{ kg/m}^3$$
$$g = 9.80 \text{ m/s}^2$$
$$P = ?$$

BASIC EQUATION:
$$P = Dhg$$

WORKING EQUATION: same

Substitution: $P = (100\overline{0} \text{ kg/m}^3)(52.0 \text{ m})(9.80 \text{ m/s}^2)$

$$= 5.10 \times 10^5 \text{ N/m}^2$$

$$= 5.10 \times 10^5 \text{ Pa}$$

$$= 51\overline{0} \text{ kPa}$$

14.2   PASCAL'S PRINCIPLE

One of the most important principles in the study of hydraulics is named after Pascal:

PASCAL'S PRINCIPLE

Pressure applied to a confined liquid is transmitted, without loss, throughout the entire liquid and to the walls of the container.

The hydraulic jack or press illustrates this principle. If we apply a force to the small piston, the pressure is transmitted, without loss, in all directions. The pressure on the large piston is the same as the pressure on the small piston. However, the *total force* on the large piston is greater because its surface area is larger.

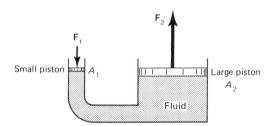

The pressures on piston 1 and piston 2 are

$$P_1 = \frac{F_1}{A_1} \quad \text{and} \quad P_2 = \frac{F_2}{A_2}$$

respectively. But, by Pascal's principle, $P_1 = P_2$. Thus,

$$\frac{F_1}{A_1} = \frac{F_2}{A_2}$$

or

$$\frac{F_1}{F_2} = \frac{A_1}{A_2} = \frac{r_1^2}{r_2^2} = \frac{D_1^2}{D_2^2}$$

where r is radius and D is the diameter of the piston. Can you show how the last two ratios follow from the first two ratios?

example   The pistons of a hydraulic press have radii of 2.00 cm and 12.0 cm.

(a) What force must be applied to the smaller piston to exert a force of $50\overline{0}0$ N on the larger?
(b) What is the pressure on each piston?
(c) What is its actual mechanical advantage?

(a) DATA:

$$F_2 = 50\overline{0}0 \text{ N}$$

$$r_2 = 12.0 \text{ cm}$$

$$F_1 = ?$$

$$r_1 = 2.00 \text{ cm}$$

BASIC EQUATION:

$$\frac{F_1}{F_2} = \frac{r_1^2}{r_2^2}$$

WORKING EQUATION:

$$F_1 = \frac{F_2 r_1^2}{r_2^2}$$

SUBSTITUTION:

$$F_1 = \frac{(50\overline{0}0 \text{ N})(2.00 \text{ cm})^2}{(12.0 \text{ cm})^2}$$

$$= 139 \text{ N}$$

(b) DATA:

$$F_2 = 50\overline{0}0 \text{ N}$$

$$r_2 = 12.0 \text{ cm} = 0.120 \text{ m}$$

$$A_2 = \pi r^2 = \pi(0.120 \text{ m})^2$$

$$P_2 = ?$$

BASIC EQUATION:

$$P_2 = \frac{F_2}{A_2}$$

WORKING EQUATION: same

SUBSTITUTION:

$$P_2 = \frac{50\overline{0}0 \text{ N}}{\pi(0.120 \text{ m})^2}$$

$$= 1.11 \times 10^5 \text{ N/m}^2$$

$$= 1.11 \times 10^5 \text{ Pa}$$

$$= 111 \text{ kPa}$$

$$P_1 = 111 \text{ kPa}   \text{(by Pascal's principle)}$$

(c) DATA:

$$F_R = 50\bar{0}0 \text{ N}$$

$$F_E = 139 \text{ N}$$

$$\text{AMA} = ?$$

BASIC EQUATION:

$$\text{AMA} = \frac{F_R}{F_E}$$

WORKING EQUATION: same

SUBSTITUTION:

$$\text{AMA} = \frac{50\bar{0}0 \text{ N}}{139 \text{ N}}$$

$$= 36.0$$

14.3 ARCHIMEDES' PRINCIPLE

One of the most important effects of pressure produced by a fluid is named after Archimedes (287–212 B.C.):

> **ARCHIMEDES' PRINCIPLE**
>
> *Any object placed in a fluid is buoyed up by a force equal to the weight of the displaced fluid.*

A floating boat displaces an amount of water equal to its weight, which equals the buoyant force of the water on the boat. A weather balloon displaces an amount of air equal to its weight, which equals the buoyant force of the air on the balloon.

A solid concrete block 15.0 cm $\times$ 20.0 cm $\times$ 10.0 cm weighs 67.6 N. When it is lowered into the water, its weight registers 38.2 N. The buoyant force is 67.6 N $-$ 38.2 N $=$ 29.4 N. The volume of the displaced water is

$$V = lwh$$

$$= 15.0 \text{ cm} \times 20.0 \text{ cm} \times 10.0 \text{ cm}$$

$$= 30\bar{0}0 \text{ cm}^3 = 3.00 \times 10^{-3} \text{ m}^3$$

The mass of the displaced water is

$$m = DV$$

$$= \left(100\bar{0} \text{ kg/m}^3\right)(3.00 \times 10^{-3} \text{ m}^3)$$

$$= 3.00 \text{ kg}$$

The weight of the displaced water is then

$$W = mg$$
$$= (3.00 \text{ kg})(9.80 \text{ m/s}^2)$$
$$= 29.4 \text{ N}$$

which equals the buoyant force.

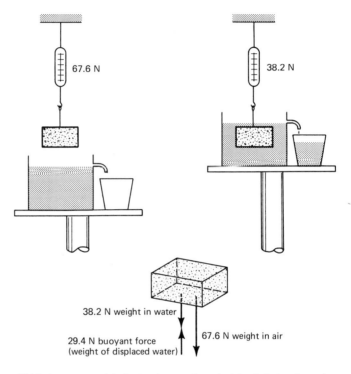

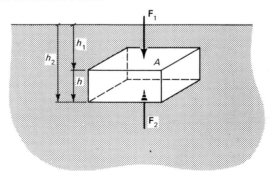

38.2 N weight in water

29.4 N buoyant force
(weight of displaced water)

67.6 N weight in air

Weight in water = weight in air − buoyant force (weight of displaced water)

To develop an equation for this buoyant force, consider the rectangular solid placed in the liquid as shown below.

The vertical surfaces of the block are in equilibrium. On the top surface, the liquid exerts a downward force of

$$F_1 = ADh_1 g$$

On the bottom surface, the liquid exerts an upward force of

$$F_2 = ADh_2 g$$

Since $h_2 > h_1$, the liquid exerts a net buoyant force upward of

$$F = F_2 - F_1$$
$$= ADh_2 g - ADh_1 g$$
$$= ADg(h_2 - h_1)$$
$$= ADgh$$
$$= VDg \quad \text{(volume of block} = Ah)$$
$$= \text{weight of the displaced liquid}$$

example A flat-bottom river barge is 12.0 m wide, 36.0 m long, and 5.00 m deep.

(a) How many m³ of water will it displace if the top stays 1.00 m above the water?
(b) What load (in N) will the barge contain under these conditions if the barge weighs 4.50×10^6 N in dry dock?

(a) DATA:

$l = 36.0$ m

$w = 12.0$ m

$h = 5.00$ m $- 1.00$ m $= 4.00$ m

$V = ?$

BASIC EQUATION:

$V = lwh$

WORKING EQUATION:

same

SUBSTITUTION:

$V = (36.0 \text{ m})(12.0 \text{ m})(4.00 \text{ m})$

$= 1730 \text{ m}^3$

(b) Next, find the mass of the displaced water:

DATA:

$D = 100\bar{0} \text{ kg/m}^3$

$V = 1730 \text{ m}^3$

$m = ?$

BASIC EQUATION:

$D = \dfrac{m}{V}$

WORKING EQUATION:

$m = DV$

SUBSTITUTION:

$m = (100\bar{0} \text{ kg/m}^3)(1730 \text{ m}^3)$

$= 1.73 \times 10^6 \text{ kg}$

Its weight is then

$$W = mg = (1.73 \times 10^6 \text{ kg})(9.80 \text{ m/s}^2)$$

$$= 1.70 \times 10^7 \text{ N}$$

That is, the buoyant force is 1.70×10^7 N.

Since 4.50×10^6 N of the buoyant force is used to buoy up the barge itself, the difference

$$1.70 \times 10^7 \text{ N} - 4.50 \times 10^6 \text{ N} = 1.25 \times 10^7 \text{ N}$$

is the weight of the load that it is carrying.

14.4 SPECIFIC GRAVITY

In Section 13.11, the specific gravity of a substance is defined as the ratio of the density of the substance to the density of water. That is,

$$\boxed{\text{sp gr} = \frac{D_{\text{substance}}}{D_{\text{water}}}}$$

Now we will discuss the methods for finding the specific gravities of various substances.

Solids: denser than water   If we consider equal volumes of an insoluble substance and water, the equation above can be written

$$\boxed{\text{sp gr} = \frac{\text{weight of substance in air}}{\text{weight of equal volume of displaced water}}}$$

Or, since the weight of the displaced water equals the buoyant force of the water, we may also write

$$\boxed{\text{sp gr} = \frac{\text{weight of substance in air}}{\text{buoyant force of water}}}$$

example 1   A piece of bronze weighs 10.4 N in air and displaces $12\bar{0}$ cm^3 of water when fully submerged. Find its specific gravity.

First, find the mass of the displaced water.

DATA:

$$D = 100\bar{0} \text{ kg/m}^3$$

$$V = 12\bar{0} \text{ cm}^3 \times \frac{1 \text{ m}^3}{10^6 \text{ cm}^3} = 1.20 \times 10^{-4} \text{ m}^3$$

$$m = ?$$

BASIC EQUATION:	$D = \dfrac{m}{V}$
WORKING EQUATION:	$m = DV$
SUBSTITUTION:	$m = (100\bar{0} \text{ kg/m}^3)(1.20 \times 10^{-4} \text{ m}^3)$
	$= 0.120 \text{ kg}$

Its weight is then

$$W = mg = (0.120 \text{ kg})(9.80 \text{ m/s}^2)$$
$$= 1.18 \text{ N}$$

Then, to find the specific gravity of the bronze:

DATA:	weight of substance in air = 10.4 N
	weight of equal volume of water = 1.18 N
	sp gr = ?
BASIC EQUATION:	sp gr = $\dfrac{\text{weight of substance in air}}{\text{weight of equal volume of water}}$
WORKING EQUATION:	same
SUBSTITUTION:	sp gr = $\dfrac{10.4 \text{ N}}{1.18 \text{ N}}$
	$= 8.81$

example 2 A brick weighs 19.1 N in air and 8.5 N in water. Find its specific gravity.

DATA:	weight of substance in air = 19.1 N
	buoyant force of water = 19.1 N $-$ 8.5 N = 10.6 N
	sp gr = ?
BASIC EQUATION:	sp gr = $\dfrac{\text{weight of substance in air}}{\text{buoyant force of water}}$
WORKING EQUATION:	same
SUBSTITUTION:	sp gr = $\dfrac{19.1 \text{ N}}{10.6 \text{ N}}$
	$= 1.80$

Solids: less dense than water In this case, the solid floats on water and the buoyant force equals the weight of the solid in air. So we could weigh the substance in air and measure the amount of displaced water and proceed as in Example 1.

example 3 A piece of wood weighs 4.00 N in air. When it is submerged in water, it displaces 785 cm^3. Find its specific gravity.

First, find the mass of the displaced water.

DATA:
$$D = 100\bar{0} \text{ kg/m}^3$$

$$V = 785 \text{ cm}^3 \times \frac{1 \text{ m}^3}{10^6 \text{ cm}^3} = 7.85 \times 10^{-4} \text{ m}^3$$

$$m = ?$$

BASIC EQUATION:
$$D = \frac{m}{V}$$

WORKING EQUATION:
$$m = DV$$

SUBSTITUTION:
$$m = (100\bar{0} \text{ kg/m}^3)(7.85 \times 10^{-4} \text{ m}^3)$$
$$= 0.785 \text{ kg}$$

Its weight is then

$$W = mg = (0.785 \text{ kg})(9.80 \text{ m/s}^2)$$
$$= 7.69 \text{ N}$$

Then, to find its specific gravity:

DATA:
weight of substance in air $= 4.00$ N

weight of displaced water $= 7.69$ N

sp gr $= ?$

BASIC EQUATION:
$$\text{sp gr} = \frac{\text{weight of substance in air}}{\text{weight of displaced water}}$$

WORKING EQUATION:
same

SUBSTITUTION:
$$\text{sp gr} = \frac{4.00 \text{ N}}{7.69 \text{ N}}$$
$$= 0.520$$

A sinker may be used to submerge the solid if the weight of the sinker is subtracted from the computations.

example 4 A chunk of cork weighs 0.80 N in air. A sinker weighs 3.50 N when submerged in water. The combined weight of the cork and sinker when both are submerged is 0.90 N. Find the specific gravity of the cork.

SKETCH:

DATA:

weight of substance in air = 0.80 N

buoyant force of water = (0.80 N + 3.50 N) − 0.90 N

$$= 3.40 \text{ N}$$

sp gr = ?

BASIC EQUATION.

$$\text{sp gr} = \frac{\text{weight of substance in air}}{\text{buoyant force of water}}$$

WORKING EQUATION:

same

SUBSTITUTION:

$$\text{sp gr} = \frac{0.80 \text{ N}}{3.40 \text{ N}}$$

$$= 0.24$$

Now, let us study three methods which compare the weight of a liquid with the weight of an equal volume of water.

Liquids: pycnometer method   The bottle shown below is called a pycnometer bottle. It has a very small hole in the stopper which allows precise filling of a

Pycnometer bottle

liquid each time it is used. By comparing the weight (or mass) of any liquid with the weight (or mass) of water, we can find the specific gravity of the liquid. That is,

$$\text{sp gr} = \frac{\text{weight (or mass) of liquid}}{\text{weight (or mass) of equal volume of water}}$$

example 5   A given pycnometer has a mass of 24.5 g. When filled with water, its mass is 73.5 g. When filled with a nitric acid solution, its mass is 95.1 g. Find the specific gravity of the acid solution.

DATA:
$$\text{mass of liquid} = 95.1 \text{ g} - 24.5 \text{ g} = 70.6 \text{ g}$$

$$\text{mass of equal volume of water} = 73.5 \text{ g} - 24.5 \text{ g} = 49.0 \text{ g}$$

$$\text{sp gr} = ?$$

BASIC EQUATION:
$$\text{sp gr} = \frac{\text{mass of liquid}}{\text{mass of equal volume of water}}$$

WORKING EQUATION: same

SUBSTITUTION:
$$\text{sp gr} = \frac{70.6 \text{ g}}{49.0 \text{ g}}$$

$$= 1.44$$

Liquids: loss of weight method   The denser a liquid, the greater its buoyant force. We can take a given solid, compare its buoyant forces in a liquid and water, and find the specific gravity of the liquid. That is,

$$\boxed{\text{sp gr} = \frac{\text{buoyant force of liquid}}{\text{buoyant force of water}}}$$

example 6   A glass ball weighs 0.75 N in air, 0.50 N in water, and 0.58 N in a liquid of unknown specific gravity. Find the specific gravity of the unknown liquid.

DATA:
$$\text{buoyant force of liquid} = 0.75 \text{ N} - 0.58 \text{ N} = 0.17 \text{ N}$$

$$\text{buoyant force of water} = 0.75 \text{ N} - 0.50 \text{ N} = 0.25 \text{ N}$$

$$\text{sp gr} = ?$$

BASIC EQUATION:
$$\text{sp gr} = \frac{\text{buoyant force of liquid}}{\text{buoyant force of water}}$$

WORKING EQUATION: same

SUBSTITUTION:
$$\text{sp gr} = \frac{0.17 \text{ N}}{0.25 \text{ N}}$$

$$= 0.68$$

Liquids: hydrometer method   A hydrometer is a sealed glass tube weighted at one end so that it floats vertically in a liquid. It sinks in the liquid until it displaces an amount of liquid equal to its own weight. The densities of the displaced liquids are inversely proportional to the depths to which the tube sinks. That is, the greater the density of the liquid, the less the tube sinks, and vice versa. A hydrometer usually has a scale inside the tube and is calibrated so that it floats in water at the 1.00 mark.

Common laboratory hydrometers for (a) light liquids, (b) for light or heavy liquids, and (c) for heavy liquids are shown below.

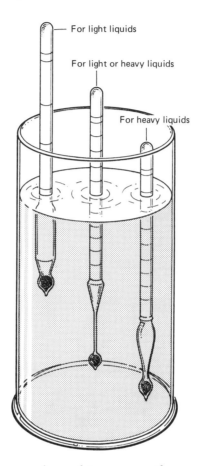

Hydrometers are commonly used to measure the specific gravities of battery acid and antifreeze in a radiator. In a lead storage battery, the electrolyte is a solution of sulfuric acid and water. The following chart gives common specific gravities of various conditions of a lead storage battery.

Condition	Specific gravity
New (fully charged)	1.3
Old (discharged)	1.15

The chart on page 314 gives various winter temperatures and the corresponding specific gravities at which the antifreeze and water solution is safe.

Temperature (°C)	Specific gravity
− 1.24	1.0049
− 2.99	1.0113
− 6.89	1.0232
− 19.82	1.0478
− 44.83	1.0738
− 51.23	1.0784

Common battery tester (a) and antifreeze tester hydrometers (b) are shown below.

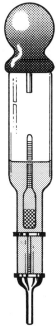
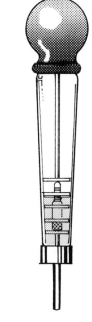

(a) Common battery tester hydrometer   (b) Antifreeze tester hydrometer

One other factor must be considered in the use of the hydrometer—that of temperature. Significant differences in readings will occur over a range of temperatures. Specific gravities of some common liquids:

Liquid	Specific gravity	Liquid	Specific gravity
Benzene	0.9	Seawater	1.025
Ethyl alcohol	0.79	Sulfuric acid	1.84
Gasoline	0.68	Turpentine	0.87
Kerosene	0.82	Water	1.000
Mercury	13.6		

14.5 FLUID FLOW

The study of the motion of fluids is very complex. Think for a minute about the motion of water flowing down a fast moving mountain stream that contains boulders and rapids and about the motion of the air during a thunderstorm or during a tornado. These types of motion are complex, indeed. We will limit our discussion to the more simple examples of fluid flow.

Streamline flow is the smooth flow of a fluid through a tube or pipe. By smooth flow we mean that each particle of the fluid follows the same uniform path as all the others. If the speed of the flow becomes too great or if the tube changes direction or diameter too abruptly, the motion of the flow is described as *turbulent*.

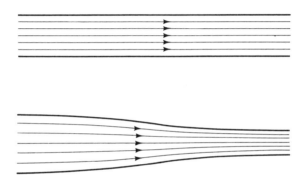

Streamline flow of a fluid through a tube or pipe

The *flow rate* of a fluid is the volume of fluid flowing past a given point in a pipe per unit time. Assume that we have a streamline flow through a straight section of pipe at speed v. During a time interval t, each particle of fluid travels a distance vt. If A is the cross-sectional area of the pipe, the volume of fluid passing a given point during time interval t is vtA. Thus, the flow rate is given by

$$Q = \frac{vtA}{t}$$

or

$$\boxed{Q = vA}$$

where Q is the flow rate,
$\qquad v$ is velocity of the fluid, and
$\qquad A$ is the cross-sectional area of the tube or pipe.

example   Water flows through a fire hose of diameter 6.4 cm at a velocity of 5.9 m/s. Find the flow rate of the fire hose in L/min.

DATA:

$$v = 5.9 \text{ m/s}$$

$$r = 3.2 \text{ cm} = 0.032 \text{ m}$$

$$A = \pi r^2 = \pi(0.032 \text{ m})^2$$

$$Q = ?$$

BASIC EQUATION: $Q = vA$

WORKING EQUATION: same

SUBSTITUTION: $Q = (5.9 \text{ m/s})\pi(0.032 \text{ m})^2 \times \dfrac{10^3 \text{ L}}{1 \text{ m}^3} \times \dfrac{60 \text{ s}}{1 \text{ min}}$

$$= 1100 \text{ L/min}$$

For an incompressible fluid, the flow rate is constant throughout the pipe. If the cross-sectional area of the pipe changes and streamline flow is maintained, the flow rate is the same all along the pipe.

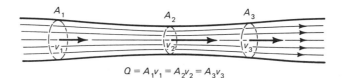

$$Q = A_1 v_1 = A_2 v_2 = A_3 v_3$$

That is, as the cross-sectional area increases, the velocity decreases, and vice versa.

14.6   BERNOULLI'S PRINCIPLE

What happens to the pressure as the cross-sectional area of the pipe changes? This concept can be illustrated by use of a Venturi meter, shown below.

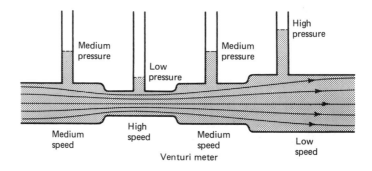

Venturi meter

Here the vertical tubes act like pressure gauges; the higher the column, the higher the pressure. As you can see, the higher the speed, the lower the pressure, and vice versa.

The explanation of this change in pressure of a fluid in streamline flow was found by Daniel Bernoulli (1700–1782).

BERNOULLI'S PRINCIPLE

For an incompressible fluid in a streamline flow, the quantity

$$P + \frac{1}{2}Dv^2 + Dgh$$

remains constant at all points along the flow.

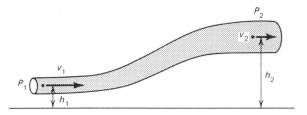

That is,

$$P_1 + \frac{1}{2}Dv_1^2 + Dgh_1 = P_2 + \frac{1}{2}Dv_2^2 + Dgh_2$$

where P is the pressure,
$\quad$ D is the density of the fluid
$\quad$ v is the velocity of the fluid,
$\quad$ h is the height above some reference line, and
$\quad$ g is the gravitational acceleration (9.80 m/s²).
If the pipe remains horizontal ($h_1 = h_2$), Bernoulli's equation becomes

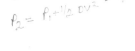

$$P_1 + \frac{1}{2}Dv_1^2 = P_2 + \frac{1}{2}Dv_2^2$$

example 1   In a Venturi meter, the water pressure is 150.0 kPa, where the velocity of the water is 5.0 m/s. If the water pressure increases to 160.0 kPa, find its velocity.

DATA:
$\qquad P_1 = 150.0 \text{ kPa} = 1.500 \times 10^5 \text{ Pa} = 1.500 \times 10^5 \text{ N/m}^2$

$\qquad v_1 = 5.0 \text{ m/s}$

$\qquad P_2 = 160.0 \text{ kPa} = 1.600 \times 10^5 \text{ Pa} = 1.600 \times 10^5 \text{ N/m}^2$

$\qquad D = 100\bar{0} \text{ kg/m}^3$

$\qquad v_2 = ?$

BASIC EQUATION: $\qquad P_1 + \frac{1}{2}Dv_1^2 = P_2 + \frac{1}{2}Dv_2^2 \qquad (h = 0)$

WORKING EQUATION: $\qquad v_2 = \sqrt{\dfrac{P_1 - P_2 + \frac{1}{2}Dv_1^2}{\frac{1}{2}D}}$

SUBSTITUTION:

$$v_2 = \sqrt{\frac{1.500 \times 10^5 \ \text{N/m}^2 - 1.600 \times 10^5 \ \text{N/m}^2 + \frac{1}{2}(100\bar{0} \ \text{kg/m}^3)(5.0 \ \text{m/s})^2}{\frac{1}{2}(100\bar{0} \ \text{kg/m}^3)}}$$

$$= \sqrt{\frac{2500 \ \text{N/m}^2}{\frac{1}{2}(100\bar{0} \ \text{kg/m}^3)}} \qquad \boxed{\left(\frac{\text{kg}}{\text{m}^3}\right)\left(\frac{\text{m}}{\text{s}}\right)^2 = \frac{(\text{kg m/s}^2)}{\text{m}^2} = \text{N/m}^2}$$

$$= \sqrt{\frac{2500 \ (\text{kg m/s}^2)/\text{m}^2}{\frac{1}{2}(100\bar{0} \ \text{kg/m}^3)}} \qquad \boxed{\sqrt{\frac{(\text{kg m/s}^2)/\text{m}^2}{\text{kg/m}^3}} = \sqrt{\frac{\text{kg m m}^3}{\text{kg s}^2 \text{m}^2}} = \sqrt{\frac{\text{m}^2}{\text{s}^2}} = \text{m/s}}$$

$$= 2.2 \ \text{m/s}$$

If the liquid in a column is stationary ($v_1 = v_2 = 0$),

$$P_1 + Dgh_1 = P_2 + Dgh_2$$

Then

$$P_1 - P_2 = Dgh_2 - Dgh_1$$

or

$$\Delta P = Dg(h_2 - h_1)$$

which is equivalent to the equation

$$\Delta P = Dg(\Delta h)$$

or

$$\Delta P = Dgh \qquad \text{(which we derived in Section 14.1)}$$

This shows that Bernoulli's principle applies in this case also.

Now consider a large tank of a liquid with a small orifice through which the liquid is free to flow (page 319). In this case, $P_1 = P_2$; that is, the pressures at both points are normal atmospheric pressures. If the orifice is small, the level of the liquid will fall very slowly as the liquid flows out the orifice. In other words, v_1 is practically zero. Then Bernoulli's equation

$$P_1 + \frac{1}{2}Dv_1^2 + Dgh_1 = P_2 + \frac{1}{2}Dv_2^2 + Dgh_2$$

becomes

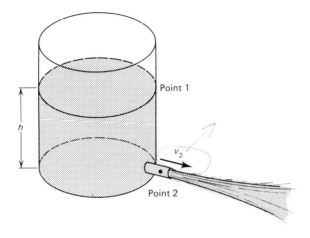

$$Dgh_1 = \tfrac{1}{2} Dv_2^2 + Dgh_2$$

$$Dg(h_1 - h_2) = \tfrac{1}{2} Dv_2^2$$

$$Dgh = \tfrac{1}{2} Dv_2^2 \qquad (h = h_1 - h_2)$$

$$2gh = v_2^2$$

or

$$\boxed{v_2 = \sqrt{2gh}}$$

The last equation is known as *Torricelli's theorem*, which relates that the velocity at which the liquid is discharged is dependent on only the height (or depth) of the liquid.

The flow rate of the liquid through the orifice is then

$$Q = vA$$

or

$$\boxed{Q = A\sqrt{2gh}}$$

example 2 A water tank of radius 8.00 m and height 12.0 m is filled with water. A hole of area 2.00 mm^2 is punctured in the side at the bottom of the tank. At what rate is the water leaking through the hole?

DATA:

$$A = 2.0 \text{ mm}^2 \times \left(\frac{1 \text{ m}}{1000 \text{ mm}}\right)^2 = 2.0 \times 10^{-6} \text{ m}^2$$

$$h = 12.0 \text{ m}$$

$$g = 9.80 \text{ m/s}^2$$

$$Q = ?$$

BASIC EQUATION:

$$Q = A\sqrt{2gh}$$

WORKING EQUATION: same

SUBSTITUTION:

$$Q = (2.0 \times 10^{-6} \text{ m}^2)\sqrt{2(9.80 \text{ m/s}^2)(12.0 \text{ m})}$$

$$= 3.1 \times 10^{-5} \text{ m}^3/\text{s}$$

$$= 3.1 \times 10^{-5} \text{ m}^3/\text{s} \times \frac{1000 \text{ L}}{1 \text{ m}^3} \times \frac{60 \text{ s}}{1 \text{ min}}$$

$$= 1.9 \text{ L/min}$$

Another application of Bernoulli's principle involves airplane travel. The figure below shows the flow of air rushing past the wing of an airplane. The velocity, v_1, of the air above the wing is greater than the velocity of the air below, v_2. Thus, the pressure at P_2 is greater, which causes a lift on the wing.

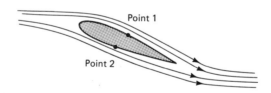

EXERCISES

Sketch	Data	Basic Equation	Working Equation	Substitution
12 cm² $h = ?$ $b = 6.0$ cm	$A = 12 \text{ cm}^2$ $b = 6.0$ cm $h = ?$	$A = bh$	$h = \dfrac{A}{b}$	$h = \dfrac{12 \text{ cm}^2}{6.0 \text{ cm}}$ $= 2.0$ cm

1. Find the pressure at the bottom of a water tower that is filled to a depth of 18.0 m.
2. A submarine is at a depth of 1.00 km in the Atlantic Ocean.
   (a) What is the pressure at this depth?
   (b) What is the total force pushing down on a circular hatch that is 80.0 cm in diameter?

3. The depth of a given point on the floor of the Pacific Ocean is 11,300 m.

(a) What is the pressure at this depth?

(b) What is the total force exerted by the seawater on a spherical instrument package of diameter 1.20 m at this depth?

4. A tank 30.0 cm × 30.0 cm × 50.0 cm deep is filled with mercury.

(a) Find the pressure at the bottom of the tank.

(b) Find the total force on the bottom of the tank.

(c) Find the total force on any side of the tank.

(d) Answer parts (a), (b), and (c) if the tank is filled with water.

(e) How do the answers compare?

5. What must the water pressure at ground level be to supply water to the third floor of a building (7.8 m high) with a pressure of 275 kPa at the third-floor level?

6. A filled cylindrical water tank of radius 10.0 m and height 40.0 m sits on the highest hill in a community.

(a) What is the total force on the bottom of the tank?

(b) What is the total force on the sides of the tank?

(c) What is the pressure on the bottom of the tank?

(d) What is the water pressure at house *A*?

(e) What is the water pressure at house *B*?

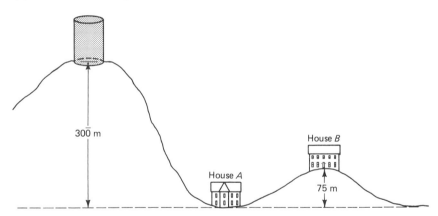

7. What is the actual mechanical advantage of a hydraulic press that produces a pressing force of 7200 N when the applied force is $6\bar{0}0$ N?

8. The AMA of a hydraulic jack is 450. What is the weight of the heaviest automobile that can be lifted by an applied force of $4\bar{0}$ N?

9. The AMA of a hydraulic jack is 360. What force must be applied to lift an automobile weighing 12,000 N?

10. The small circular piston of a hydraulic press has an area of 8.00 cm². If the applied force is 25.0 N, what must the area of the large piston be to exert a pressing force of $36\bar{0}0$ N?

11. The pistons of a hydraulic jack have radii of 5.00 cm and 30.0 cm.

(a) What force must be applied to the smaller piston to exert a force of $50\bar{0}0$ N on the larger piston?

(b) What is the pressure on the larger piston?

(c) What is the pressure on the smaller piston?

(d) What is the actual mechanical advantage of the jack?

12. A force of 60.0 N is applied to the smaller piston of radius 2.50 cm on a hydraulic press.

    (a) What must be the radius of the larger piston to lift a weight of 16,800 N?

    (b) What is the pressure on the smaller piston?

    (c) What is the pressure on the larger piston?

    (d) What is the actual mechanical advantage of the press?

13. A rock weighs 25.7 N in air and 18.9 N in water. What is the buoyant force of the water?

14. A metal casting displaces 327 cm^3 of water. What is the buoyant force of the water?

15. A flat-bottom barge is 12.0 m wide, 30.0 m long, and 6.00 m deep.

    (a) How many m^3 of water will it displace if the top stays 1.00 m above the water?

    (b) What load (in N) will the barge contain under these conditions if the barge weighs 3.6×10^6 N in dry dock?

16. A sunken barge having a total weight of 1.6×10^7 N is to be raised using inflated plastic bags. What volume of air is needed to raise the barge

    (a) in a freshwater lake?

    (b) in a seawater gulf?

17. What percent (by volume) of an iceberg is below the water level in the ocean?

18. A white pine raft is 1.00 m $\times$ 4.00 m $\times$ 10.0 cm. What load will it support without sinking?

19. A piece of lead weighs 25.0 N in air and 22.8 N in water. Find its specific gravity.

20. A chunk of iron has a mass of 11.5 kg in air. When placed under water, it displaces 1470 cm^3 of water. Find its specific gravity.

21. An aluminum rod 18.0 cm long and 3.00 cm in diameter has a mass of 344 g. Find its specific gravity.

22. A wooden block weighs 9.63 N in air. When submerged in water, it displaces 1640 cm^3. Find its specific gravity.

23. A piece of cork weighs 0.55 N in air. A sinker weighs 2.70 N when submerged in water. The combined weight of cork and sinker when both are submerged is 0.63 N. Find the specific gravity of the cork.

24. A maple plaque contains some embedded pure silver handicraft work. Its total weight is 2.98 kg. When immersed in water, it displaces 2500 cm^3.

    (a) What is the mass (in grams) of silver contained in the plaque? (The density of maple is $60\overline{0}$ kg/m^3 and of silver is 10,500 kg/m^3.)

    (b) What is its apparent weight while submerged?

25. A given pycnometer has a mass of 24.5 g. When filled with water, its mass is 73.5 g. When filled with gasoline, its mass is 58.3 g. Find the specific gravity of the gasoline.

26. A glass ball weighs 0.75 N in air, 0.50 N in water, and 0.30 N in a sulfuric acid solution. Find the specific gravity of the sulfuric acid solution.

27. Water flows through a fire hose of diameter 3.8 cm at a velocity of 5.0 m/s. Find the flow rate of the fire hose in L/min.

28. What pressure (in kPa) is required in a fire hose to have the stream of water reach a height of 25 m?

29. Water flows from a pipe at 650 L/min.

(a) What is the diameter of the pipe if the velocity of the water is 1.5 m/s?

(b) What is the velocity of the water if the diameter of the pipe is 20.0 cm?

30. In a horizontal water pipe, the pressure is $8\overline{0}$ kPa when the velocity is 2.7 m/s.

(a) If the water pressure decreases to 65 kPa, find its velocity.

(b) If the velocity changes to 1.5 m/s, what is the water pressure?

31. Water flows though a horizontal pipe 20.0 cm in diameter at a velocity of 4.00 m/s.

(a) What is its velocity after it enters a 10.0-cm-diameter pipe?

(b) What is the flow rate (in L/min)?

32. Water flows from a 15.0-cm-diameter main at a pressure of $35\overline{0}$ kPa at a velocity of 6.00 m/s into a secondary pipe of diameter 10.0 cm to a building on a hill that is 25.0 m above the main.

(a) What is the pressure of the water as it enters the building?

(b) What is the flow rate (in L/min)?

33. A gasoline tank of radius 2.00 m and height 5.00 m is filled. A 1.5-mm^2 hole is punctured in the side at the bottom of the tank.

(a) At what velocity is the gasoline leaking from the tank?

(b) At what rate is the gasoline leaking from the tank (in L/min)?

15.1 TEMPERATURE

Basically, temperature is a measure of the hotness or coldness of an object. Temperature could be measured in a simple way by using your hand to sense the hotness or coldness of an object. However, the range of temperatures that your hand can withstand is too small; and your hand is not precise enough to measure temperature. Therefore, other methods are used for measuring temperature.

A property of matter that can be used to accurately measure common temperatures is that the volume of a liquid or solid changes uniformly as its temperature changes. The liquid in glass thermometers is an example. This type of thermometer consists of a hollow glass bulb and a hollow glass tube joined together as shown in the top left figure, page 325. A small amount of liquid mercury or alcohol is placed in the bulb. The air is then removed from the tube. When the liquid is heated, it expands and rises up the glass tube. The height to which the liquid rises indicates the temperature.

The Celsius thermometer is standardized by marking two points on the glass which indicate the liquid level at two known temperatures. The temperatures used are the melting point of ice (0°C) and the boiling point of water at sea level (100°C). The distance between these marks is then divided into 100 equal segments called degrees.

The figure on the right, page 325, shows some Celsius temperature readings.

In some technical work, it is necessary to use the absolute temperature scale, which is called the Kelvin scale. This is called an absolute scale because 0 K refers

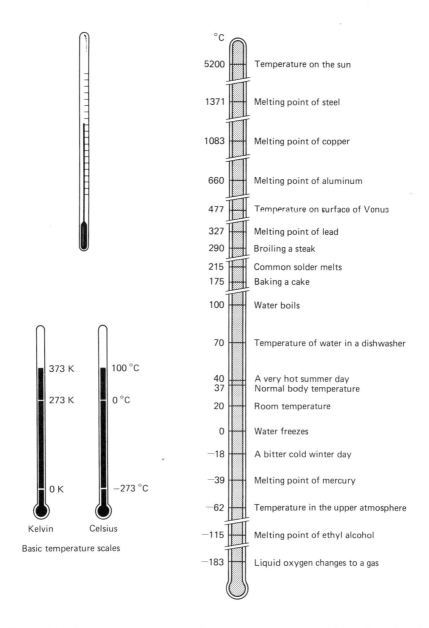

°C

5200	Temperature on the sun
1371	Melting point of steel
1083	Melting point of copper
660	Melting point of aluminum
477	Temperature on surface of Venus
327	Melting point of lead
290	Broiling a steak
215	Common solder melts
175	Baking a cake
100	Water boils
70	Temperature of water in a dishwasher
40	A very hot summer day
37	Normal body temperature
20	Room temperature
0	Water freezes
−18	A bitter cold winter day
−39	Melting point of mercury
−62	Temperature in the upper atmosphere
−115	Melting point of ethyl alcohol
−183	Liquid oxygen changes to a gas

373 K   100 °C

273 K   0 °C

0 K   −273 °C

Kelvin   Celsius

Basic temperature scales

to the theoretical lowest temperature, the temperature at which all molecular motion is assumed to stop and molecular energy is a minimum. A temperature of absolute zero can never be actually reached. The reasons are discussed in the next section. There is no such limit for high temperatures.

The relationship between the Celsius and Kelvin scales is

$$K = °C + 273.15°$$

That is,

$$0 \ K = -273.15°C$$

Since three-significant-digit accuracy is most common, we will use the following relationships in our calculations:

$$K = °C + 273°$$
$$0 \text{ K} = -273°C$$

Note that the degree symbol (°) is used when a Celsius temperature is given but is *not* used when a Kelvin temperature is given. That is,

$$100°C = 373 \text{ K}$$
$$50°C = 323 \text{ K}$$
$$-25°C = 248 \text{ K}$$

Since 0°C (273 K) is commonly used in many calculations, this temperature is also often referred to as *standard temperature*.

We could also measure temperature by using the fact that when objects are heated they give off light of different colors. When an object is heated, in the absence of chemical reactions, it first gives off red light. As it is heated more, it appears white (see the table below).

METALLURGY AND HEAT TREATMENT

Temperatures, steel colors, and related processes

Colors		°C	Processes
	White	1400°	Welding
		1300°	High speed steel hardening
	Yellow-white		(1180–1340 °C)
		1200°	
	Yellow	1100°	
		1000°	
	Orange-red		Alloy tool steel hardening
Heat colors		900°	(820–1070 °C)
	Light cherry red	800°	
			Carbon tool steel hardening
	Cherry red	700°	(730–840 °C)
	Dark red		
		600°	High speed steel tempering
	Very dark red	500°	(540–590 °C)
	Black red in dull light or darkness	400°	
			Carbon tool steel tempering
	Pale blue (310 °C)	300°	(150–570 °C)
	Violet (285 °C)		
Temper colors	Purple (275 °C)		
	Yellowish-brown (255 °C)	200°	
	Dark straw (240 °C)		
	Light straw (220 °C)	100°	
		0°	

Chemical reactions sometimes cause different colors to be given off. When carbon steel is heated and exposed to air, several colors are given off before it appears red. This is due to a chemical reaction involving the carbon. If we could measure the color of the light given off, we could then determine the temperature. Although this works only for high temperatures, it is used in the production of metal alloys. The temperature of hot molten metals is determined this way.

15.2   HEAT

When a machinist drills a hole in a metal block, it becomes very hot. As the drill does mechanical work on the metal, the temperature of the metal increases. How can we explain this? We need to look at the difference between the metal at low temperatures and at high temperatures. At high temperatures the molecules in the metal vibrate more rapidly than at low temperatures. Their velocity is higher at high temperatures, and thus their kinetic energy ($KE = \frac{1}{2}mv^2$) is greater.

To raise the temperature of a material, we must speed up the molecules; that is, we must add energy to them. *Heat is the name given to this energy which is being added to or taken from a material.*

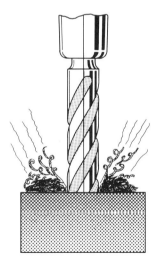

Drilling a hole in a metal block causes a temperature increase. As the drill turns, it collides with molecules of the metal, causing them to speed up. This mechanical work done on the metal has caused an increase in the energy (speed) of the molecules. For this reason, any friction between two surfaces results in a temperature rise of the materials.

Since heat is a form of energy, we should measure it in joules, the SI energy unit. However, it was not always known that heat was a form of energy, and special metric units for heat were developed and are still in use. These units are the calorie and the kilocalorie. The calorie (cal) is the amount of heat (energy) needed to raise the temperature of 1 gram of water 1°C. The kilocalorie (kcal) is the amount of heat (energy) needed to raise the temperature of 1 kilogram of water 1°C. By international agreement, the calorie and kilocalorie are defined in terms of joules as follows:

$$1 \text{ cal} = 4.18605 \text{ J}$$
$$1 \text{ kcal} = 4186.05 \text{ J}$$

For most calculations, we round these definitions to three significant digits.

We said in the last section that temperatures below absolute zero cannot be reached. Now we can see the reason for this. To lower the temperature of a substance, we need to remove some of the energy of motion of the molecules (heat). When we have removed all the heat possible (when the molecules are moving as slowly as possible), we have reached the lowest possible temperature called absolute zero. Lower temperatures cannot be reached because all the heat has been removed. However, there is no upper limit on temperature because we can always add more heat (energy) to a substance to increase its temperature.

Technically, what is the difference between temperature and heat? Temperature is a measure of the average velocity of the molecules in a substance. Then, heat is the total thermal energy (kinetic and potential) which is added to or taken away from the molecules in a substance.

There are basically two ways of changing the temperature of an object:

1. By doing work *on* the substance, such as the work done by the drill on the metal block discussed earlier in this section.
2. By supplying energy *to* the object, such as mechanical, chemical, or electrical.

15.3   HEAT TRANSFER

The movement of heat from a hot engine to the air is necessary to keep the engine from overheating. The heat produced by a furnace is transferred to the various rooms in a house. The movement of heat is a major technical problem.

The net transfer of heat from one object to another is always from the warmer object to the colder one or from the warmer part of an object to a colder part.

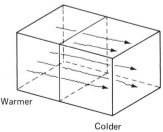

Warmer

Colder

There are three methods of heat transfer: conduction, convection, and radiation. The usual method of heat transfer in solids is *conduction*. When one end of a metal rod is heated, the molecules in that end move faster than before. These molecules collide with other molecules and cause them to move faster also. In this way the heat is transferred from one end of the metal to the other. Another example of conduction is the transferring of the excess heat produced in the combustion chamber of an engine through the engine block in the water coolant.

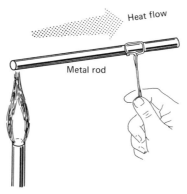

Heat conduction

The conduction of heat through some materials is better than through others. A poor conductor of heat is called an *insulator*. A list of several good conductors and poor conductors is given below.

Good heat conductors	*Poor heat conductors*
Aluminum	Air
Copper	Asbestos
Steel	Glass
	Wood

Another method of heat transfer is called *convection* (page 330). This is the movement of warm gases or liquids from one place to another. The wind carries heat along with it. The water coolant in an engine carries hot water from the engine block to the radiator by a convection process. The wind is a natural convection process. The engine coolant is a forced convection process because it depends on a pump.

The natural convection process occurs because warm air is lighter (less dense) than cool air. The warm air rises and the cool air falls.

The third method of heat transfer is *radiation*. Put your hand several inches from a hot iron. The heat you feel is not transferred by conduction because air is a poor conductor. It is not transferred by convection because hot air rises. This kind

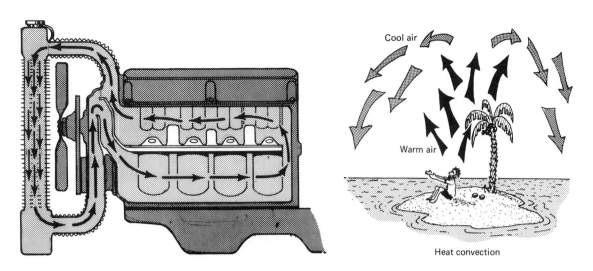

Heat convection

of heat transfer is called radiation. This radiant heat is similar to light and passes through air, glass, and the vacuum of space. The energy that comes to us from the sun is in the form of radiant energy.

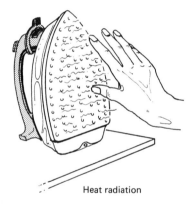

Heat radiation

Dark objects absorb radiant heat and light objects reflect radiant heat. This is why we feel cooler on a hot day when we wear light-colored clothing than when we wear dark clothing.

15.4 HEAT CAPACITY

If we place a piece of steel and a pan of water in the direct sunlight, we would notice that the water becomes only slightly warmer while the steel gets quite hot. Why should one get so much hotter than the other? If equal masses of steel and water were placed over the same flame for 1 min, the temperature of the steel would increase almost 10 times more than that of the water. The water has a greater capacity to absorb heat.

Consider 250-g blocks of aluminum, copper, iron, lead, and zinc, each of the same cross-sectional area. The heights will vary because of their different densities. We shall heat them to the same temperature in boiling water and then place the five metal blocks on a smooth block of paraffin. As the diagram below shows, the aluminum melts by far the most paraffin, followed by iron. Copper and zinc seem to melt the same amount, and lead melts the least amount.

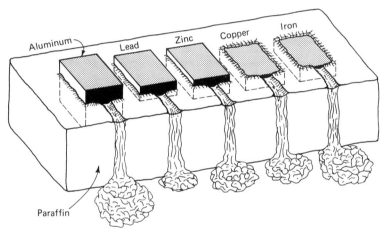

These five metal blocks of equal mass and equal cross section have been heated to the same temperature. Because they have different heat capacities, they melt the paraffin to different depths and thus melt different amounts of paraffin.

This experiment illustrates that different metals absorb and then give off different amounts of heat even though they go through the same temperature change and are of equal mass. We say that these metals differ in heat capacity. Those materials with high heat capacity warm more slowly because they can absorb a great amount of heat/C° while warming. Similarly, they cool more slowly because they give off more heat/C° while cooling. The *heat capacity* of an object is the amount of heat needed to raise its temperature 1C°. That is,

$$\text{heat capacity} = \frac{Q}{\Delta T}$$

where Q is the amount of heat and
ΔT is the change in temperature.

15.5   SPECIFIC HEAT

Heat capacity does not take into account the mass of the material, so that different amounts of different materials can be compared.

The *specific heat* of a material is the amount of heat necessary to change the temperature of a unit of mass of the material 1C°. That is,

$$c = \frac{Q}{m\,\Delta T}$$

or

$$\boxed{Q = mc\,\Delta T}$$

where c is the specific heat,
Q is the amount of heat,
m is the mass of the material, and
ΔT is the change in temperature.
The following table gives the specific heats of some common materials.

Material	Specific heat	
	cal/g C° or kcal/kg C°	J/kg C°
Air	0.24	1̄000
Alcohol, ethyl	0.58	2400
Aluminum	0.22	920
Brass (40% zinc)	0.092	390
Copper	0.092	390
Glass	0.21	880
Gold	0.031	130
Ice	0.51	2100
Iron (steel)	0.115	481
Lead	0.031	130
Mercury	0.033	140
Silver	0.056	230
Soil (dry)	0.20	840
Steam	0.48	2̄000
Water	1.00	4190
Wood	0.42	1800
Zinc	0.092	390

example 1   How many joules of heat must be absorbed to cool $25\bar{0}$ kg of aluminum from $15\bar{0}°C$ to $2\bar{0}°C$?

DATA:

$$m = 25\bar{0} \text{ kg}$$

$$c = 920 \text{ J/kg C}°$$

$$\Delta T = 15\bar{0}°C - 2\bar{0}°C = 13\bar{0}C°$$

$$Q = ?$$

BASIC EQUATION:

$$Q = mc\,\Delta T$$

WORKING EQUATION: same

SUBSTITUTION:

$$Q = (25\bar{0} \text{ kg})(920 \text{ J/kg C}°)(13\bar{0}C°)$$

$$= 3.0 \times 10^7 \text{ J} \quad \text{or} \quad 3\bar{0} \text{ M J}$$

example 2   How many kilocalories of heat must be added to 10.0 kg of steel to raise its temperature from $20°C$ to its white-hot temperature of $1370°C$?

DATA:

$$m = 10.0 \text{ kg}$$

$$c = 0.115 \text{ kcal/kg C}°$$

$$\Delta T = 1370°C - 20°C = 1350C°$$

$$Q = ?$$

BASIC EQUATION:

$$Q = mc\,\Delta T$$

WORKING EQUATION: same

SUBSTITUTION:

$$Q = (10.0 \text{ kg})(0.115 \text{ kcal/kg C}°)(1350C°)$$

$$= 1550 \text{ kcal}$$

When two substances at different temperatures are "mixed" together, heat flows from the warmer body to the cooler body until they reach the same temperature. Part of the heat lost by the warmer body is transferred to the cooler body and part is lost to the surrounding objects or the air. In many cases almost all the heat is transferred to the cooler body. We will assume that all the heat lost by the warmer body is transferred to the cooler body. That is, the heat lost by the warmer body equals the heat gained by the cooler body. The amount of heat lost or gained by a body is

$$Q = mc\,\Delta T$$

By formula,

$$c_\ell \left(m_g (T_\ell - T_f) \right)$$

$$
\boxed{
\begin{array}{c}
Q_{lost} = Q_{gained} \\
m_\ell c_\ell (T_\ell - T_f) = m_g c_g (T_f - T_g)
\end{array}
}
$$

where the subscript ℓ refers to the warmer body, which *loses* heat; the subscript g refers to the cooler body, which *gains* heat; and T_f is the final temperature of the mixture. This principle is commonly called the *method of mixtures*.

example 3 If 5.00 kg of brass at 275°C is placed in 20.0 kg of water at 15°C, what is the final temperature of the mixture?

DATA:

$m_\ell = 5.00$ kg	$m_g = 20.0$ kg
$c_\ell = 390$ J/kg C°	$c_g = 4190$ J/kg C°
$T_\ell = 275$°C	$T_g = 15$°C
	$T_f = ?$

BASIC EQUATION: $m_\ell c_\ell (T_\ell - T_f) = m_g c_g (T_f - T_g)$

WORKING EQUATION: In this type of problem, it is easier to substitute directly into the basic equation.

SUBSTITUTION:

$$(5.00 \text{ kg})(390 \text{ J/kg C°})(275°C - T_f) = (20.0 \text{ kg})(4190 \text{ J/kg C°})(T_f - 15°C)$$

$$5.4 \times 10^5 \text{ J} - (2.0 \times 10^3 \text{ J/C°})T_f = (8.38 \times 10^4 \text{ J/C°})T_f - 1.3 \times 10^6 \text{ J}$$

$$1.8 \times 10^6 \text{ J} = (8.58 \times 10^4 \text{ J/C°})T_f$$

$$21°C = T_f$$

example 4 If $20\bar{0}$ g of steel at $22\bar{0}$°C is placed in $50\bar{0}$ g of water at $1\bar{0}$°C, what is the final temperature of the mixture?

DATA:

$m_\ell = 20\bar{0}$ g	$m_g = 50\bar{0}$ g
$c_\ell = 0.115$ cal/g C°	$c_g = 1.00$ cal/g C°
$T_\ell = 22\bar{0}$°C	$T_g = 1\bar{0}$C°
	$T_f = ?$

BASIC EQUATION: $m_\ell c_\ell (T_\ell - T_f) = m_g c_g (T_f - T_g)$

SUBSTITUTION:

$$(20\bar{0} \text{ g})(0.115 \text{ cal/g C°})(22\bar{0}°C - T_f) = (50\bar{0} \text{ g})(1.00 \text{ cal/g C°})(T_f - 1\bar{0}°C)$$

$$5060 \frac{\text{cal °C}}{\text{C°}} - 23.0 \frac{\text{cal}}{\text{C°}} T_f = 50\bar{0} \frac{\text{cal}}{\text{C°}} T_f - 5\bar{0}00 \frac{\text{cal °C}}{\text{C°}}$$

$$10,100 \text{ cal} = 523 \ \frac{\text{cal}}{\text{C}°} \ T_f$$

$$\frac{10,100 \text{ cal}}{523 \text{ cal}/\text{C}°} = T_f$$

$$19.3°\text{C} = T_f$$

example 5   The following data were collected in the laboratory to determine the specific heat of an unknown metal:

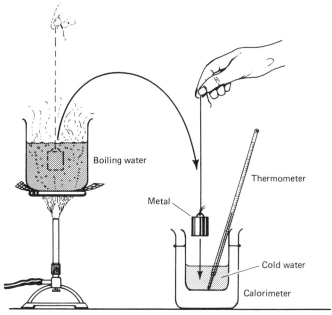

Apparatus for measuring the specific heat of a metal by the method of mixtures.

Mass of aluminum calorimeter*	123 g	m_c
Specific heat of calorimeter	0.22 cal/g C°	c_c
Mass of water	$25\overline{0}$ g	m_w
Specific heat of water	1.00 cal/g C°	c_w
Mass of metal	196 g	m_m
Initial temperature of water and calorimeter	15.0°C	T_w, T_c
Initial temperature of metal	99.3°C	T_m

*Note: A calorimeter is usually a metal cup inside another metal cup, which is insulated by the air between them.

Final temperature of calorimeter,
water, and metal   21.1°C T_f

BASIC EQUATION: $Q_{lost} = Q_{gained}$

The metal loses heat while the calorimeter and water gain heat. That is,

$$Q_m = Q_c + Q_w$$

WORKING EQUATION: $m_m c_m(T_m - T_f) = m_c c_c(T_f - T_c) + m_w c_w(T_f - T_w)$

SUBSTITUTION: $(196 \text{ g})c_m(99.3°C - 21.1°C) =$

$(123 \text{ g}) (0.22 \text{ cal/g C°})(21.1°C - 15.0°C) + (25\bar{0} \text{ g})(1.00 \text{ cal/g C°})(21.1°C - 15.0°C)$

$(1.53 \times 10^4 \text{ g C°}) c_m = 1700 \text{ cal}$

$c_m = 0.11 \text{ cal/g C°}$

15.6   MECHANICAL EQUIVALENT OF HEAT

The temperature of two blocks being rubbed together, of a saw blade sawing through a board, and of a drill bit drilling a hole in a metal rises as long as the frictional forces are applied. Thus, heat and work are somehow related.

James Prescott Joule (1818–1889), an English scientist, determined the relationship between heat and work by performing the following famous experiment. He used an apparatus similar to the one below.

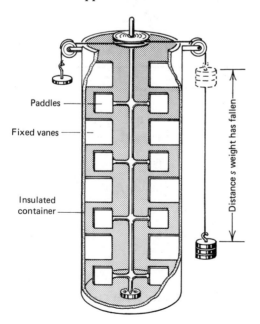

Paddles

Fixed vanes

Insulated
container

Distance s weight has fallen

A simplified model of the apparatus used by Joule
to determine the mechanical equivalent of heat

The falling mass, m, causes the paddles to rotate, which in turn stirs the liquid. The stirring is done against the frictional resistance (viscosity) of the liquid, which produces heat and raises the temperature of the liquid. The work done on the liquid equals the loss in potential energy of the mass as it falls a distance s minus its kinetic energy as it reaches the end of its fall. (In Joule's experiments v was small.) Thus, the work done on the liquid by the falling mass is

$$W = mgs - \tfrac{1}{2}mv^2$$

The amount of heat gain by the liquid is

$$Q = m_{\text{liq}} c_{\text{liq}} (T_f - T_i)$$

No matter what liquid is used, Joule found that the work done on the liquid was directly proportional to the heat gained by the liquid. That is,

$$W = kQ$$

In addition, Joule found that k, the proportionality constant, was the same for all liquids: namely,

$$k = 4.19 \text{ J/cal}$$

That is,

1 cal of heat is produced by 4.19 J of work

or   1 kcal of heat is produced by 4190 J of work

These last two relationships are known as the *mechanical equivalent of heat*. And

4.19 J of work produces 1 cal of heat

4190 J of work produces 1 kcal of heat

The opposite effect of Joule's experiment occurs in the gasoline engine. As the piston moves down, air and gasoline (vapor) enter the cylinder through the valves. As the piston moves up, the valves close, trapping and compressing the air and gasoline (vapor) to a small fraction of its original volume. The compression, in turn, causes the temperature of these gases to rise. Then a spark causes a chemical reaction between the oxygen and gas vapors producing heat, which increases the pressure of the gases inside the cylinder. The heated expanding gases do work by pushing on the piston. Thus, some of the heat is changed into mechanical energy and the temperature in the cylinder decreases. Some of the heat escapes through the exhaust and through the engine block. Thus, heat energy is changed to mechanical energy.

example   Find the temperature increase of water as it falls from the top to the bottom of a waterfall of height h.

As a mass m of water falls a height h, its potential energy loss is

$$PE = mgh$$

If we assume that all the potential energy loss is converted into kinetic energy and thus in turn is converted to and dissipated as heat to the falling water, the heat gain is

$$Q = mc \, \Delta T$$

Thus,

$$PE = Q$$

or

$$mgh = mc \, \Delta T$$

Then

$$\Delta T = \frac{gh}{c}$$

EXERCISES

Sketch	Data	Basic Equation	Working Equation	Substitution
$A = 12 \text{ cm}^2$, $h = ?$, $b = 6.0 \text{ cm}$	$A = 12 \text{ cm}^2$ $b = 6.0 \text{ cm}$ $h = ?$	$A = bh$	$h = \dfrac{A}{b}$	$h = \dfrac{12 \text{ cm}^2}{6.0 \text{ cm}}$ $= 2.0 \text{ cm}$

1. Change each of the following melting point temperatures to Kelvin.
   (a) aluminum, 660°C
   (b) lead, 327°C
   (c) iron, 1535°C
   (d) mercury, −39°C
   (e) ice, 0°C

2. Change each of the following Kelvin temperatures to °C.
   (a) 55 K
   (b) 230 K
   (c) 303 K
   (d) 565 K
   (e) 1200 K
   (f) −32 K

3. How many calories of heat must be added to $75\overline{0}$ g of steel to raise its temperature from 75°C to $30\overline{0}$°C?

4. How many kilocalories of heat must be added to $120\overline{0}$ kg of copper to raise its temperature from 25°C to 275°C?

5. How many joules of heat are given off by $50\overline{0}$ kg of aluminum when it cools from $35\overline{0}$°C to $2\overline{0}$°C?

6. How many joules of heat are absorbed by an electric freezer in lowering the temperature of $185\overline{0}$ g of water from $8\overline{0}$°C to $1\overline{0}$°C?

7. A $52\overline{0}$-kg steam boiler is made of steel and contains $30\overline{0}$ kg of water at $4\overline{0}$°C. Assuming that 75 percent of the heat is delivered to the boiler and water, how many calories are required to raise the temperature of both the boiler and water to $10\overline{0}$°C?

8. Assume that 2.5 kg of steel are dropped into 11.0 kg of water at 21°C. The final temperature is 28°C. What was the initial temperature of the steel?

9. If 5.00 kg of water at 93°C are mixed with 7.00 kg of water at 18°C, what is the final temperature of the mixture?

10. A 25$\bar{0}$-g piece of tin at 99°C is dropped in 10$\bar{0}$ g of water at 1$\bar{0}$°C. If the final temperature of the mixture is 20°C, what is the specific heat of the tin in cal/g°C?

11. How many grams of water at 2$\bar{0}$°C must be added to 8$\bar{0}$0 g of water at 9$\bar{0}$°C to have a mixture whose final temperature is 5$\bar{0}$°C?

12. Assume that 125 kg of aluminum at 275°C are dropped into 40$\bar{0}$ kg of water at 15°C. What is the final temperature?

13. Assume that 45 kg of lead at 315°C are dropped into 10$\bar{0}$ kg of water at 2$\bar{0}$°C. What is the final temperature of the mixture?

14. If 125$\bar{0}$ g of copper at 2$\bar{0}$°C are mixed with 50$\bar{0}$ g of water at 95°C, what is the final temperature of the mixture?

15. If 50$\bar{0}$ g of brass at 20$\bar{0}$°C and 30$\bar{0}$ g of steel at 15$\bar{0}$°C are added to 90$\bar{0}$ g of water in a 150-g aluminum pan and both are at 2$\bar{0}$°C, what is the final temperature of this mixture, assuming that no heat is lost to the surroundings?

16. The following data were collected in the laboratory to determine the specific heat of an unknown metal.

Mass of copper calorimeter	153 g
Specific heat of calorimeter	0.092 cal/g C°
Mass of water	275 g
Specific heat of water	1.00 cal/g C°
Mass of metal	236 g
Initial temperature of water and calorimeter	16.2°C
Initial temperature of metal	99.6°C
Final temperature of calorimeter, water, and metal	22.7°C

Find the specific heat of the unknown metal.

17. The following data were collected in the laboratory to determine the specific heat of an unknown metal.

Mass of aluminum calorimeter	132 g
Specific heat of calorimeter	920 J/kg C°
Mass of water	285 g
Specific heat of water	4190 J/kg C°
Mass of metal	215 g

Initial temperature of water

and calorimeter   12.6°C

Initial temperature of metal   99.1°C

Final temperature of calorimeter,

water, and metal   18.6°C

Find the specific heat of the unknown metal.

18. What is the mechanical work equivalent of 3640 cal?

19. What is the mechanical work equivalent of 84.7 kcal?

20. What is the heat equivalent (in cal) of 265 J?

21. What is the heat equivalent (in kcal) of 8.52×10^5 J?

22. Gasoline has a mass density of 0.700 g/cm^3 and gives off 1.15×10^4 cal/g when it burns.

   (a) How many joules of work can be obtained by burning 1.00 L of gasoline?

   (b) If the gasoline is burned in an engine that is $2\overline{0}$ percent efficient, how many joules of work can be obtained?

23. Natural gas gives off 1.00×10^5 cal/g when it burns in a turbine engine. Natural gas is burned in the turbine at the rate of 2.50 g/s. If the turbine is 25 percent efficient, what is the power output in kilowatts?

24. The water going over Niagara Falls drops 50.6 m. Find the rise in temperature as it falls from the top to the bottom of the Falls.

25. A lead bullet of mass 5.00 g falls a height of $25\overline{0}$ m. If half of the energy of the bullet is converted into internal heat energy, find its rise in temperature.

THERMAL EXPANSION OF SOLIDS AND LIQUIDS

16.1 EXPANSION OF SOLIDS

Most solids expand when heated and contract when cooled. They expand or contract in all three dimensions—length, width, and thickness.

When a solid is heated, the expansion is due to the increased length of the vibrations of the atoms and molecules. This results in the solid expanding in all directions. This increase in volume results in a decrease in density. Engineers, technicians, and designers must know the effects of thermal expansion.

You have no doubt heard of highway pavements buckling on a hot summer day.

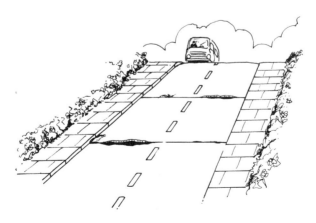

Bridges are built with special joints that allow for expansion and contraction of the bridge deck.

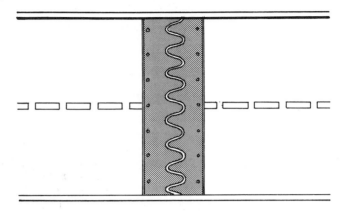

The clicking noise of a train's wheels passing over the rails can be heard more in winter than in summer. The space between rails is larger in winter than in summer. If the rails were placed snugly end to end in the winter, they would buckle in the summer.

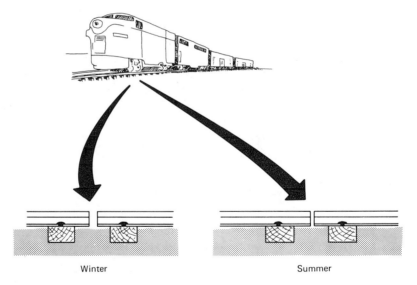

Winter   Summer

Similarly, pipelines and buildings must be designed and built to allow for this expansion and contraction.

There are some advantages to solids expanding. A bimetal bar is made by fusing two different metals together side by side. When heated, the brass expands more than the steel, which makes the bar curve (page 343). The thermostat operates on this principle.

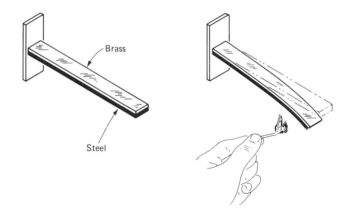

As shown below, the basic parts of a thermostat are a bimetal strip on the right and a regular metal strip on the left. The bimetal strip of brass and steel bends with the temperature. The regular metal strip is moved by hand to set the temperature desired.

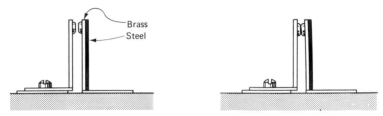

This particular bimetal bar is made and placed so that it bends to the left when cooled. As a result, when it comes in contact with the strip on the left, it completes a circuit, which turns on the furnace. When the room warms to the desired temperature, the bimetal bar moves back to the right, which opens the contacts and shuts off the heat.

Linear expansion   The amount that a solid expands depends on the following:

1. The material—different materials expand at different rates. Steel expands at a rate less than that of brass.
2. The length of the solid—the longer the solid, the larger the expansion. A $2\bar{0}$-cm steel rod will expand twice as much as a $1\bar{0}$-cm steel rod.
3. The amount of change in temperature—the greater the change in temperature, the greater the expansion.

These factors can be written as a formula:

$$\Delta l = \alpha l \, \Delta T$$

where Δl is the change in length,

α is the constant called the coefficient of linear expansion,*

l is the original length or any other linear measurement, and

ΔT is the change in temperature.

The following table lists the coefficients of linear expansion for some common solids.

Material	α (per C°)
Aluminum	2.3×10^{-5}
Brass	1.9×10^{-5}
Concrete	1.1×10^{-5}
Copper	1.7×10^{-5}
Glass	9.0×10^{-6}
Pyrex	4.0×10^{-6}
Steel	1.3×10^{-5}
Zinc	2.6×10^{-5}

example 1 A steel railroad rail is 20.000 m long at $-2\bar{0}°C$. How much will it expand in summer when the temperature is $3\bar{0}°C$?

DATA:

$$l = 20.000 \text{ m}$$

$$\Delta T = 3\bar{0}°C - (-2\bar{0}°C) = 5\bar{0}C°$$

$$\alpha = 1.3 \times 10^{-5}/C°$$

$$\Delta l = ?$$

BASIC EQUATION:

$\Delta l = \alpha l \, \Delta T$

WORKING EQUATION:

same

SUBSTITUTION:

$$\Delta l = (1.3 \times 10^{-5}/C°)(20.000 \text{ m})(5\bar{0}C°)$$

$$= 0.013 \text{ m} \quad \text{or} \quad 13 \text{ mm}$$

Pipes that undergo large temperature changes are usually installed with a loop as shown below to allow for expansion and contraction.

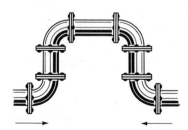

*Defined as the change in the unit length of a solid when its temperature changes 1°.

example 2   What allowance for expansion must be made for a steel pipe 120.00 m long which handles coolants and must undergo temperature changes of $20\overline{0}C°$?

DATA:	$\alpha = 1.3 \times 10^{-5}/C°$
	$l = 120.00$ m
	$\Delta T = 20\overline{0}C°$
	$\Delta l = ?$
BASIC EQUATION:	$\Delta l = \alpha l\ \Delta T$
WORKING EQUATION:	same
SUBSTITUTION:	$\Delta l = (1.3 \times 10^{-5}/C°)(120.00\text{ m})(20\overline{0}C°)$
	$= 0.31$ m or 31 cm

Area expansion   Surfaces expand in length and width when heated. This relationship can be written

$$\Delta A = \alpha' A\ \Delta T$$

where ΔA is the change in area,
  α' is the coefficient of area expansion,*
  A is the original surface area, and
  ΔT is the change in temperature.

The coefficient of area expansion, α', is approximately equal to twice the coefficient of linear expansion, α. Thus, the formula for area expansion becomes

$$\Delta A = 2\alpha A\ \Delta T$$

example 3   The top of a circular copper dish has an area of 38.50 cm^2 at $-5°C$. What is the change in area when its temperature is increased to $12\overline{0}°C$?

DATA:	$A = 38.50$ cm^2
	$\alpha = 1.7 \times 10^{-5}/C°$
	$\Delta T = 12\overline{0}°C - (-5°C) = 125C°$
BASIC EQUATION:	$\Delta A = 2\alpha A\ \Delta T$
WORKING EQUATION:	same
SUBSTITUTION:	$\Delta A = 2(1.7 \times 10^{-5}/°C)(38.50\text{ cm}^2)(125C°)$
	$= 0.16$ cm^2

*Defined as the change in the unit surface area of a solid when its temperature changes 1°.

Volume expansion Similarly, solids expand in length, in width, and in thickness when heated. This relationship can be written

$$\Delta V = \alpha'' V \,\Delta T$$

where ΔV is the change in volume,
  α'' is the coefficient of volume expansion,*
  V is the original volume, and
  ΔT is the change in temperature.

The coefficient of volume expansion, α'', is approximately equal to three times the coefficient of linear expansion, α. Thus, the formula for volume expansion becomes

$$\Delta V = 3\alpha V \,\Delta T$$

example 4 A section of concrete measures 7.00 m $\times$ 15.0 m $\times$ 30.0 m at $5\bar{0}°$C. What allowance for change in volume is necessary for a temperature of $-25°$C?

DATA: $V = 7.00$ m $\times$ 15.0 m $\times$ 30.0 m $= 3150$ m^3

  $\alpha = 1.1 \times 10^{-5}/\text{C}°$

  $\Delta T = 5\bar{0}°\text{C} - (-25°\text{C}) = 75\text{C}°$

BASIC EQUATION: $\Delta V = 3\alpha V \,\Delta T$

WORKING EQUATION: same

SUBSTITUTION: $\Delta V = 3(1.1 \times 10^{-5}/\text{C}°)(3150 \text{ m}^3)(75\text{C}°)$

  $= 7.8$ m^3

16.2 EXPANSION OF LIQUIDS

Similarly, liquids generally expand when heated and contract when cooled. The thermometer is made using this principle.

When a thermometer is placed under your tongue, the heat from your mouth causes the mercury in the bottom of the thermometer to expand (page 347). Mercury is then forced to rise up the thin calibrated tube. The formula for volume expansion of liquids is

$$\Delta V = \beta V \,\Delta T$$

*Defined as the change in unit volume of a solid when its temperature changes 1°.

where β is the coefficient of volume expansion for liquids.

The following table lists the coefficients of volume expansion for some common liquids. Note that the coefficients for volume expansion of liquids are generally larger than for solids.

Liquid	β (per $C°$)
Acetone	1.49×10^{-3}
Alcohol, ethyl	1.12×10^{-3}
Carbon tetrachloride	1.24×10^{-3}
Gasoline	1.08×10^{-3}
Mercury	1.8×10^{-4}
Petroleum	9.6×10^{-4}
Turpentine	9.7×10^{-4}
Water	2.1×10^{-4}

example   If petroleum at $0°C$ occupies $25\bar{0}$ L, what is its volume at $5\bar{0}°C$?

DATA:
$$\beta = 9.6 \times 10^{-4}/C°$$
$$V = 250 \text{ L}$$
$$\Delta T = 5\bar{0}\,C°$$
$$\Delta V = ?$$

BASIC EQUATION:
$$\Delta V = \beta V \,\Delta T$$

WORKING EQUATION: same

SUBSTITUTION:
$$\Delta V = (9.6 \times 10^{-4}/C°)(25\bar{0} \text{ L})(5\bar{0}\,C°)$$
$$= 12 \text{ L}$$

$$\text{volume at } 5\bar{0}°C = 25\bar{0} \text{ L} + 12 \text{ L} = 262 \text{ L}$$

16.3 EXPANSION OF WATER

Water is unusual in its expansion characteristics. Most of us have seen the mound in the middle of each ice cube in ice-cube trays. This evidence shows the expansion of water during its change of state from liquid to solid form. Nearly all liquids are densest at their lowest temperature before a change of state to become solids. As the temperature drops, the molecular motion slows and the substance becomes denser.

Water does not follow the general rule stated above. Because of its unusual structural characteristics, water is densest at 4°C instead of 0°C. A graph of its change in density with an increase in temperature appears below.

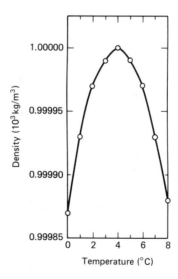

When ice melts at 0°C, the water formed *contracts* as the temperature is raised to 4°C. Then it begins to *expand*, as do most other liquids.

EXERCISES

Sketch	Data	Basic Equation	Working Equation	Substitution
12 cm² $h = ?$ $b = 6.0$ cm	$A = 12$ cm² $b = 6.0$ cm $h = ?$	$A = bh$	$h = \dfrac{A}{b}$	$h = \dfrac{12 \text{ cm}^2}{6.0 \text{ cm}}$ $= 2.0$ cm

1. Find the increase in length of a copper pipe 25.000 m long at 0°C when it is heated to $15\bar{0}$°C.

2. Find the increase in length of a zinc rod 50.00 m long at 15°C when it is heated to $13\bar{0}$°C.

3. A steel pipe 125.00 m long is installed at $2\bar{0}$°C. Find the decrease in length when coolants at $-9\bar{0}$°C pass through the pipe.

4. Compute the increase in length of 300.00 m of copper wire when its temperature changes from 14°C to 34°C.

5. A steel tape measures 200.00 m at $1\bar{0}$°C. What is its length at $5\bar{0}$°C?

6. A brass rod 20.000 cm long expands 0.008 cm when it is heated. Find the temperature change.

7. The road bed on a bridge 215.00 m long is made of concrete. What allowance is needed for temperatures of $-4\bar{0}$°C in winter and 75°C in summer?

8. An aluminum plug has a diameter of 10.003 cm at $4\bar{0}$°C. At what temperature will it fit exactly into a hole of constant diameter 10.000 cm?

9. The diameter of a steel drill bit at $1\bar{0}$°C is 2.000 cm. Find its diameter at 175°C.

10. A brass ball has a diameter 12.000 cm and is 0.011 cm too large to pass through a hole in a copper plate when the ball and plate are at a temperature of $2\bar{0}$°C.

    (a) What is the temperature of the ball when it will just pass through the plate, assuming that the temperature of the plate does not change?

    (b) What is the temperature of the plate when the ball will just pass through, assuming that the temperature of the ball does not change?

11. A brass cylinder has a cross-sectional area of 74.8 cm^2 at $-1\bar{0}$°C. Find its change in area when heated to $12\bar{0}$°C.

12. The volume of the cylinder in Exercise 11 is 237 cm^3 at $24\bar{0}$°C. Find its change in volume when cooled to $-6\bar{0}$°C.

13. An aluminum rod has a radius of 10.50 cm at $1\bar{0}$°C. What is its cross-sectional area when heated to $15\bar{0}$°C?

14. A steel rod has a radius of 12.00 cm at $8\bar{0}$°C. What is its cross-sectional area when cooled to 5°C?

15. A glass plug has a volume of 60.00 cm^3 at 12°C. What is its volume at 76°C?

16. A section of concrete dam is a rectangular solid 6.00 m by 16.0 m by 25.0 m at 35°C. What allowance for change in volume is necessary for a temperature of $-2\bar{0}$°C?

17. A quantity of carbon tetrachloride has a volume of 625 L at 12°C. What is its volume at 48°C?

18. Some mercury has a volume of 2.500 L at $-3\bar{0}$°C. What is the change in volume when heated to $5\bar{0}$°C?

19. Some petroleum has a volume of 9.00 m^3 at $-2\bar{0}$°C. What is its volume at 25°C?

20. What is the increase in volume of 35.00 L of acetone when heated from 28°C to 38°C?

21. Some water at 85°C has a volume of 1.25×10^5 L. What is its volume at $1\bar{0}$°C?

22. A $120\bar{0}$-L tank of gasoline is completely filled at $1\bar{0}$°C. How much spills over if the temperature rises to 45°C?

23. Calculate the increase in volume of 215.00 cm^3 of mercury when its temperature increases from $1\bar{0}$°C to 25°C.

24. What is the drop in volume of alcohol in a railroad tank car which contains 6.2×10^4 L if the temperature drops from 25°C to $1\bar{0}$°C?

25. A gasoline service station owner receives a 33,00$\bar{0}$-L truckload of gasoline, which is at 32°C. It cools to 15°C in his underground tank. At 40 cents/L, how much money does he lose due to the contraction of the gasoline?

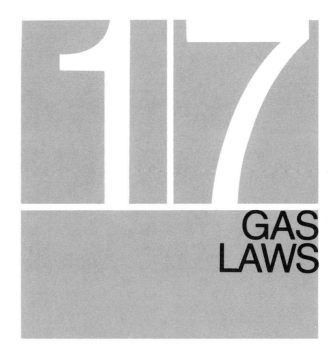

GAS LAWS

17.1 GAS PRESSURE—A MOLECULAR VIEW

When air—or any gas—is blown into a balloon, what causes the balloon to inflate?

A gas is composed of molecular particles that are in constant motion. Collisions between molecules are perfectly elastic; that is, their kinetic energies and momentums are totally conserved. Any gas expands to completely fill its container. The motion of the molecules is continual and at random. As the air is blown into the balloon above, the many billions of air molecules continually bombard each other and the sides of the balloon, producing pressure on the inside surface of the balloon. The pressure may be increased by blowing more air into the balloon. The additional molecules produce more collisions and thus more pressure on the balloon's inside surface.

Of course, there are also air molecules bombarding the outside surface of the balloon. The balloon's inflation becomes constant when the inside molecular bombarding forces equal the outside molecular bombarding forces and the elastic forces of the balloon material itself.

If we put a typical inflated toy balloon in a vacuum, it would probably burst because the inside molecular bombarding forces would overcome the elastic forces of the balloon material in the absence of the outside molecular bombarding forces. The sum of these inside molecular bombarding forces can be very large. The lifts in many automotive service bays operate under air pressure. That is, the sum of these inside molecular bombarding forces can lift an automobile. Air bearings, a thin air space under pressure, allow one to horizontally move heavy equipment.

17.2   ATMOSPHERIC PRESSURE

Since air is composed of molecules, air has mass and therefore exerts pressure. Thus, the atmosphere exerts pressure on objects on the surface of the earth. This atmospheric pressure can be illustrated by using a bell jar with a hole in the top over which a thin rubber membrane is stretched. As air is pumped out of the bell jar, the inside air pressure is reduced by removing a number of air molecules. Thus, there are fewer molecular bombardments on the inside surface of the rubber membrane than there are on its outside surface. The outside air pressure, now being greater than the inside air pressure, pushes the rubber membrane down into the bell jar.

Air pressure

Thin rubber
membrane

To vacuum pump

When a straw is used to drink, the air pressure inside the straw is reduced. As a result, the outside air pressure is higher than the pressure in the straw, which forces the fluid up the straw.

Air pressure pushes down on fluid with one atmosphere pressure.

Air pressure is reduced

Earlier we saw that the pressure on a submerged body increases as the body goes deeper into the liquid. Some animals live near the bottom of the ocean, where the pressure of the water above them is so great that it would collapse any human body and most submarines. But through the process of evolution, the animals have adapted to this tremendous pressure.

Similarly, we on Earth live at the bottom of a fluid, the atmosphere, which is several kilometres deep. We do not feel the pressure from this fluid because it is the same from all directions and also because our bodies have become accustomed to it. When the air pressure becomes unequal, its force becomes quite evident in the form of wind. This wind may be a cool summer breeze or the tremendous concentrated force of a tornado.

What is the pressure of our atmosphere equivalent to? Experiments have shown that the atmosphere at sea level supports a column of water 10.34 m high in a tube in which the air has been removed. The atmosphere supports 76.0 cm of mercury in a similar tube (see page 354). This is not surprising since mercury is 13.6 times as dense as water and

$$\frac{1}{13.6} \times 10.34 \text{ m} \times \frac{100 \text{ cm}}{1 \text{ m}} = 76.0 \text{ cm}$$

The pressure of the atmosphere (a column of air equal to the height of the atmosphere) can be expressed in terms of the height of an equivalent column of mercury, which is 76.0 cm at sea level. The pressure can then be determined by

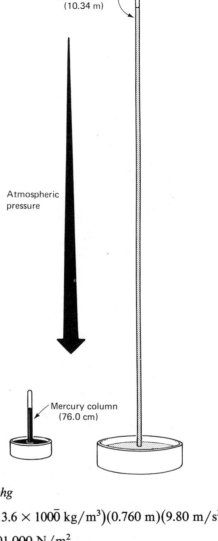

Water column
(10.34 m)

Atmospheric
pressure

Mercury column
(76.0 cm)

$$P = Dhg$$

$$= (13.6 \times 100\bar{0} \text{ kg/m}^3)(0.760 \text{ m})(9.80 \text{ m/s}^2)$$

$$= 101{,}000 \text{ N/m}^2$$

$$= 101{,}000 \text{ Pa}$$

$$= 101 \text{ kPa}$$

Normal atmospheric pressure at sea level has been more accurately determined to be 101.32 kPa. This value is also called *standard pressure*. The figure on page 355 illustrates the normal range of atmospheric pressures.

17.3 GAUGE PRESSURE

Most gas pressures, such as gas for cooking or the amount of air in automobile tires, are measured by gauges. Most gauges are made to show a reading of zero at normal atmospheric pressure. The actual or total pressure is called the *absolute*

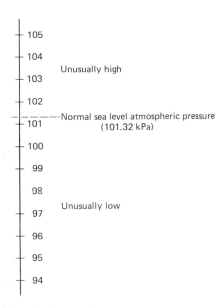

— 105
— 104
— 103   Unusually high
— 102
– – – – – – Normal sea level atmospheric pressure
— 101   (101.32 kPa)
— 100
— 99
— 08
— 97   Unusually low
— 96
— 95
— 94

Normal range of atmospheric pressures

pressure, which is found as follows:

$$\text{absolute pressure} = \text{gauge pressure} + \text{atmospheric pressure}$$

Thus, a tire whose gauge pressure reads $20\overline{0}$ kPa has an absolute pressure of $20\overline{0}$ kPa $+ 101$ kPa, or 301 kPa. This is normally written

$$P_{ga} = 20\overline{0} \text{ kPa}$$

$$P_{atm} = 101 \text{ kPa}$$

$$P_{abs} = 301 \text{ kPa}$$

Thus,

$$\boxed{P_{abs} = P_{ga} + P_{atm}}$$

17.4 VOLUME AND PRESSURE

The behavior or nature of a gas is dependent upon three factors: its pressure, its volume, and its temperature. Since *all* gases behave almost identically (when not near their liquification points) with corresponding changes in these factors, we can study the behavior of *all* gases at the same time. This study is generally categorized under the heading of "the gas laws."

The first gas law we present here is called *Boyle's law*, which states: *If the temperature of a gas is constant, its absolute pressure is inversely proportional to its volume.* That is,

$$P = \frac{k}{V} \qquad \text{where } k \text{ is a proportionality constant}$$

355

If we consider the absolute pressure and volume at various readings, Boyle's law becomes

$$P_1 V_1 = P_2 V_2 \quad (= P_3 V_3 = P_4 V_4 = \cdots = k)$$

The graph of Boyle's law is shown below.

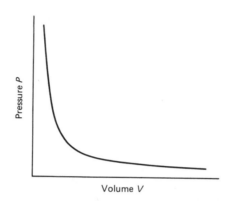

That is, as the volume of a confined gas decreases, its pressure increases, and vice versa.

example 1   A given amount of oxygen occupies 750 cm³ at an absolute pressure of 290 kPa. What is the volume if the pressure is decreased to 240 kPa?

DATA:

$$P_1 = 290 \text{ kPa}$$
$$V_1 = 750 \text{ cm}^3$$
$$P_2 = 240 \text{ kPa}$$
$$V_2 = ?$$

BASIC EQUATION:

$$P_1 V_1 = P_2 V_2$$

WORKING EQUATION:

$$V_2 = \frac{P_1 V_1}{P_2}$$

SUBSTITUTION:

$$V_2 = \frac{(290 \text{ kPa})(750 \text{ cm}^3)}{240 \text{ kPa}}$$
$$= 910 \text{ cm}^3$$

example 2   A cylindrical tube at normal atmospheric pressure (101 kPa) is fitted with a sealed plunger so that no air leaks out as the plunger is moved. The plunger is then moved down, which compresses the air to one-fifth of its original volume.

(a) Find the absolute pressure of the compressed air.
(b) Find the gauge pressure of the compressed air.

(a) DATA:
$$P_1 = 101 \text{ kPa}$$
$$V_2 = \tfrac{1}{5} V_1$$
$$P_2 = ?$$

BASIC EQUATION:
$$P_1 V_1 = P_2 V_2$$

WORKING EQUATION:
$$P_2 = \frac{P_1 V_1}{V_2}$$

SUBSTITUTION:
$$P_2 = \frac{(101 \text{ kPa})(V_1)}{\tfrac{1}{5} V_1}$$
$$= 505 \text{ kPa (absolute)}$$

(b) DATA:
$$P_{\text{abs}} = 505 \text{ kPa}$$
$$P_{\text{atm}} = 101 \text{ kPa}$$
$$P_{\text{ga}} = ?$$

BASIC EQUATION:
$$P_{\text{abs}} = P_{\text{ga}} + P_{\text{atm}}$$

WORKING EQUATION:
$$P_{\text{ga}} = P_{\text{abs}} - P_{\text{atm}}$$

SUBSTITUTION:
$$P_{\text{ga}} = 505 \text{ kPa} - 101 \text{ kPa}$$
$$= 404 \text{ kPa}$$

If the pressure of a given amount of gas is increased, its density increases as the gas molecules are forced closer together. Also, if the pressure is decreased, the density decreases. That is, the density of a gas is directly proportional to its pressure. In equation form,

$$\boxed{\frac{D_1}{D_2} = \frac{P_1}{P_2}}$$

example 3 Ammonia has a density of 0.76 kg/m^3 at 0°C at normal atmospheric pressure. Find its density when the pressure is increased to 275 kPa at the same temperature.

DATA:
$$D_1 = 0.76 \text{ kg/m}^3$$
$$P_1 = 101 \text{ kPa}$$
$$P_2 = 275 \text{ kPa}$$
$$D_2 = ?$$

BASIC EQUATION:
$$\frac{D_1}{D_2} = \frac{P_1}{P_2}$$

WORKING EQUATION:

$$D_2 = \frac{D_1 P_2}{P_1}$$

SUBSTITUTION:

$$D_2 = \frac{(0.76 \text{ kg/m}^3)(275 \text{ kPa})}{101 \text{ kPa}}$$

$$= 2.1 \text{ kg/m}^3$$

17.5   VOLUME AND TEMPERATURE

The second gas law to be considered here is called *Charles' law*, which states: *If the pressure on a gas is constant, its volume is directly proportional to its absolute temperature.* That is,

$$V = kT \qquad \text{where } k \text{ is a proportionality constant}$$

If we consider the volume and the absolute temperature at various readings, Charles' law becomes

$$\frac{V_1}{T_1} = \frac{V_2}{T_2} \qquad \left(= \frac{V_3}{T_3} = \frac{V_4}{T_4} = \cdots = k \right)$$

The graph of Charles' law is shown below.

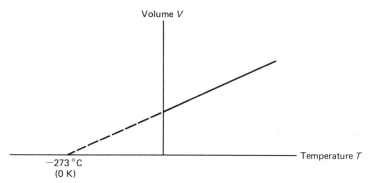

The plot of the equation is a straight line because the volume is directly proportional to its absolute temperature. Note that if the straight line is extended, it intercepts the T-axis at $-273°C$ or 0 K. If we let V_0 be the volume of the gas at $0°C$ (273 K), we find that $k = V_0/273$, which is also the slope of the line. This means that for each increase or decrease in temperature of $1°C$, the volume increases or decreases $\frac{1}{273}$ of its volume at $0°C$. Then, one could reason that the volume of a gas at $-273°C$ (0 K) is zero. But Charles' law, as all gas laws, applies only within normal temperatures well above the temperatures at which the gas would liquify.

In Chapter 16 we found that the volume of a solid changes as its temperature changes by

$$\Delta V = 3\alpha V \,\Delta T$$

while the volume of a liquid changes by

$$\Delta V = \beta V \,\Delta T$$

During his work, Charles discovered that the coefficient of volume expansion is $\beta = \frac{1}{273}$ per C° *for all gases*, using the reasoning discussed above.

example 1 A given amount of helium occupies 1750 cm³ at $3\bar{0}$°C. What will its volume be at $-1\bar{0}$°C?

DATA: $V_1 = 1750 \text{ cm}^3$

$T_1 = 3\bar{0}°\text{C} + 273° = 303 \text{ K}$

$T_2 = -1\bar{0}°\text{C} + 273° = 263 \text{ K}$

$V_2 = ?$

BASIC EQUATION: $\dfrac{V_1}{T_1} = \dfrac{V_2}{T_2}$

WORKING EQUATION: $V_2 = \dfrac{V_1 T_2}{T_1}$

SUBSTITUTION: $V_2 = \dfrac{(1750 \text{ cm}^3)(263 \text{ K})}{303 \text{ K}}$

$= 1520 \text{ cm}^3$

example 2 A gas occupies 25.0 m³ at $4\bar{0}$°C. At what temperature will the gas occupy 30.0 m³?

DATA: $V_1 = 25.0 \text{ m}^3$

$T_1 = 4\bar{0}°\text{C} + 273° = 313 \text{ K}$

$V_2 = 30.0 \text{ m}^3$

$T_2 = ?$

BASIC EQUATION: $\dfrac{V_1}{T_1} = \dfrac{V_2}{T_2}$

WORKING EQUATION: $T_2 = \dfrac{V_2 T_1}{V_1}$

SUBSTITUTION: $T_2 = \dfrac{(30.0 \text{ m}^3)(313 \text{ K})}{25.0 \text{ m}^3}$

$= 376 \text{ K} = 103°\text{C}$

17.6 IDEAL GAS LAW

We find it very difficult to keep either the pressure constant or the temperature constant. We find that Boyle's and Charles' laws when combined are much more common in everyday applications. Boyle's and Charles' laws combined is sometimes called the *ideal gas law*, which states: *The volume of an ideal gas is directly proportional to its absolute temperature and inversely proportional to its absolute pressure.* That is,

$$V = \frac{kT}{P} \quad \text{where } k \text{ is a proportionality constant}$$

Then, if we consider these quantities at various readings, the ideal gas law becomes

$$\boxed{\frac{V_1 P_1}{T_1} = \frac{V_2 P_2}{T_2} \quad \left(= \frac{V_3 P_3}{T_3} = \frac{V_4 P_4}{T_4} = \cdots = k\right)}$$

An ideal gas is considered to be composed of infinitely small, perfectly elastic molecules which exert no attractive or repulsive forces on each other except during contact. Real gases that are within normal ranges of temperature and pressure very closely conform to the behavior of an ideal gas. However, at temperatures or pressures near condensation or liquification, the ideal gas laws do not apply and significant deviations occur.

example We have 125 m³ of acetylene at $2\bar{0}°C$ at 13.8 MPa. What is its pressure if its volume is changed to $10\bar{0}$ m³ at 35°C?

DATA:

$$V_1 = 125 \text{ m}^3$$

$$P_1 = 13.8 \text{ MPa}$$

$$T_1 = 2\bar{0}°C + 273° = 293 \text{ K}$$

$$V_2 = 10\bar{0} \text{ m}^3$$

$$T_2 = 35°C + 273° = 308 \text{ K}$$

$$P_2 = ?$$

BASIC EQUATION:

$$\frac{V_1 P_1}{T_1} = \frac{V_2 P_2}{T_2}$$

WORKING EQUATION:

$$P_2 = \frac{V_1 P_1 T_2}{V_2 T_1}$$

SUBSTITUTION:

$$P_2 = \frac{(125 \text{ m}^3)(13.8 \text{ MPa})(308 \text{ K})}{(10\bar{0} \text{ m}^3)(293 \text{ K})}$$

$$= 18.1 \text{ MPa}$$

A commonly used reference in gas laws is called *standard temperature and pressure* (STP). Standard temperature is the freezing point of water, 0°C. Standard pressure is equivalent to atmospheric pressure, 101.32 kPa.

EXERCISES

Sketch	Data	Basic Equation	Working Equation	Substitution
$h = ?$ $b = 6.0$ cm	$A = 12$ cm^2 $b = 6.0$ cm $h = ?$	$A = bh$	$h = \dfrac{A}{b}$	$h = \dfrac{12 \text{ cm}^2}{6.0 \text{ cm}}$ $= 2.0$ cm

1. What is the pressure (in kPa) of

(a) 2 atm?

(b) 5 atm?

(c) $\frac{1}{5}$ atm?

2. A backpacker in the Rocky Mountains reads his barometer as $52\overline{0}$ mm of mercury. Find this pressure in kPa.

3. A tire gauge reads 195 kPa. What is the absolute pressure?

4. Some air at 120 kPa absolute pressure occupies 1400 cm^3. What is its volume at 150 kPa?

5. Some gas at standard pressure occupies 18.5 m^3. What is its absolute pressure if its volume is 22.5 m^3?

6. Some gas at 95.0 kPa absolute pressure occupies 65.0 L. What is its volume at 75.0 kPa absolute pressure?

7. Some gas has a density of 2.75 kg/m^3 at an absolute pressure of 108 kPa. What is its density if the pressure is decreased to 96 kPa?

8. A gas has a density of 1.75 kg/m^3 at standard pressure. What is its absolute pressure (in kPa) when the density is 1.45 kg/m^3?

9. Some methane at $35\overline{0}$ kPa gauge pressure occupies 750 m^3. What is its gauge pressure if its volume is $5\overline{0}0$ m^3?

10. Some helium at 103 kPa gauge pressure occupies 20.0 m^3. What is its volume at 138 kPa gauge pressure?

11. Some nitrogen at $5\overline{0}0$ kPa gauge pressure occupies 13.0 m^3. What is its volume at $35\overline{0}$ kPa gauge pressure?

12. A gas at 2.1 MPa gauge pressure occupies 40.0 m^3. What is its gauge pressure if its volume is

(a) doubled?

(b) tripled?

(c) halved?

13. A gas tank gauge registers $100\overline{0}$ kPa. After some gas has been used, the gauge registers $5\overline{0}0$ kPa.

(a) What percent of the gas remains in the tank?

(b) What percent of the gas that remains in the tank can be used?

14. A $100\bar{0}$-kL water tank is half-filled to a depth of 10.0 m under normal atmospheric conditions. Then $25\bar{0}$ kL of water is pumped into the tank while no air is allowed to escape. What is the water pressure at the bottom of the tank?

15. Some gas occupies a volume of 325 cm³ at 41°C. What is its volume at 94°C?

16. Some oxygen occupies 275 m³ at -15°C. What is its volume at 35°C?

17. Some methane occupies 1500 L at 45°C. What is its volume at $2\bar{0}$°C?

18. Some helium occupies $120\bar{0}$ m³ at $2\bar{0}$°C. At what temperature will its volume be $60\bar{0}$ m³?

19. An automobile tire has a gauge pressure of $20\bar{0}$ kPa at 0°C. Assuming no air leaks and no change in the volume of the tire, what will the gauge pressure be at 35°C?

20. We have $60\bar{0}$ L of oxygen at 10.3 MPa gauge pressure at 15°C. What is the volume at 8.28 MPa gauge pressure at 35°C?

21. We have $80\bar{0}$ m³ of natural gas at a pressure of $120\bar{0}$ mm of mercury at $3\bar{0}$°C. What is the temperature if the volume is changed to $120\bar{0}$ m³ at $150\bar{0}$ mm of mercury?

22. We have $140\bar{0}$ L of nitrogen at 167 kPa gauge pressure at 53°C. What is the temperature if the volume changes to $80\bar{0}$ L at 208 kPa gauge pressure?

23. An acetylene welding tank has a gauge pressure of 13.8 MPa at 5°C. If the temperature rises to $3\bar{0}$°C, what is its new gauge pressure?

24. What is the new gauge pressure in Exercise 23 if the temperature falls to $-3\bar{0}$°C?

25. An ideal gas occupies a volume of 5.00 L at STP. What is its gauge pressure if the volume is doubled and its temperature is increased to $20\bar{0}$°C?

26. An ideal gas occupies a volume of 5.00 L at STP.

    (a) What is its temperature if its volume is halved and its absolute pressure is doubled?

    (b) What is its temperature if its volume is halved and its gauge pressure is doubled?

18

CHANGE OF STATE

18.1 FUSION

Many industries are concerned with a change of state in the materials they use. In foundries the principal activity is to change the state of solid metals to liquid, pour the liquid into molds, and allow it to become solid again.

This change of state from solid to liquid is called *melting* or *fusion*. The change from the liquid to solid state is called *freezing* or *solidification*. Most solids have a crystal structure and a definite melting point at any given pressure. Fusion and solidification of these substances occur at the same temperature. For example, water at 0° Celsius changes to ice and ice changes to water at the same temperature. There is no temperature change during change of state. Ice at 0°C changes to water at 0°C. Some substances, such as butter and glass, have no particular melting temperature but change state gradually.

Although there is no temperature change during a change of state, *there is a transfer of heat*. A melting solid *absorbs* heat and a solidifying liquid *gives off* heat. When 1 g of ice at 0°C melts, it absorbs 8$\bar{0}$ cal of heat. Similarly, when 1 g of water freezes at 0°C, ice at 0°C is produced and 8$\bar{0}$ cal of heat is released.

When 1 kg of ice at 0°C melts, it absorbs 8$\bar{0}$ kcal of heat. Similarly, when 1 kg of water freezes at 0°C, ice at 0°C is produced and 8$\bar{0}$ kcal of heat is released.

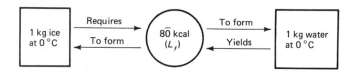

Or, when 1 kg of ice at 0°C melts, it absorbs 335 kJ of heat. Then, when 1 kg of water freezes at 0°C, ice at 0°C is produced and 335 kJ of heat is released.

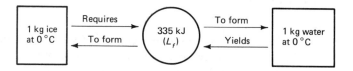

The amount of heat required to melt 1 g or 1 kg of a solid is called its heat of fusion. That is,

$$L_f = \frac{Q}{m}$$

where L_f is the heat of fusion,
  Q is the amount of heat, and
  m is the mass.

example 1   If 6.6 kcal of heat is required to melt 1.2 kg of lead at 327°C (melting point of lead), what is the heat of fusion of lead?

DATA:

$$Q = 6.6 \text{ kcal}$$

$$m = 1.2 \text{ kg}$$

$$L_f = ?$$

BASIC EQUATION: $$L_f = \frac{Q}{m}$$

WORKING EQUATION: same

SUBSTITUTION: $$L_f = \frac{6.6 \text{ kcal}}{1.2 \text{ kg}}$$

$$= 5.5 \text{ kcal/kg}$$

example 2 An aluminum calorimeter of mass 145 g contains 288 g of water at 45.0°C. Then 115 g of ice at 0.0°C is added to the water. After the ice is completely melted, the temperature of the water is 12.0°C. Find the heat of fusion of the ice.

Here again the basic equation is $Q_{\text{lost}} = Q_{\text{gained}}$. Heat is lost by the water and the calorimeter. Heat is gained by the melting ice and by the resulting ice water. Let us rewrite the data as follows:

Mass of aluminum calorimeter	145 g	m_c
Specific heat of calorimeter	0.22 cal/gC°	c_c
Mass of water	288 g	m_w
Specific heat of water	1.00 cal/gC°	c_w
Initial temperature of water and calorimeter	45.0°C	T_w, T_c
Mass of ice	115 g	m_i
Mass of resulting ice water	115 g	m_{iw}
Specific heat of ice water	1.00 cal/gC°	c_{iw}
Initial temperature of ice water	0.0°C	T_{iw}
Final temperature of water and calorimeter	12.0°C	T_f

Then

BASIC EQUATION: $$Q_{\text{lost}} = Q_{\text{gained}}$$

WORKING EQUATION: $$m_w c_w (T_w - T_f) + m_c c_c (T_c - T_f) = m_i L_f + m_{iw} c_{iw} (T_f - T_{iw})$$

SUBSTITUTION:

$$(288 \text{ g})(1.00 \text{ cal/gC°})(45.0°C - 12.0°C) + (145 \text{ g})(0.22 \text{ cal/gC°})(45.0°C - 12.0°C)$$
$$= (115 \text{ g})L_f + (115 \text{ g})(1.00 \text{ cal/gC°})(12.0°C - 0.0°C)$$

$$9.50 \times 10^3 \text{ cal} + 1.1 \times 10^3 \text{ cal} = (115 \text{ g})L_f + 1.38 \times 10^3 \text{ cal}$$

$$9.2 \times 10^3 \text{ cal} = (115 \text{ g})L_f$$

$$8\bar{0} \text{ cal/g} = L_f$$

In general, for heat of fusion of water, use

$$\boxed{L_f = 8\bar{0} \text{ cal/g} = 8\bar{0} \text{ kcal/kg} = 335 \text{ kJ/kg}}$$

18.2 VAPORIZATION

Steam heating systems in homes and factories are important applications of the principles of change of state from liquid to the gaseous or vapor state. This change of state is called *vaporization*. Boiling water shows this change of state. The reverse process (change from gas to liquid) is called *condensation*. As steam condenses in radiators, large amounts of heat are released.

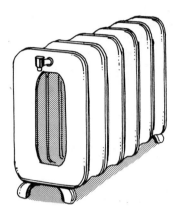

While a liquid is boiling, the temperature of the liquid does not change. There is, however, a transfer of heat. A liquid being vaporized (boiling) *absorbs* heat. As a vapor condenses, heat is given off. When 1 g of water at $10\overline{0}°C$ changes to steam at $10\overline{0}°C$, it absorbs $54\overline{0}$ cal; similarly, when 1 g of steam at $10\overline{0}°C$ condenses to water at $10\overline{0}°C$, $54\overline{0}$ cal of heat is given off.

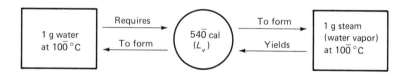

When 1 kg of water at $10\overline{0}°C$ changes to steam at $10\overline{0}°C$, it absorbs $54\overline{0}$ kcal of heat. Similarly, when 1 kg of steam at $10\overline{0}°C$ condenses to water at $10\overline{0}°C$, $54\overline{0}$ kcal of heat is given off.

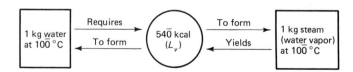

Or, when 1 kg of water at $10\bar{0}°C$ changes to steam at $10\bar{0}°C$, it absorbs 2.26 MJ of heat. Then when 1 kg of steam at $10\bar{0}°C$ condenses to water at $10\bar{0}°C$, 2.26 MJ of heat is given off.

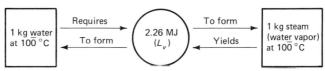

The amount of heat required to vaporize 1 g or 1 kg of a liquid is called its *heat of vaporization*. That is,

$$L_v = \frac{Q}{m}$$

where L_v is the heat of fusion,
    Q is the amount of heat, and
    m is the mass.

example 1 If 21 MJ of heat is required to vaporize 25 kg of ethyl alcohol at 78.5°C (boiling point of ethyl alcohol), what is the heat of vaporization of ethyl alcohol?

DATA:

$$Q = 21 \text{ MJ}$$

$$m = 25 \text{ kg}$$

$$L_v = ?$$

BASIC EQUATION:

$$L_v = \frac{Q}{m}$$

WORKING EQUATION:

same

SUBSTITUTION:

$$L_v = \frac{21 \text{ MJ}}{25 \text{ kg}}$$

$$= 0.84 \text{ MJ/kg} \quad \text{or} \quad 840 \text{ kJ/kg}$$

example 2 A copper calorimeter of mass 179.9 g contains 305.0 g of water at 12.0°C. Then 45.0 g of steam at 100.0°C is added to the water. After the steam condenses, the temperature of the water is 89.2°C. Find the heat of vaporization of the steam.

Again, the basic equation is $Q_{lost} = Q_{gained}$. Heat is lost by the steam in condensing and by the resulting steam water. Heat is gained by the water and the calorimeter. Let us rewrite the data as follows:

Mass of copper calorimeter	179.9 g	m_c
Specific heat of calorimeter	0.092 cal/gC°	c_c
Mass of water	305.0 g	m_w
Specific heat of water	1.00 cal/gC°	c_w

Initial temperature of water and calorimeter	12.0°C	T_w, T_c
Mass of steam	45.0 g	m_s
Mass of resulting steam water	45.0 g	m_{sw}
Specific heat of steam water	1.00 cal/gC°	c_{sw}
Initial temperature of steam water	100.0°C	T_{sw}
Final temperature of water and calorimeter	89.2°C	T_f

Then

BASIC EQUATION: $$Q_{lost} = Q_{gained}$$

WORKING EQUATION: $$m_s L_v + m_{sw} c_{sw} (T_{sw} - T_f) = m_w c_w (T_f - T_w) + m_c c_c (T_f - T_c)$$

SUBSTITUTION:

$$(45.0 \text{ g}) L_v + (45.0 \text{ g})(1.00 \text{ cal/gC}°)(100.0°C - 89.2°C)$$

$$= (305.0 \text{ g})(1.00 \text{ cal/gC}°)(89.2°C - 12.0°C)$$

$$+ (179.9 \text{ g})(0.092 \text{ cal/gC}°)(89.2°C - 12.0°C)$$

$$(45.0 \text{ g}) L_v + 486 \text{ cal} = 2.35 \times 10^4 \text{ cal} + 1.3 \times 10^3 \text{ cal}$$

$$(45.0 \text{ g}) L_v = 2.43 \times 10^4 \text{ cal}$$

$$L_v = 54\bar{0} \text{ cal/g}$$

In general, for heat of vaporization of water, use

$$\boxed{L_v = 54\bar{0} \text{ cal/g} = 54\bar{0} \text{ kcal/kg} = 2.26 \text{ MJ/kg}}$$

When vaporization occurs from a liquid to a gas or vapor, we call it *evaporation*. When vaporization occurs directly from a solid to a gas or a vapor without passing through the liquid phase, we call it *sublimation*. When dry ice is left at room temperature, the direct change from a solid to a gas is an example of sublimation. The rate of evaporation (or sublimation) increases at higher temperatures because the energy of the molecules is larger at the higher temperatures.

The table on page 369 shows some common substances and their heat characteristics.

From the table we may calculate the amount of heat released when a given quantity of steam is cooled through a change of state. We will use the heat of vaporization and the method of mixtures. We know that the heat lost during cooling is given by $Q = mc \, \Delta T$. The total amount of heat released by changing steam to water equals the amount of heat lost by the steam during cooling: $(m_{steam} c_{steam} \, \Delta T_{steam})$ plus the quantity of heat lost during condensation $(m_{steam} L_v)$, plus the heat lost by the water during cooling $(m_{water} c_{water} \, \Delta T_{water})$. Therefore,

$$Q = (m_{steam})(c_{steam})(\Delta T_{steam}) + (m_{steam}) L_v + (m_{water})(c_{water})(\Delta T_{water})$$

Heat constants

	Melting point (°C)	Boiling point (°C)	Specific heat (cal/g C° or kcal/kgC°)	Specific heat (J/kg C°)	Heat of fusion (cal/g or kcal/kg)	Heat of fusion (J/kg)	Heat of vaporization (cal/g or kcal/kg)	Heat of vaporization (J/kg)
Alcohol, ethyl	−117	78.5	0.58	2400	24.9	1.04×10^5	204	8.54×10^5
Aluminum	660		0.22	920	76.8	3.21×10^5		
Brass	840		0.092	390				
Copper	1083		0.092	390	49.0	2.05×10^5		
Glass			0.21	880				
Ice	0		0.51	2100	$\overline{80}.$	3.35×10^5		
Iron (steel)	1540		0.115	481	7.89	3.30×10^4		
Lead	327		0.031	130	5.86	2.45×10^4		
Mercury	−38.9	357	0.033	140	2.82	1.18×10^4	65.0	2.72×10^5
Silver	961		0.056	230	26.0	1.09×10^5		
Steam			0.48	$2\overline{0}00$				
Water (liquid)	0	$10\overline{0}$	1.00	4190			$54\overline{0}$	2.26×10^6
Zinc	419		0.092	390	23.0	9.63×10^4		

example 3   How many calories are released when 275 g of steam at $15\bar{0}°C$ is cooled to water at $3\bar{0}°C$?

DATA:
$$m = 275 \text{ g}$$
$$T_i \text{ of steam} = 15\bar{0}°C$$
$$T_f \text{ of steam (water)} = 3\bar{0}°C$$
$$Q = ?$$

BASIC EQUATION: $Q = (m_{steam})(c_{steam})(\Delta T_{steam}) + (m_{steam})L_v + (m_{water})(c_{water})(\Delta T_{water})$

WORKING EQUATION: same

SUBSTITUTION: $Q = (275 \text{ g})(0.48 \text{ cal/gC°})(15\bar{0}°C - 10\bar{0}°C) + (275 \text{ g})(54\bar{0} \text{ cal/g})$
$$+ (275 \text{ g})(1.00 \text{ cal/gC°})(10\bar{0}°C - 3\bar{0}°C)$$
$$= 6600 \text{ cal} + 149{,}000 \text{ cal} + 19{,}000 \text{ cal}$$
$$= 175{,}000 \text{ cal}   \text{ or }   175 \text{ kcal}$$

The following graph shows the heat gained (in cal) by 1 g of ice at $-2\bar{0}°C$ as it warms to 0°C, melts at 0°C, warms as water to $10\bar{0}°C$, vaporizes at $10\bar{0}°C$, and warms as steam to $12\bar{0}°C$. Note that during changes of state there are no temperature changes.

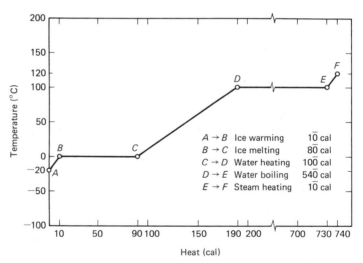

Heat gained by one gram of ice at $-2\bar{0}$ °C as it is converted to steam at $12\bar{0}$ °C.

The following graph shows the heat gained (in J) by 1 kg of ice at $-2\bar{0}°C$ as it warms to 0°C, melts at 0°C, warms as water to $10\bar{0}°C$, vaporizes at $10\bar{0}°C$, and warms as steam to $12\bar{0}°C$.

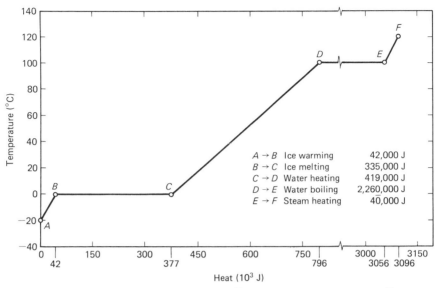

$A \rightarrow B$	Ice warming	42,000 J
$B \rightarrow C$	Ice melting	335,000 J
$C \rightarrow D$	Water heating	419,000 J
$D \rightarrow E$	Water boiling	2,260,000 J
$E \rightarrow F$	Steam heating	40,000 J

Heat gained by one kilogram of ice at $-2\overline{0}\,°C$ as it is converted to steam at $12\overline{0}\,°C$

18.3 EFFECTS OF PRESSURE AND IMPURITIES ON CHANGE OF STATE

Automobile cooling systems present important problems concerning change of state. Most substances contract on solidifying. Water and a few other substances, however, expand. The tremendous force exerted by this expansion is shown by the number of cracked automobile blocks and burst radiators suffered by careless motorists every winter.

Impurities in water tend to *lower* the freezing point. Alcohol has a lower freezing point than water and is used in some types of antifreeze. By mixing antifreeze with water in the cooling system, the freezing point of the water may be lowered to avoid freezing in winter. Automobile engines may also be ruined in

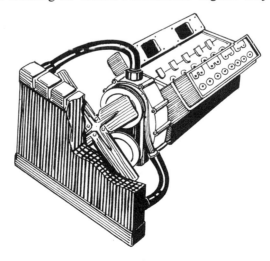

winter by overheating if the water in the radiator is frozen, preventing the engine from being cooled by circulation in the system.

An increase in the pressure on a liquid *raises* the boiling point. Automobile manufacturers utilize this fact by pressurizing their cooling systems and thereby raising the boiling point of the coolant used.

A decrease in the pressure on a liquid *lowers* the boiling point. Frozen concentrated orange juice is produced by subjecting the pure juice to very low pressures at which the water in the juice is evaporated. The consumer must restore the lost water before serving the juice.

18.4   EVAPORATION AS A COOLING PROCESS

Your body uses evaporation to help control its temperature. When you become too warm, your sweat glands produce water, which evaporates from your skin. As the water evaporates, the body loses heat at the rate of $Q/m = L_v$. When you are in a cool breeze, the perspiration evaporates more rapidly, which cools you faster. When the weather is hot and humid, you tend to remain hot because the perspiration does not vaporize as quickly under these conditions.

When a person, especially a child, has a very high body temperature, the common medical practice is to rub the body with rubbing alcohol because it readily evaporates from the skin. As it evaporates, it removes heat from the body.

During evaporation the molecules of a liquid are continually leaving the surface of the liquid. Some have enough energy to leave, freeing themselves from the liquid's surface. The rest, not having enough energy to leave, remain as part of the liquid.

The rate of evaporation of a liquid depends on the following:

1. The amount of surface area exposed. The larger the surface area exposed, the more molecules that have a chance to escape from the surface.

2. The temperature. The higher the temperature, the higher the molecular energy of the molecules, which allows more molecules to escape.

3. Air currents. Air currents blowing over the liquid's surface remove many of the molecules that have been evaporated before they may fall back into the liquid. This is why a cool summer breeze "feels so good."

4. The *volatility* of a liquid is a measure of its ability to vaporize. Examples of highly volatile liquids are rubbing alcohol and gasoline. The more volatile the liquid, the greater its rate of evaporation.

5. The air pressure on or above the liquid. The lower the pressure, the greater the rate of evaporation. Under a partial vacuum, there are fewer air molecules around with which the liquid molecules may collide, allowing for a higher rate of escape and a higher rate of evaporation.

EXERCISES

Sketch	Data	Basic Equation	Working Equation	Substitution
$h = ?$ $b = 6.0$ cm 12 cm²	$A = 12$ cm² $b = 6.0$ cm $h = ?$	$A = bh$	$h = \dfrac{A}{b}$	$h = \dfrac{12 \text{ cm}^2}{6.0 \text{ cm}}$ $= 2.0$ cm

1. How many calories of heat is needed to melt 125 g of ice at 0°C?

2. How many kilograms of ice at 0°C can be melted by the addition of 635 kcal of heat?

3. How many joules of heat is needed to melt 15.0 kg of ice at 0°C?

4. How many kilograms of ice at 0°C can be melted by the addition of 4.00 MJ of heat?

5. How many joules of heat is needed to melt 25.0 kg of silver at 961°C (melting point of silver)?

6. If 906 kcal of heat is needed to melt 18.5 kg of copper at 1083°C (melting point of copper), what is the heat of fusion of copper (in kcal/kg)?

7. How many calories of heat is needed to melt 325 g of ice at 0°C and to raise the temperature of the melted ice to 35°C?

8. How many joules of heat need to be removed from 10.0 L of water at 25.0°C to freeze it to ice at 0°C?

9. An aluminum calorimeter of mass 165 g contains $35\bar{0}$ g of water at 50.0°C. Then 125 g of ice at 0.0°C is added to the water. What is the final temperature of the mixture?

10. A copper calorimeter of mass 175 g contains 375 g of water at 45.0°C. How much ice at 0.0°C must be added to the water to make the temperature of the final mixture 20.0°C?

11. How many kilocalories of heat is needed to vaporize 3.5 kg of water at $10\bar{0}$°C?

12. How many grams of steam at $10\bar{0}$°C can be condensed to water with the removal of 120 kcal of heat?

13. How many joules of heat need to be removed to condense 1.5 kg of steam at $10\bar{0}$°C?

14. How many litres of water at $10\bar{0}$°C are vaporized by the addition of 5.00 MJ of heat?

15. How many joules of heat need to be removed from 1.25 kg of steam at 115°C to condense it to water and cool the water to $5\bar{0}$°C?

16. How many kilocalories of heat is needed to vaporize 5.00 kg of water at $10\bar{0}$°C and raise the temperature of the steam to 145°C?

17. A copper calorimeter of mass 225.0 g contains 275.0 g of water at 5.0°C. Then 39.5 g of steam at 100.0°C condenses and is added to the water. What is the final temperature of the mixture?

18. An aluminum calorimeter of mass 275.0 g contains 450.0 g of water at 7.0°C. How much steam at 100.0°C must be condensed and added to the water to make the final temperature of the mixture 18.0°C?

19. How many calories of heat is liberated when $20\bar{0}$ g of steam at $12\bar{0}$°C is changed to ice at -12°C?

20. How many joules of heat is needed to change 20.0 kg of ice at -15°C to steam at 115°C?

.51

heat ice

$m c_{ice} (-15° - 0°) +$

melt ice

$m h_F$ 80

+

heat water

1.0

$m c_{water} (100 - 0) +$

boil

540

$m h_v$

+

.48

heat steam

$m c_{steam} (115° - 100°)$

19.1 PROPERTIES OF A WAVE

A *wave* is a disturbance propagated through a medium or through space. The disturbance may be a displacement of atoms from their equilibrium positions in an elastic medium, a variation of pressure in a gas, a pulse sent along a rope, or a change in electric flux density in empty space. In each case there is a transfer of energy in the direction of the propagation of the disturbance or wave. Waves may be one-dimensional, as in the case of a violin string; two-dimensional, as in the case of a drumhead; or three-dimensional, as in the case of sound.

A wave may be a single-pulse disturbance which travels out from a source, as is the case when one cracks a whip; or it may be a wave train which is generated when the source of the disturbance is periodic. A wave train generated by a simple harmonic source will be used as our model wave. A simple harmonic wave traveling in the positive x-direction is represented mathematically by the expression

$$y = A \sin\left[2\pi\left(\frac{x}{\lambda} - \frac{t}{T}\right) + \phi\right]$$

where y is the displacement,
A is the amplitude,
x is the position in the direction of wave propagation,
λ is the wavelength,

t is the independent variable time,

T is the period of the wave, and

ϕ is the phase angle.

The *displacement*, y, represents the disturbance as a function of position and time, x and t. Displacement is the variation of the physical quantity for a specific wave from its equilibrium value. The amplitude of the wave, A, is the maximum displacement possible.

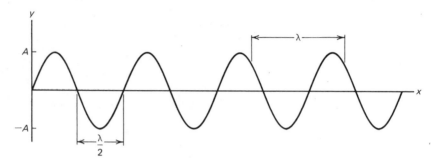

A *wavelength*, λ, is the length of the period of the disturbance pattern. The period, T, is the time required for a wavelength to pass a point on the x-axis. The reciprocal of the period is called the frequency, f, and is the number of complete oscillations per unit of time. An oscillation per second is called a hertz (Hz). The term ϕ is a phase angle and permits the wave to be in any stage of its oscillation at $x = 0$, $t = 0$.

The velocity, v, of propagation of a wave is the wavelength, λ, divided by the time, T, required for that wavelength to travel through a point, or

$$v = \frac{\lambda}{T} = \lambda f$$

This speed is the speed at which the disturbance is moving along the x-axis; it is also the speed that energy is being transmitted along the x-axis; it is *not* the speed with which the disturbance is moving away from or toward its equilibrium position. It should be emphasized that no medium is being transferred by the wave. Matter may oscillate about an equilibrium point, but there is no net transfer of matter by a wave.

The expression $2\pi(x/\lambda - t/T)$ may be written $(2\pi/\lambda)(x - vt)$, since $v = \lambda/T$, and now the displacement using functional notation is given by $y = f(x - vt)$. It can be seen by inspection that as time, t, increases, x must increase in order for $f(x - vt)$ to have the same value. Accordingly, the wave, as presented, corresponds to a wave traveling to the right. A wave traveling to the left would have an argument $2\pi(x/\lambda + t/T)$.

The wave equation presented is one-dimensional, simple-harmonic, and extends over the entire *x*-axis. Although the equation is restricted, it is also general in purpose, since no qualification of physical quantity has been necessary. This wave model will be used to illustrate some properties of waves.

$$y = A \sin\left[2\pi\left(\frac{x}{\lambda} - \frac{t}{T}\right) + \phi\right]$$

example   A wave is represented by the equation

$$y - 3.0 \sin \pi(4.0x - 1\overline{0}00t)$$

where *y* and *x* are in metres and *t* is in seconds.

(a) What is the period of this wave?
(b) What is this wave's frequency of oscillation?
(c) What is the wavelength?
(d) What are the speed and the direction of propagation?
(e) What is the amplitude of this wave?
(f) What is the displacement of the wave at the position $x = 3.0$ m when *t* is zero?

Compare the expression for this wave with the model expression on page 375.

(a) The coefficient of *t* in this wave's argument is compared to the model expression.

$$\frac{2\pi}{T} = \pi 1\overline{0}00$$

or

$$T = 0.0020 \text{ s}$$

(b) The frequency of oscillation, *f*, is the reciprocal of the period.

$$f = \frac{1}{T} = \frac{1}{0.0020 \text{ s}}$$

$$f = 5\overline{0}0 \text{ Hz}$$

(c) The coefficient of *x* in the argument to that in the model expression is

$$\frac{2\pi}{\lambda} = \pi(4.0)$$

$$\lambda = 0.50 \text{ m}$$

(d) The velocity of the wave is to the right because the coefficient of *t* is negative. The speed is the ratio of the wavelength to the period:

$$v = \frac{\lambda}{T} = \frac{0.50 \text{ m}}{0.0020 \text{ s}}$$

$$v = 250 \text{ m/s}$$

(e) By direct comparison, the amplitude is given by

$$A = 3.0 \text{ m}$$

(f) Substituting the given values of *x* and *t* in the wave expression gives

$$y = 3.0 \sin \pi[4.0(3.0) - 1000(0)]$$

$$y = 0$$

Although the following discussion will use material examples to illustrate waves, the description applies also to electromagnetic waves.

19.2 TRANSVERSE AND LONGITUDINAL WAVES

Waves in nature can generally be divided into two categories, transverse and longitudinal. A *transverse wave* is one in which the displacement is perpendicular to the direction of propagation of the wave. In the figure below, notice that each particle is oscillating in simple harmonic motion about its equilibrium position. The phase difference in the motion of each particle is such that the wave form and the energy move to the right. An example of a transverse wave is water. A buoy on water moves up and down as the wave passes by it, illustrating the transverse up-and-down motion of the water. There is also no transfer of water in the direction of propagation.

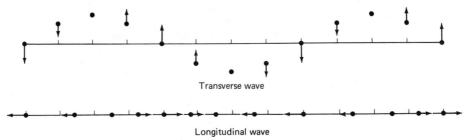

Transverse wave

Longitudinal wave

A *longitudinal wave* is one in which the displacement is along the line of the wave propagations and its direction of energy transfer. The figure showing a longitudinal wave has the particles vibrating in harmonic motion along the direction of the wave velocity. Each particle is in a different phase, with the result that the wave and energy move to the right in the illustration. Attention is called again to the fact that there is no net transfer of material by this wave. An example of a longitudinal wave is sound.

A solid can transmit energy by both transverse and longitudinal modes; however, a fluid can support only longitudinal waves because fluids have no sheer elasticity.

19.3 SUPERPOSITION OF WAVES

Waves exhibit a quality called *superposition*. When two waves of the same type pass through the same region of space, they superimpose on each other in such a manner that their displacements add together to form a new wave in that region. The figure on page 379 shows two waves of equal amplitude and wavelength, but different in phase by ϕ.

$$y_1 = A \, \sin\left[2\pi\left(\frac{x}{\lambda} + \frac{t}{T}\right)\right]$$

$$y_2 = A \, \sin\left[2\pi\left(\frac{x}{\lambda} + \frac{t}{T}\right) + \phi\right]$$

and their resultant wave

$$y = y_1 + y_2 = 2A \, \cos\frac{\phi}{2} \sin\left[2\pi\left(\frac{x}{\lambda} + \frac{t}{T}\right) + \frac{\phi}{2} \right]$$

(See trigonometric identity 27 on page 592.)

is the two waves superimposed. Note that the period and wavelength of the new wave are the same as the period and wavelength of the component waves. The amplitude of the new wave is $2A \cos \phi/2$, where ϕ is the phase difference between the two waves. If $\phi = 0, 2\pi, 4\pi, \dots, 2n\pi$, the new wave has an amplitude that is twice the amplitude of the component waves. If $\phi = \pi, 3\pi, \dots, (2n + 1)\pi$, the new wave has an amplitude of zero. In the figure below, a phase of $\phi = \pi/4$ is used. Two waves may have a difference of phase because of a difference in path traveled. A path difference of one wavelength would correspond to a phase difference of 2π and would produce constructive interference. A path difference of a half-wavelength would correspond to a phase difference of π and produce destructive interference.

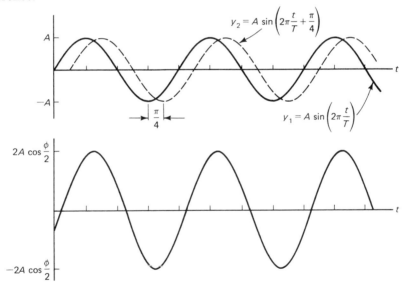

example   Two wave sources are separated by a distance of 15.0 m in water. The sources are synchronized so that they emit waves in unison which are 3.00 m long.

(a) Find the positions of maximum displacement on the line between the sources.
(b) Find the positions of minimum displacement on the line between the sources.

(a) A path difference between two sources can be related to the phase difference by

$$\phi = \left(\frac{2\pi}{\lambda}\right) \text{(path difference)}$$

The condition for a maximum is that ϕ be an integer multiple of 2π.

$$\phi = n2\pi = \left(\frac{2\pi}{\lambda}\right) \text{(path difference)}$$

or that the path difference must be an integer multiple of wavelength. For this example, referring to the figure, one sees that

$$\text{path difference} = (15.0 - x) - x = n\lambda$$

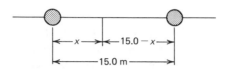

Solving for x and replacing λ by its value, 3.00 m.

$$15.0 - 2x = n(3.00)$$

$$x = 7.50 - 1.50n \qquad \text{(in metres)}$$

So the positions of maximum displacement between the two sources are at distances of 7.50 m, 6.00 m, 4.50 m, 3.00 m, and 1.50 m from each source, corresponding to $n = 0$, 1, 2, 3, and 4.

(b) The criterion for a minimum displacement is that ϕ be $(2n + 1)\pi$.

$$\left(\frac{2\pi}{\lambda}\right)(\text{path difference}) = (2n + 1)\pi$$

$$\text{path difference} = (2n + 1)\left(\frac{\lambda}{2}\right)$$

Placing the path difference equal to an odd multiple of the half-wavelength gives

$$15.0 - 2x = (2n + 1)\left(\frac{3.00}{2}\right)$$

or

$$x = 7.50 - (2n + 1)(0.75) \qquad \text{(in metres)}$$

The distances for minimum displacement between the two sources are 6.75 m, 5.25 m, 3.75 m, 2.25 m, and 0.75 m from each source, corresponding to $n = 0, 1, 2, 3$, and 4.

19.4   WAVE INTENSITY

The *intensity* of a wave, I, is the rate at which energy is transmitted through an area perpendicular to the direction of wave propagation.

$$\boxed{I = \frac{E}{At}}$$

where I is the intensity in W/m^2,
E is the energy in joules, J,
A is the area in m^2, and
t is the time in s.

The intensity of a wave at a position in space is proportional to the square of the amplitude of the wave. From the preceding section, the resultant of the two waves that were in phase had twice the amplitude of a single wave but four times the intensity, whereas the resultant formed by the waves that were π out of phase had no amplitude and no intensity.

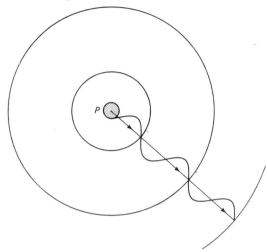

In the figure above, a point source of power P is sending out waves in all directions, so that any sphere with the source at the center would have the same displacement at all points on its surface. We would refer to this as a spherical wave front. The source is emitting energy at a constant rate so that the rate that energy passes through any sphere is the same and the intensity at any distance r from the point source is $I = P/A = P/(4\pi r^2)$, or

$$\frac{I_1}{I_2} = \frac{r_2^2}{r_1^2}$$

The intensity of waves from a point source is inversely proportional to the square of the distance from the source.

example   The intensity from a point source of waves is found to be 12 W/m² at a distance of 2.0 m from the source. What is the intensity at a distance of 5.0 m from the source?

DATA:

$$I_1 = 12 \text{ W/m}^2$$

$$r_1 = 2.0 \text{ m}$$

$$r_2 = 5.0 \text{ m}$$

$$I_2 = ?$$

The intensity is inversely proportional to the distance from the source.

BASIC EQUATION:
$$\frac{I_1}{I_2} = \frac{r_2^2}{r_1^2}$$

WORKING EQUATION:
$$I_2 = \frac{r_1^2}{r_2^2} I_1$$

SUBSTITUTION:
$$I_2 = \frac{(2.0 \text{ m})^2}{(5.0 \text{ m})^2}(12 \text{ W/m}^2)$$
$$= 1.9 \text{ W/m}^2$$

19.5   REFLECTION OF WAVES

When a wave is traveling from one medium to a second medium, part of the energy of the wave may be reflected and the remainder transmitted into the new medium. The transmitted wave is always transmitted in phase; however, the reflected wave may be in phase or π out of phase. If the wave is traveling from a medium that is denser to a less dense medium, the reflected wave is in the same phase as the incident wave; whereas if the wave is traveling from a less dense to a denser medium, the reflected wave is π out of phase with the incident wave. An example of this phenomenon can be demonstrated with a rope. If one end of the rope is left free and a pulse is sent along the rope, the free end will respond with an increased displacement. When the pulse arrives at the free end, the reflected wave is in phase with the incident wave and their displacements add in superposition. If the end of the rope is secured to a heavy chain or a heavier rope or a fixed ring, the reflected wave will be π out of phase with the incident wave and there will be no displacement at the end point.

19.6   STANDING WAVES

Consider two waves of equal wavelength and amplitude; however, one moves to the right and the other moves to the left. These waves are represented by the expressions

$$y_1 = A \sin 2\pi\left(\frac{x}{\lambda} - \frac{t}{T}\right)$$
$$y_2 = A \sin 2\pi\left(\frac{x}{\lambda} + \frac{t}{T}\right)$$

When these waves are superimposed, the resultant wave is given by

$$y = y_1 + y_2 = 2A \sin 2\pi\frac{x}{\lambda} \cos 2\pi\frac{t}{T}$$

(See trigonometric identity 27 on page 592.)

The first factor $2A \sin 2\pi(x/\lambda)$ is spatial and gives an amplitude envelope that is a function of position. The second factor is a simple harmonic function of time. The resultant wave has each point along the x-axis vibrating in simple harmonic motion within the amplitude envelope of $\pm 2A \sin 2\pi(x/\lambda)$, as shown below. At $x = \lambda/2$, λ, $3\lambda/2$, and so on, are positions of no displacement called nodes. Midway between the nodes are positions of maximum amplitude called antinodes.

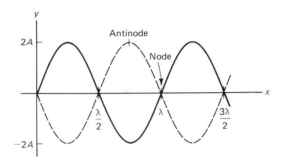

This wave is called a *standing wave* and is the superposition of two similar waves traveling in opposite directions. There is no net propagation of energy. The energy is in the harmonic oscillations of the medium and changes back and forth from kinetic energy to potential energy with the oscillations.

19.7   VIBRATING STRING

The speed of a wave traveling along a string is dependent upon the linear mass density of the string and the tension with which the string is stretched. The greater the mass per unit length, the slower the wave travels while the speed of the wave is increased with tension. Let P be the tension on the string in newtons and w be the mass per unit length in kilograms per metre. The velocity, v, of the wave along the string is given by

$$v = \sqrt{\frac{P}{w}}$$

A string on an instrument is fixed at both ends and a wave reflected from the end is π out of phase with the incident wave. A necessary end condition for standing waves on a stretched string is that a node exists at the end points or that the length of the string be a multiple of $\lambda/2$.

$$l = \frac{n\lambda}{2} \qquad \text{or} \qquad \lambda = \frac{2l}{n}$$

Recall that the velocity of propagation of a wave is λf; then

$$\lambda f = v = \sqrt{\frac{P}{w}}$$

$$f = \frac{1}{\lambda}\sqrt{\frac{P}{w}} = \frac{n}{2l}\sqrt{\frac{P}{w}}$$

$$\boxed{f = nf_0 \quad \text{where} \quad f_0 = \frac{1}{2l}\sqrt{\frac{P}{w}}}$$

This equation gives the condition for standing waves to exist in a stretched string. Only discrete values of frequencies are possible; indeed, they are integer multiples of a fundamental frequency f_0. The lowest frequency is called the fundamental frequency or first harmonic; the next frequency when $n = 2$ is called the second harmonic, and so on.

> **example**   A string of linear density 0.00080 kg/m is stretched over a 0.250-m span. It is desired to tune the string so that its fundamental frequency is 256 Hz. What tension must be applied to the string?
>
> DATA:
> $$l = 0.250 \text{ m}$$
> $$w = 8.0 \times 10^{-4} \text{ kg/m}$$
> $$f_0 = 256 \text{ Hz}$$
> $$P = ?$$
>
> BASIC EQUATION:
> $$f_0 = \frac{1}{2l}\sqrt{\frac{P}{w}}$$
>
> WORKING EQUATION:
> $$P = 4f_0^2 l^2 w$$
>
> SUBSTITUTION:
> $$P = 4(256 \text{ Hz})^2(0.250 \text{ m})^2(8.0 \times 10^{-4} \text{ kg/m})$$
>
> $$\boxed{\text{Hz}^2 \text{ m}^2 \text{ kg/m} = \frac{1}{\text{s}^2} \cdot \text{m}^2 \cdot \frac{\text{kg}}{\text{m}} = \text{kg m/s}^2 = \text{N}}$$
>
> $$P = 13 \text{ N}$$

EXERCISES

Sketch	Data	Basic Equation	Working Equation	Substitution
12 cm² $h = ?$ $b = 6.0$ cm	$A = 12 \text{ cm}^2$ $b = 6.0 \text{ cm}$ $h = ?$	$A = bh$	$h = \dfrac{A}{b}$	$h = \dfrac{12 \text{ cm}^2}{6.0 \text{ cm}}$ $= 2.0 \text{ cm}$

1. A wave is represented by the equation

$$y = 2.5 \sin(4.87x + 0.025t + 0.79),$$

where y and x are in metres and t is in seconds.

(a) What is the period of the wave?

(b) What is the frequency of the wave's source?

(c) What is the wavelength?

(d) What are the speed and the direction of propagation of this wave?

(e) What is the amplitude of this wave?

(f) What is the displacement of this wave at the position $x = 3.0$ m when t is zero?

2. Write an expression to represent a wave whose frequency is 384 Hz and speed of propagation is 343 m/s. Let this wave have an amplitude of 0.0050 m and travel along the $+x$-axis. (Let $\phi = 0$.)

3. Two sources produce water waves that are 3.00 m long. These sources are π out of phase and separated by a distance of 10.0 m. At what positions will maximum values of displacements be found?

4. Refer to Exercise 3 and determine the positions where minimum values of displacement will be found.

5. Two sources emitting 2.00-m waves in phase are separated by 8.00 m. Locate the positions of the antinodes along the line between the sources. (An antinode is at each source.)

6. Locate the position of the displacement nodes in Exercise 5.

7. A point source produces 2.0 W of light.

(a) What is the intensity at a point 0.50 m from the source?

(b) What is the intensity at a point 2.0 m from the source?

8. Two similar waves with the same frequency and speed differ in phase by $\pi/4$.

(a) In terms of the amplitude A of a single wave, what is the maximum amplitude the resultant wave of these two can have?

(b) In terms of the intensity from a single wave, what would be the intensity of the resultant wave?

9. A string whose linear density is 0.00200 kg/m is stretched to produce a wave speed of 125 m/s. What tension was applied to the string?

10. A string whose linear density is 0.0050 kg/m is stretched to produce a wave speed of 200 m/s.

(a) What length is necessary to produce a fundamental frequency of 320 Hz?

(b) What tension was applied to stretch the string?

SOUND

20.1 NATURE OF SOUND

A vibratory disturbance with frequency in the approximate range between 20 and 20,000 Hz produces a longitudinal wave in an elastic medium called *sound*. If the frequency of vibration is greater than 20,000 Hz, the wave is called *ultrasonic*. For frequencies less than about 20 Hz, the wave is *infrasonic*. The approximate range 20 to 20,000 Hz is the audible frequency range for human beings. Most adult people have a frequency response range which is somewhat abbreviated on each end of this spectrum, but still spans about eight octaves. Sound may travel through any medium, solid or fluid; however, our daily experience with it is almost always through air.

20.2 PHYSIOLOGICAL AND PHYSICAL PROPERTIES

Our normal atmospheric pressure is 101 kPa and the variation of pressure produced in person-to-person conversation is of the order of 10^{-4} Pa. This relatively small signal-to-background ratio is not only detected but is distinguishable as to pitch and quality. There are physical terms used to describe sound waves, and there are physiological terms used to describe the sensations produced by sound. These terms are related to each other but are not the same. *Intensity* is a physical term which gives the energy per unit time through a unit area by the wave; *loudness* is a term that describes how faint or strong the sensation of sound seems to the observer. The sound must reach a certain physical intensity before the observer hears any sound at all. After this threshold of audibility has been reached, the sensation of loudness increases as the common logarithm of the intensity. Frequency and pitch are very closely related; however, pitch has a small dependence on loudness and a sound of the same frequency may have a different pitch when loudness is at a higher level. Timbre and wave form are closely related. Most stereo enthusiasts know that the quality of sound from a given speaker depends upon the sound box in which it is installed. The shape, size, and finish of the sound box determine which harmonics will be enhanced and diminished to give the quality of sound from the system.

20.3   SPEED OF SOUND

Sound is transmitted by solids, liquids, and gases. The speed at which sound travels through any specific medium depends upon the elastic constants and density of that medium. The speed of sound in air depends upon barometric pressure and temperature. It is generally accepted that the speed of sound at 1 atm pressure at 0°C is 331 m/s and increases with temperature. The following equation gives the speed of sound as a function of temperature:

$$v = 331 \text{ m/s} + (0.61 \text{ m/s/°C})T$$

where T is expressed in degrees Celsius.

example   Determine the speed of sound when the air temperature is 23°C.

DATA: $T = 23°C$

                               $v = ?$

BASIC EQUATION: $v = 331 \text{ m/s} + (0.61 \text{ m/s/°C})T$

WORKING EQUATION: same

SUBSTITUTION: $v = 331 \text{ m/s} + (0.61 \text{ m/s/°C})(23°C)$

                               $= 345 \text{ m/s}$

20.4   INTERFERENCE OF SOUND WAVES

Sound waves, like other waves, superimpose on each other. In regions where two or more sound waves are passing, their amplitudes add together to give a resultant wave that has an amplitude equal to the algebraic sum of the individual waves. In public places, especially those which were constructed for athletic events but because of capacity get used as auditoriums, there exist regions where it is impossible to hear some sounds from the podium, and other regions where the same sounds seem to be clear and amplified. There are many factors that make for good acoustics in an auditorium and one of these is the interference phenomenon. The attempt here is to present interference in its basic nature without the complexities of multifrequency sounds and sound damping and sound delaying systems.

It should be emphasized that while calculations deal with amplitudes and combinations of amplitudes, it is the intensities that are detected. Intensity (Section 19.4) is directly proportional to the square of the amplitude.

example 1 Two loudspeakers from the same source are separated from each other by a distance of 4.000 m.

(a) Determine at what frequencies of sound the intensity would be enhanced at a position 10.000 m in front of one speaker and on a line perpendicular to the line through the two speakers.
(b) Determine at what frequencies the intensity would be reduced.
Assume the speed of sound to be 343 m/s on this day.

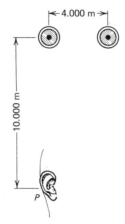

(a) Refer to the figure above and notice that the path from one speaker to the *point P* is 10.000 m and $\sqrt{(10.000 \text{ m})^2 + (4.000 \text{ m})^2}$ is the path from the other speaker to *point P*.

$$\text{path difference} = \sqrt{(10.000 \text{ m})^2 + (4.000 \text{ m})^2} - 10.000 \text{ m}$$
$$= 0.770 \text{ m}$$

In Section 19.3, we saw that to have maximum constructive interference, the path difference must be an integer multiple of the wavelength, or

$$n\lambda = 0.770 \text{ m} \qquad \text{or} \qquad \lambda = \frac{0.770}{n} \text{ m}$$

The speed of a wave is the product of its frequency and wavelength.

$$v = f\lambda$$
$$f = \frac{v}{\lambda} = \frac{v}{0.770 \text{ m}/n} = \frac{vn}{0.770 \text{ m}}$$
$$f = \frac{343 \text{ m/s}}{0.770 \text{ m}} n = 445n \text{ Hz}$$

The maximum value of n would be the greatest integer value that gives a frequency less than 20,000 Hz.

$$n_{max} = \text{*INT}\left[\frac{20,000}{445}\right] = 44$$

Maximum values of loudness would be found for

$$f = 445n \text{ Hz for } n = 1, 2, \ldots, 44$$

(b) In Section 19.3, it was shown that two similar waves interfere destructively if their path difference is $(2n + 1)\lambda/2$.

$$\text{path difference} = 0.770 \text{ m} = (2n + 1)\frac{\lambda}{2}$$

$$v = f\lambda = f\left(\frac{2(0.770 \text{ m})}{2n + 1}\right)$$

$$f = \frac{(2n + 1)v}{2(0.770 \text{ m})}$$

$$f = \frac{(2n + 1)(343 \text{ m/s})}{2(0.770 \text{ m})} = (2n + 1)223 \text{ Hz}$$

The maximum value of f can be determined by finding the maximum integer value of n that satisfies the inequality

$$(2n + 1)(223 \text{ Hz}) \leqslant 20,000$$

$$n_{max} = \text{INT}\left[\frac{1}{2}\left(\frac{20,000}{223} - 1\right)\right] = 44$$

The frequencies for which these would be minimum in loudness are

$$f = (2n + 1)(223 \text{ Hz}), \text{ for}$$

$$n = 0, 1, 2, \ldots, 44$$

If two sound waves of the same amplitude but slightly different frequencies are present in the same region of space, they superimpose and give a resultant wave whose frequency is the average of the original frequencies and whose intensity increases and decreases. Each intensity maximum is called a *beat*.

Let the two waves be represented by

$$y_1 = A \sin 2\pi f_1 t$$

$$y_2 = A \sin 2\pi f_2 t$$

$$y = y_1 + y_2 = 2A\left[\cos 2\pi\left(\frac{f_1 - f_2}{2}\right)t\right]\left[\sin 2\pi\left(\frac{f_1 + f_2}{2}\right)t\right]$$

(See trigonometric identity 27 on page 592.)

---

*This notation refers to finding the largest integer less than or equal to the quantity in the brackets.

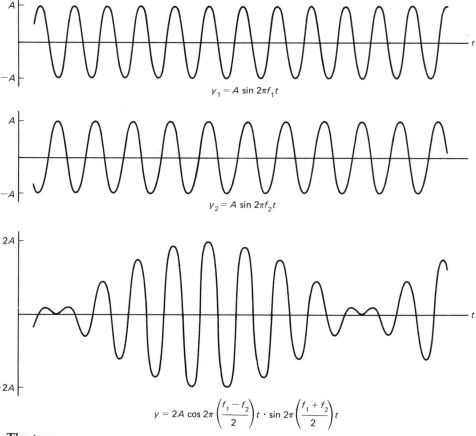

$$y_1 = A \sin 2\pi f_1 t$$

$$y_2 = A \sin 2\pi f_2 t$$

$$y = 2A \cos 2\pi \left(\frac{f_1 - f_2}{2} \right) t \cdot \sin 2\pi \left(\frac{f_1 + f_2}{2} \right) t$$

The term

$$\sin 2\pi \left(\frac{f_1 + f_2}{2} \right) t$$

of course, is the continuing harmonic oscillation at the average frequency. The term

$$2A \cos 2\pi \left(\frac{f_1 - f_2}{2} \right) t$$

may be thought of as an amplitude term (see the figure above) whose value varies from 0 to $2A$, giving a warbling effect called beats. A point where the amplitude is zero is a node, and a point where the amplitude is a maximum is an antinode. The distance between two successive nodes is one-half wavelength. The antinodes are midway between the nodes. Since an antinode of displacement or a maximum of intensity occurs twice in each complete oscillation, $|f_1 - f_2|$ is called the *beat frequency* and corresponds to the number of beats heard each second. Actually, a person cannot distinguish more than about six discrete intensity maximums per second and cannot detect higher beat frequencies with his or her ear.

example 2 A tuning fork of unknown frequency was found to produce four beats per second when sounded simultaneously with a fork of known frequency 256 Hz. A piece of molding clay is then added to the tongs of the unknown fork. The two forks are then sounded in unison and now produce a single beat per second. What is the frequency of the unknown fork?

Since the unknown fork produces a 4-Hz beat frequency with the known fork, it must have a frequency of (256 ± 4) Hz. The addition of molding clay to the unknown fork increases its inertia without changing its elastic properties, thus lowering its natural frequency. That is, if the frequency of the unknown fork had been 252 Hz, the beat frequency would have increased. Because it was not, the frequency of the unknown fork must have been $26\bar{0}$ Hz.

20.5 RESONANCE

Just as the vibrating string had certain discrete frequencies that produced standing waves, cavities support certain discrete sound waves. When a source of sound produces a standing wave in a cavity, this is called *resonance*. Most musical instruments are designed to produce resonance for certain sets of sounds. The typical example is the pipe organ. Standing waves in both open and closed pipes will be presented.

Consider a pipe that is open on both ends. In this case, the reflected wave is reflected at each end of the pipe in phase with the incident wave and, consequently, an antinode at each end of the pipe is a necessary condition for standing waves to exist in the pipe. In the figure below, one sees that the antinode condition can be satisfied by a pipe whose length is $\lambda/2$, $2(\lambda/2)$, $3(\lambda/2)$, ... , or

$$l = n\frac{\lambda}{2}$$

Pipe open at both ends

$\ell = \frac{\lambda}{2}$ $\quad \ell = 2\left(\frac{\lambda}{2}\right)$ $\quad \ell = 3\left(\frac{\lambda}{2}\right)$ $\quad \ell = 4\left(\frac{\lambda}{2}\right)$

The resonant frequencies for the pipe open at both ends would be

$$f_n = \frac{n}{2l} v$$

where v is the speed of sound,

  l is the length of the pipe, and

  n is an integer that corresponds to the harmonic of the fundamental frequency of resonance $v/2l$.

  If we have a pipe that is open on one end and closed on the other end, the conditions for a standing wave demand that the open end be an antinode and the closed end a node. The figure below shows several means to meet these end conditions. The length of the pipe must be $\lambda/4$, $3(\lambda/4)$, . . . , $(2n + 1)(\lambda/4)$, or

$$l = (2n + 1)\frac{\lambda}{4}$$

The resonant frequencies f_{2n+1}, for the pipe closed at one end would be

$$f_{2n+1} = \frac{(2n + 1)v}{4l}$$

where v is the speed of sound,

  l is the length of the pipe, and

  $2n + 1$ is the corresponding harmonic of the fundamental frequency $v/4l$.

  It will be noted that for the closed pipe, the even-numbered harmonics are not permitted.

Pipe closed at one end

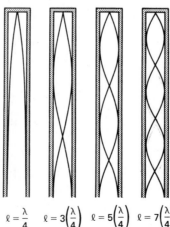

$$\ell = \frac{\lambda}{4} \qquad \ell = 3\left(\frac{\lambda}{4}\right) \qquad \ell = 5\left(\frac{\lambda}{4}\right) \qquad \ell = 7\left(\frac{\lambda}{4}\right)$$

example

(a) What would be the length of the pipe for a 384-Hz sound if the organ design is open-end?

(b) What would be the length for a closed-end organ design?

Assume the speed of sound to be 343 m/s.

(a) The fundamental frequency is the lowest frequency and corresponds to $f_1 = v/2l$.

DATA: $\quad\quad\quad\quad\quad\quad\quad\quad f_n = f_1 = 384 \text{ Hz}$

$$v = 343 \text{ m/s}$$

$$n = 1$$

$$l = ?$$

BASIC EQUATION: $\quad\quad\quad\quad f_n = \dfrac{nv}{2l}$

WORKING EQUATION: $\quad\quad l = \dfrac{nv}{2f_n}$

SUBSTITUTION: $\quad\quad\quad l = \dfrac{(1)(343 \text{ m/s})}{(2)(384 \text{ Hz})}$ $\boxed{\dfrac{\text{m/s}}{\text{Hz}} = \dfrac{\text{m/s}}{1/\text{s}} = \text{m}}$

$$= 0.447 \text{ m} \quad \text{or} \quad 44.7 \text{ cm}$$

(b) The fundamental frequency for a closed pipe corresponds to an n value of 0.

DATA: $\quad\quad\quad\quad\quad\quad\quad\quad f_{2n+1} = f_1 = 384 \text{ Hz}$

$$v = 343 \text{ m/s}$$

$$n = 0$$

$$l = ?$$

BASIC EQUATION: $\quad\quad\quad\quad f_{2n+1} = \dfrac{(2n+1)v}{4l}$

WORKING EQUATION: $\quad\quad l = \dfrac{(2n+1)v}{4(f_{2n+1})}$

SUBSTITUTION: $\quad\quad\quad l = \dfrac{[2(0)+1](343 \text{ m/s})}{4(384 \text{ Hz})}$

$$= 0.223 \text{ m} \quad \text{or} \quad 22.3 \text{ cm}$$

It should be recognized that the open pipe can also support all harmonics of 384 Hz, whereas the closed pipe is resonant only with the odd harmonics of the fundamental frequency.

20.6   THE DOPPLER EFFECT

At one time or other most people observe that some familiar and continuous sounds appear to change pitch with relative motion. At a railroad crossing, the whistle of the train appears to be shriller as the train approaches than it is as the train leaves the crossing. At an air show, as the plane streaks across the field, the sound of its engine appears to go to a lower pitch as the plane passes overhead. This effect is called the *Doppler effect* and for sound is given by

$$f' = f\frac{v \pm v_o}{v \mp v_s}$$

where f is the true frequency of the source,
  f' is the apparent frequency heard by the observer,
  v is the speed of sound in air,
  v_o is the speed of the observer, and
  v_s is the speed of the source.

The speeds, v_o and v_s, must be relative to the medium transmitting the sound. The signs $\pm$ and $\mp$ in the numerator and denominator, respectively, need discussion. The top signs in the numerator and the denominator are used when the motion of the observer is toward the source or when the motion of the source is toward the observer. This says that when relative motion of source and observer is toward the other, the apparent frequency heard is increased. The lower signs are used when the motion of the observer is away from the source or the motion of the source is away from the observer. If the relative motion of source and observer is away from the other, the apparent pitch heard by the observer decreases. The four combinations of relative motions of observer and source and their corresponding frequencies are given in the table below.

Source	Observer	Formula
$v_s \rightarrow$	$\leftarrow v_o$	$f' = f\left(\dfrac{v + v_o}{v - v_s}\right)$
$\leftarrow v_s$	$v_o \rightarrow$	$f' = f\left(\dfrac{v - v_o}{v + v_s}\right)$
$v_s \rightarrow$	$v_o \rightarrow$	$f' = f\left(\dfrac{v - v_o}{v - v_s}\right)$
$\leftarrow v_s$	$\leftarrow v_o$	$f' = f\left(\dfrac{v + v_o}{v + v_s}\right)$

The figure below shows a source moving to the left and two stationary observers: one on the left, which the source moves toward, and one on the right, which the source moves away from. Successive wave fronts emitted as the source moves are shown.

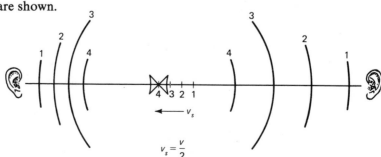

$$v_s = \frac{v}{2}$$

Note that the waves coming toward the observer on the left are crowded together and would appear to be of a higher frequency, while those approaching the observer on the right are spread apart and appear to be of a lower frequency to that observer.

example   At an amusement park, a mother and her daughter board a tram to survey the park while the father waits at the station to meet friends who are joining them at the park. As the tram is departing at a rate of 15.0 m/s, it blows its whistle, whose frequency is $42\bar{0}$ Hz.

(a) What frequency is heard by the mother and daughter riding the tram?
(b) What frequency is heard by the father waiting at the station?
(c) What frequency is heard by the family friends who are approaching the station from the opposite direction in their car at 30.0 m/s?
(Assume the speed of sound to be 343 m/s.)

(a) DATA: $f = 42\bar{0}$ Hz

$v = 343$ m/s

$v_s = 15.0$ m/s away from observer

$v_o = 15.0$ m/s toward source

$f' = ?$

BASIC EQUATION: $f' = f\left(\dfrac{v + v_o}{v + v_s}\right)$

WORKING EQUATION: same

SUBSTITUTION: $f' = (42\bar{0}\text{ Hz})\left(\dfrac{343\text{ m/s} + 15.0\text{ m/s}}{343\text{ m/s} + 15.0\text{ m/s}}\right)$

$= 42\bar{0}$ Hz

The people riding on the tram would hear the frequency of the tram's whistle without enhancement.

(b) DATA:
$$f = 42\bar{0} \text{ Hz}$$
$$v = 343 \text{ m/s}$$
$$v_s = 15.0 \text{ m/s away from observer}$$
$$v_o = 15.0 \text{ m/s}$$
$$f' = ?$$

BASIC EQUATION:
$$f' = f\left(\frac{v - v_o}{v + v_s}\right)$$

WORKING EQUATION: same

SUBSTITUTION:
$$f' = (42\bar{0} \text{ Hz})\left(\frac{343 \text{ m/s} - 0 \text{ m/s}}{343 \text{ m/s} + 15.0 \text{ m/s}}\right)$$
$$= 402 \text{ Hz}$$

The father at the station hears a lower frequency as the tram departs.

(c) DATA:
$$f = 42\bar{0} \text{ Hz}$$
$$v = 343 \text{ m/s}$$
$$v_s = 15.0 \text{ m/s away from observer}$$
$$v_o = 30.0 \text{ m/s toward the source}$$
$$f' = ?$$

BASIC EQUATION:
$$f' = f\left(\frac{v + v_o}{v + v_s}\right)$$

WORKING EQUATION: same

SUBSTITUTION:
$$f' = (42\bar{0} \text{ Hz})\left(\frac{343 \text{ m/s} + 30.0 \text{ m/s}}{343 \text{ m/s} + 15.0 \text{ m/s}}\right)$$
$$= 438 \text{ Hz}$$

The family friends hear a higher pitch because they are approaching the source faster than the source moves away from them.

EXERCISES

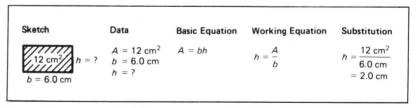

1. What is the speed of sound when the temperature is 5°C?
2. What is the speed of sound when the temperature is −5°C?

3. What is the speed of sound in a warm house where the temperature is 26°C?

4. A public address system has two speakers fed by the same amplifier. These speakers are separated by 5.000 m. What frequencies will have maximum values of displacement at a point 25.000 m directly in front of one of the speakers? Use 343 m/s for the value of the speed of sound.

5. In Exercise 4, what frequencies will have minimum values of displacement?

6. Two tuning forks of frequency $51\overline{0}$ Hz and 512 Hz are sounded simultaneously.

    (a) What frequency of sound would be heard?

    (b) What beat frequency would be heard?

7. A 384-Hz and a $36\overline{0}$-Hz tuning fork are sounded simultaneously. What beat frequency would an observer hear?

8. A tuning fork of unknown frequency produced two beats when sounded with a fork of $42\overline{0}$ Hz. The unknown fork was then loaded with molding clay on its tongs and, when sounded in unison with the standard fork, produced three beats. What is the frequency of the unknown fork?

9. What length of open pipe is needed to give resonance to a 256-Hz sound? Assume the speed of sound is 343 m/s.

10. What length of closed pipe will produce resonance with a 512-Hz sound?

11. (a) What length of closed pipe has the same frequency for its fundamental as a 1.00-m open pipe has for its second harmonic?

    (b) What is this frequency?

12. A train traveling at a speed of 80.0 km/h approaches a station and sounds a $50\overline{0}$-Hz whistle. What frequency will be heard by an observer at the station? (Assume the speed of sound is 343 m/s.)

13. A fire engine leaving the fire station traveling at $10\overline{0}$ km/h sounds its $45\overline{0}$-Hz siren. What frequency will the cook back at the station hear?

14. A police car traveling south at $10\overline{0}$ km/h sounds its siren, which has a frequency of $48\overline{0}$ Hz. What frequency will a person in front of the police car, traveling at 60.0 km/h in the same direction, hear?

15. Refer to Exercise 14. What frequency will a person in front of the police car, traveling at 60.0 km/h toward the police car, hear?

16. Refer to Exercise 14. What frequency will a person behind the police car, traveling at 60.0 km/h in the same direction, hear?

17. Refer to Exercise 14. What frequency will a person behind the police car, traveling at 60.0 km/h in the opposite direction, hear?

18. A police car traveling south at $10\overline{0}$ km/h sounds a siren with a frequency of $48\overline{0}$ Hz. There is a strong wind blowing north to south at 30.0 km/h. What frequency will a person in front of the police car, traveling at 60.0 km/h toward the police car, hear?

19. Refer to Exercise 18. What frequency will a person behind the police car, traveling at 60.0 km/h in the same direction, hear?

20. Refer to Exercise 18. What frequency will a person behind the police car, traveling at 60.0 km/h in the opposite direction, hear?

21. A 256-Hz source moves away from a stationary observer toward a wall at a speed of 2.0 m/s. Two waves reach the observer, one directly from the source and one reflected from the wall to the observer. Describe the sound the observer hears.

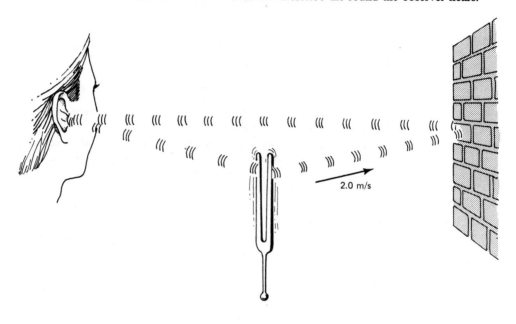

21
STATIC ELECTRICITY

21.1 ELECTRIC CHARGE

Most persons have experiences with static electricity early in their lives. Walking across a carpet on a dry day builds up a charge which discharges from the person when he or she touches an object or another person. A static charge can be built up on a comb or other plastic object by rubbing it with a cloth. The charged comb will

attract pieces of paper and other small objects. Some clothing, especially synthetic materials, come from the clothes dryer with a heavy static charge and crackle when straightened out. In fact, sparks can be seen as the clothes are discharged. If not discharged, they tend to cling to people wearing them. Trucks that carry flammable liquids prevent a buildup of static charge by dragging a chain on the pavement. Otherwise, a spark from a discharge could cause an explosion.

There are two different types of materials—one in which electric charges are free to move and the other in which the charges cannot migrate. Materials in which the charges remain in the specific region of the object where they are placed are called *insulators*. Materials where the charges are free to move are called *conductors* or *metals*. The negative charge is the more mobile charge. Throughout this discussion we will assume that the negative charge moves and the positive charge remains relatively fixed in conductors.

An electroscope can be used to illustrate some of the properties of electric charges. The electroscope consists of a metal body (ball with a stem), two gold leaves, and a case. The two gold leaves are very thin and delicate; they are very light and can be separated by small forces. A transparent case provides support for the electroscope body and protects the delicate gold leaves from convection air currents, which could easily damage them. The case is insulated from the body of the electroscope.

The electroscope can be used to demonstrate that like charges repel each other and that unlike charges attract each other. A charged rubber rod is brought into the vicinity of the ball of an electroscope. The negative charges on the rubber rod, an insulator, are locked in position; but, the negative charges on the electroscope, a conductor, are free to move. The free negative charges on the electroscope are repelled by the negatively charged rod and migrate to the gold leaves.

The gold leaves now repel each other and spread apart as in the preceding figure. If the ball of the electroscope is touched with a finger, negative charge will flow off the electroscope onto the finger, leaving the electroscope neutral except in the vicinity of the rubber rod, where the ball is positively charged because of the enforced shortage of negative charge in that region.

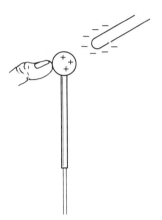

The gold leaves will hang down free of electrical forces, as shown above.

Next, the finger and then the rod are removed, leaving the electroscope positively charged or short of negative charge. The negative charge redistributes itself as equitably as possible over the metal electroscope, leaving a net positive charge along the entire surface of the electroscope. The gold leaves will spread apart as shown below. If the charged rubber rod is brought back into the vicinity of the electroscope, it drives the negative charges back into the gold leaves, neutralizing the positive charge, and the leaves come back together again.

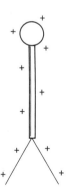

In the preceding illustration, two types of charges were used, positive and negative, and two types of materials, conductors and insulators. The unit of electrical charge in the SI system is the coulomb, C. The smallest value of charge possible is 1.6×10^{-19} C. The smallest value of negative charge has the same numerical value as the smallest positive charge and is referred to as the *charge* of an electron and is assigned the symbol *e*. The charge of an electron is negative, whereas the charge on an atomic nucleus (composed of protons and neutrons) is positive and an integer multiple of *e*. In fact, any charge, *q*, must be an integer multiple of *e*, or

$$q = ne$$

where *n* is an integer. Because of this property, the charge is said to be *quantized*.

We have seen that all matter is made up of atoms. These atoms are made of electons, protons, and neutrons. Each proton has one unit of positive charge, and each electron has one unit of negative charge. The neutron has no charge. The protons and neutrons are tightly packed into what is called the nucleus. Electrons may be thought of as small "planets" that orbit the nucleus.

An atom normally has the same number of electrons as protons and thus is uncharged. If an electron is removed, the atom is left with a positive charge. If an extra electron is added, the atom has a negative charge. When two materials are rubbed together, the atoms on the two surfaces move across each other and brush off electrons. The electrons are transferred from one surface to the other. One surface is then left with a positive charge and the other is negative. This is the process we call *electrification*.

Electrons are much less massive and hence more mobile than are atomic nuclei. Conductors have electrons that are not bound to individual atoms. These electrons are called conduction electrons and are free to move throughout the conductor much as the atoms of gas are free to move in a container.

21.2 COULOMB'S LAW

When two point electric charges are in the vicinity of each other, action and reaction forces exist between the charges. If the charges are like, either both positive or both negative, the forces are repulsive and act along the line joining the two charges, tending to spread the charges farther apart. If the charges are unlike, one negative and the other positive, the forces are attractive along the line between the charges, tending to pull the charges together.

The force between the two point charges is proportional to the product of their charges and inversely proportional to the square of the distance between the charges. This is called *Coulomb's law* and is expressed as

$$F = \frac{kq_1q_2}{r^2}$$

where *F* is the electrostatic force in newtons,
  q_1 and q_2 are the charges in coulombs, and
  r is the distance between the charges in metres.

Then k, the constant of proportionality, is given as

$$k = 9.00 \times 10^9 \text{ Nm}^2/\text{C}^2$$

This constant serves as a bridge between mechanical and electrical quantities and is written as

$$k = \frac{1}{4\pi\epsilon_0}$$

where ϵ_0 is called the permittivity constant and has a value of 8.85×10^{-12} C^2/Nm^2. This value is a measured value and establishes the relationship between electrical units and mechanical units.

Coulomb's law is given by the expression

$$F = \frac{1}{4\pi\epsilon_0} \frac{q_1 q_2}{r^2}$$

example 1 Two charges 8.0×10^{-7} C and -7.0×10^{-8} C are separated by a distance of 15 cm. What force exists between those charges?

DATA:
$$q_1 = 8.0 \times 10^{-7} \text{ C}$$

$$q_2 = -7.0 \times 10^{-8} \text{ C}$$

$$\frac{1}{4\pi\epsilon_0} = 9.00 \times 10^9 \text{ Nm}^2/\text{C}^2$$

$$r = 15 \text{ cm} = 0.15 \text{ m}$$

$$F = ?$$

BASIC EQUATION:
$$F = \frac{1}{4\pi\epsilon_0} \frac{q_1 q_2}{r^2}$$

WORKING EQUATION: same

SUBSTITUTION:
$$F = (9.00 \times 10^9 \text{ Nm}^2/\text{C}^2) \left[\frac{(8.0 \times 10^{-7} \text{ C})(-7.0 \times 10^{-8} \text{ C})}{(0.15 \text{ m})^2} \right]$$

$$\left(\frac{\text{Nm}^2}{\text{C}^2} \right) \left(\frac{(\text{C})(\text{C})}{\text{m}^2} \right) = \text{N}$$

$$= -2.2 \times 10^{-2} \text{ N}$$

(*Note:* The negative sign signifies that the force is attractive; that is, q_1 and q_2 are attracted toward each other.)

example 2 Find the repulsive force that exists between two protons separated by 5.0×10^{-15} m in a nucleus.

DATA:
$$q_1 = q_2 = 1.6 \times 10^{-19} \text{ C}$$
$$r = 5.0 \times 10^{-15} \text{ m}$$
$$\frac{1}{4\pi\epsilon_0} = 9.00 \times 10^9 \text{ Nm}^2/\text{C}^2$$
$$F = ?$$

BASIC EQUATION:
$$F = \frac{1}{4\pi\epsilon_0} \frac{q_1 q_2}{r^2}$$

WORKING EQUATION: same

SUBSTITUTION:
$$F = (9.00 \times 10^9 \text{ Nm}^2/\text{C}^2)\left[\frac{(1.6 \times 10^{-19} \text{ C})^2}{(5.0 \times 10^{-15} \text{ m})^2}\right]$$
$$= 9.2 \text{ N}$$

Note that this is a sizable force, and as most nuclei have more than two protons, we must conclude that there exists a very strong binding force in the nucleus which decreases quickly as the distance increases.

21.3   ELECTRIC FIELD OF A POINT CHARGE

A fixed electric charge produces a condition at each point in the space about it in such a way that if a positive test charge, q_0, were placed at a point, it would be acted upon by a force, F. This environment produced in space by the fixed charge is called an *electric field*, E, and may be expressed as

$$\boxed{E = \frac{F}{q}}$$

The electric field is a vector whose magnitude at each point in space is the force per unit charge. The field direction is the direction a positive charge would tend to move if placed at that point. The value of the field, E, at a radial distance, r, from the point charge, q, is

$$E = \frac{1}{4\pi\epsilon_0} \frac{q}{r^2}$$

and is directed radially away from the charge if the charge is positive and toward the charge if it is negatively charged.

example   Two charges, $q_1 = 8.0 \times 10^{-7}$ C and $q_2 = -7.0 \times 10^{-8}$ C, are separated by a distance of 15 cm.

(a) Find the field at q_2 due to q_1.
(b) Find the field at q_1 due to q_2.
(c) Find the forces on each charge.

(a) DATA:

$$q_1 = 8.0 \times 10^{-7} \, C$$

$$r = 0.15 \, m$$

$$\frac{1}{4\pi\epsilon_0} = 9.00 \times 10^9 \, Nm^2/C^2$$

$$E_1 = ?$$

BASIC EQUATION:

$$E = \frac{1}{4\pi\epsilon_0} \frac{q}{r^2}$$

WORKING EQUATION: same

SUBSTITUTION:

$$E_1 = (9.00 \times 10^9 \, Nm^2/C^2)\left[\frac{8.0 \times 10^{-7} C}{(0.15 \, m)^2}\right]$$

$$\boxed{\frac{Nm^2}{C^2} \cdot \frac{C}{m^2} = \frac{N}{C}}$$

$$= 3.2 \times 10^5 \, N/C \text{ away from the charge } q_1$$

(b) DATA:

$$q_2 = -7.0 \times 10^{-8} \, C$$

$$r = 0.15 \, m$$

$$\frac{1}{4\pi\epsilon_0} = 9.00 \times 10^9 \, Nm^2/C^2$$

$$E_2 = ?$$

BASIC EQUATION:

$$E = \frac{1}{4\pi\epsilon_0} \frac{q}{r^2}$$

WORKING EQUATION: same

SUBSTITUTION:

$$E_2 = (9.00 \times 10^9 \, Nm^2/C^2)\left[\frac{-7.0 \times 10^{-8} \, C}{(0.15 \, m)^2}\right]$$

$$= -2.8 \times 10^4 \, N/C \text{ toward the charge } q_2$$

(c) To find the force on q_2 due to the field of q_1:

DATA:

$$E_1 = 3.2 \times 10^5 \, N/C$$

$$q_2 = -7.0 \times 10^{-8} \, C$$

$$F_{21} = ?$$

BASIC EQUATION:

$$E = \frac{F}{q}$$

WORKING EQUATION: $F = qE$

SUBSTITUTION:

$$F_{21} = (-7.0 \times 10^{-8} \, C)(3.2 \times 10^5 \, N/C)$$

$$= -2.24 \times 10^{-2} \, N \text{ from } q_2 \text{ toward } q_1$$

To find the force on q_1 due to the field of q_2:

DATA:

$$E_2 = -2.8 \times 10^4 \text{ N/C}$$

$$q_1 = 8.0 \times 10^{-7} \text{ C}$$

$$F_{12} = ?$$

BASIC EQUATION:

$$E = \frac{F}{q}$$

WORKING EQUATION:

$$F = qE$$

SUBSTITUTION:

$$F_{12} = (8.0 \times 10^{-7} \text{ C})(-2.8 \times 10^4 \text{ N/C})$$

$$= -2.24 \times 10^{-2} \text{ N from } q_1 \text{ toward } q_2$$

Note: $\mathbf{F}_{12} = -\mathbf{F}_{21}$. These two forces are action–reaction forces.

21.4   LINES OF FORCE

Some visualization of electric fields can be made by use of lines of force. The positive charges would be considered as sources of the lines of force and negative charges as sinks of the lines of forces. The larger the charge, the greater the number of lines of force associated with it. The area density of the lines of forces corresponds to field strength; that is, the closer together the lines are, the stronger the forces on a charge placed in that region. The tangent to a force line at any point is the direction of the field. The figure below shows an isolated positive charge, serving as a source with lines of force emerging from it. The next figure below shows an isolated negative charge with lines of force converging into it or the charge is serving as a sink. The lower figure shows equal positive and negative charges separated by a short distance and isolated from other charges. Note that the lines of forces originate at $+q$ and terminate at $-q$ and at no place do they cross each other.

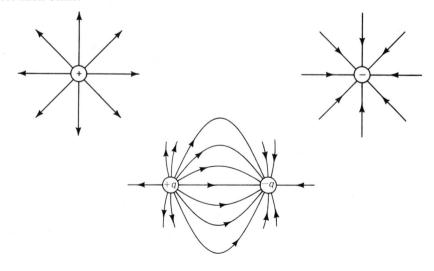

21.5 ELECTRIC FIELD OF A DISTRIBUTION OF CHARGES

If several different point charges are placed at several different points in space, then for any specific point in space, the field due to each of the several point charges can be calculated. The vector sum of the individual electric fields is the total electric field at that point.

example Two charges 6.0×10^{-8} C each are located on two vertices of an equilateral triangle 0.50 m on a side.

(a) What is the field at the third vertex?
(b) What force would act on a third charge of 6.0×10^{-8} C placed on this third vertex?

SKETCH:

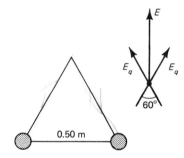

(a) DATA: The field due to a single charge:

$$q = 6.0 \times 10^{-8} \text{ C}$$

$$r = 0.50 \text{ m}$$

$$\frac{1}{4\pi\epsilon_0} = 9.00 \times 10^9 \text{ Nm}^2/\text{C}^2$$

$$E_q = ?$$

BASIC EQUATION:

$$E_q = \frac{1}{4\pi\epsilon_0} \frac{q}{r^2}$$

WORKING EQUATION: same

SUBSTITUTION:

$$E_q = (9.00 \times 10^9 \text{ Nm}^2/\text{C}^2) \left[\frac{6.0 \times 10^{-8} \text{ C}}{(0.50 \text{ m})^2} \right]$$

$$= 2.2 \times 10^3 \text{ N/C}$$

Referring to the figure, E is the vector sum of the E_q's.

$$E = 2E_q(\cos 30°) = 2(2.2 \times 10^3 \text{ N/C})(\cos 30°)$$

$$= 3.8 \times 10^3 \text{ N/C}$$

E is the field at the third vertex and it is directed away from the midpoint of the subtended side.

(b) DATA:

$$q = 6.0 \times 10^{-8} \, \text{C}$$

$$E = 3.8 \times 10^3 \, \text{N/C}$$

$$F = ?$$

BASIC EQUATION:

$$E = \frac{F}{q}$$

WORKING EQUATION:

$$F = qE$$

SUBSTITUTION:

$$F = (6.0 \times 10^{-8} \, \text{C})(3.8 \times 10^3 \, \text{N/C})$$

$$= 2.3 \times 10^{-4} \, \text{N away from the center of triangle}$$

In general, charges are not at discrete points but are distributed over an area. A metal spherical shell, pictured below, will have a charge uniformly distributed over its surface. It can be shown that for all space outside the uniformly charged sphere, the field will have radial symmetry and the same value as if the total charge were placed at the center of the shell. Inside the spherical shell, there will be no field at all. An infinite sheet of charge shown below has a uniform field perpendicular to the sheet.

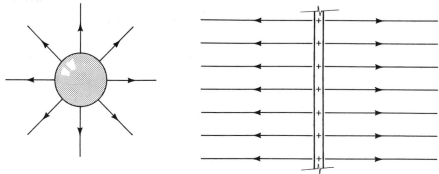

21.6 POTENTIAL DIFFERENCE

In a region of uniform field, such as is shown in the figure below, a point charge q_0 is moved from point A to point B by an external force. The amount of work done by this force is W_{AB}. The work by the external force in moving a unit charge from point A to point B is defined as the *potential difference* between A and B and may be expressed as

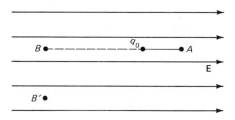

$$\boxed{V_B - V_A = \frac{W_{AB}}{q_0}} \qquad \boxed{V = \frac{J}{C}}$$

where the unit of potential difference $V_B - V_A$ is volts, W_{AB} is in joules, and q_0 is in coulombs. If, instead, the external force had moved the test charge q_0 from point A to point B', the potential difference

$$V_{B'} - V_A = \frac{W_{AB'}}{q_0}$$

But the amount of work necessary to move q_0 from A to B' is exactly the same as the work required to move the charge from A to B; that is,

$$W_{AB'} = W_{AB}$$

(see Section 7.3); consequently,

$$V_{B'} - V_A = V_B - V_A$$

or

$$V_{B'} = V_B$$

In fact, there will be families of surfaces perpendicular to the lines of force which are called *equipotential surfaces*. A charge can be moved from one place on the equipotential surface to another place without doing any work. Exactly the same work per unit charge is required to move a charge from any point on one equipotential surface to any point on a second equipotential surface. Equipotential surfaces are always perpendicular to the lines of forces, because if there were a line of force component in the plane of the surface, work would be required to move a charge.

A point charge is shown in the figure below. The lines of forces are symmetrical and uniformly distributed; thus, the equipotential surfaces are concentric spheres. The difference in potential between the sphere of radius r_B and a sphere of radius r_A is

$$V_B - V_A = \frac{q}{4\pi\varepsilon_0}\left(\frac{1}{r_B} - \frac{1}{r_A}\right)$$

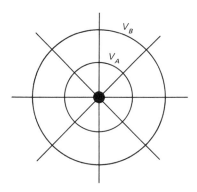

Only potential difference has been discussed and, indeed, it is only the difference that can be measured; however, it is convenient to speak of the potential at a point. If an arbitrary value of potential is assigned at one point, then the potential of all other points is defined. For the case of the point charge above, choose V_A to be zero where its field is zero: namely, where r_A approaches infinity. Drop the subscript B and then the expression for the potential at a distance r from the charge q is

$$V = \frac{1}{4\pi\epsilon_0}\frac{q}{r}$$

It should be noted that if the value of q is positive, then V is positive and work is needed to bring a positive charge to this point. Also, if all constraints were removed, the test charge would separate from the point charge. If, however, the value of q is negative, a positive charge is attracted to its charge and the two are bound.

Potentials are scalars. The potential at a point due to a distribution of charge is simply the algebraic sum of the potentials of the individual charges:

$$V = V_1 + V_2 + \cdots + V_n$$

example   Charges of 1.0×10^{-8} C, -3.0×10^{-8} C, and 2.0×10^{-8} C are located at successive corners of a $2\bar{0}$-cm square.

(a) What is the electric potential of each charge at the uncharged corner?
(b) What is the potential at the uncharged corner?

SKETCH:

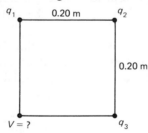

(a) DATA:

$q_1 = 1.0 \times 10^{-8}$ C

$r_1 = 0.20$ m

$q_2 = -3.0 \times 10^{-8}$ C

$r_2 = \sqrt{(0.20\text{ m})^2 + (0.20\text{ m})^2}$ (Pythagorean theorem)

$= 0.28$ m

$q_3 = 2.0 \times 10^{-8}$ C

$r_3 = 0.20$ m

$$\frac{1}{4\pi\epsilon_0} = 9.00 \times 10^9 \text{ Nm}^2/\text{C}^2$$

$$V_1 = ?$$

$$V_2 = ?$$

$$V_3 = ?$$

BASIC EQUATION: $\quad V = \dfrac{1}{4\pi\epsilon_0} \dfrac{q}{r}$

WORKING EQUATION: $\quad$ same

SUBSTITUTIONS: $\quad V_1 = 9.00 \times 10^9 \dfrac{\text{Nm}^2}{\text{C}^2} \left(\dfrac{1.0 \times 10^{-8} \text{ C}}{0.20 \text{ m}} \right)$

$$\boxed{\frac{\text{Nm}^2}{\text{C}^2} \cdot \frac{\text{C}}{\text{m}} = \frac{\text{Nm}}{\text{C}} = \frac{\text{J}}{\text{C}} = \text{V}}$$

$$= 450 \text{ V}$$

$$V_2 = 9.00 \times 10^9 \frac{\text{Nm}^2}{\text{C}^2} \left(\frac{-3.0 \times 10^{-8} \text{ C}}{0.28 \text{ m}} \right)$$

$$= -960 \text{V}$$

$$V_3 = 9.00 \times 10^9 \frac{\text{Nm}^2}{\text{C}^2} \left(\frac{2.0 \times 10^{-8} \text{ C}}{0.20 \text{ m}} \right)$$

$$= 9\overline{0}0 \text{ V}$$

(b) DATA: $\quad V_1 = 450 \text{ V} \qquad$ [from part (a)]

$$V_2 = -960 \text{ V}$$

$$V_3 = 9\overline{0}0 \text{ V}$$

BASIC EQUATION: $\quad V = V_1 + V_2 + \cdots + V_n$

WORKING EQUATION: $\quad$ same

SUBSTITUTION: $\quad V = 450 \text{ V} + (-960 \text{ V}) + 9\overline{0}0 \text{ V}$

$$= 390 \text{ V}$$

In the practical cases of an automobile or an instrument, a ground is used as an infinite source or sink of charge and is chosen as zero potential. Points in the circuit are given potential values which are the difference in potential between the point in the circuit and ground.

21.7 CAPACITANCE OF OBJECTS

A metal sphere with a charge, q, placed on it would have the same potential as the equipotential surface of radius r from a point charge q, namely:

$$V = \frac{1}{4\pi\epsilon_0} \frac{q}{r}$$

The constant $1/4\pi\epsilon_0$ is from Coulomb's law and r is a geometrical dimension (radius) of the sphere. We can write that the ratio of charge to the potential of a conducting sphere as a constant.

$$\frac{q}{V} = C \quad \text{(a constant)}$$

$$\frac{C}{V} = F$$

This constant C is defined as the capacitance of the sphere and has the unit farad, F. It should be noted that capacitance is a property of the sphere and depends only upon the sphere itself and is independent of whether the sphere is charged or not. The value of the capacitance of the sphere may be written

$$C = 4\pi\epsilon_0 r$$

In fact, any configuration of conductors may be considered as a capacitor and has a capacitance which is a property of that particular object. The capacitance of a number of different configurations may be calculated, but the value of a parallel-plate capacitor will be presented here and used in discussions whose results apply to capacitors in general.

    A parallel-plate capacitor is shown in the figure below.

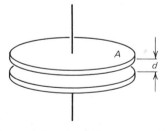

It consists of two conducting plates of surface area A separated uniformly by a distance d. The material between the plates is empty space or air and the capacitance of a parallel plate air capacitor is

$$C_0 = \frac{\epsilon_0 A}{d}$$

If some insulating material besides air is used between the plate, the capacitance of the arrangement is increased by a factor K and the capacitance is then

$$C = \frac{K\epsilon_0 A}{d}$$

The constant K is a property of the insulator between the conducting plates and is called the *dielectric constant* of the material. Even in this case the value of the capacitance, C, is a property of the object and depends upon the insulating material and geometry of the object.

Energy is required to charge a capacitor. As the charge on the capacitor increases, more work per unit charge is required to add more charge. A capacitor serves as a means to store energy. When a capacitor of C farads has been charged to a potential of V volts, the energy U stored in joules is

$$U = \tfrac{1}{2} C V^2$$

or, recalling that $q = CV$,

$$U = \tfrac{1}{2} C V^2 = \frac{\tfrac{1}{2} q^2}{C} = \tfrac{1}{2} q V$$

example

(a) If the earth were a metallic sphere, what would be the value of its capacitance?
(b) What potential would be required to place a 1.0-C charge on this capacitor?
(c) What energy would be stored on this charged capacitor?
   The radius of the earth is 6.4×10^6 m.

(a) DATA:

$$r = 6.4 \times 10^6 \text{ m}$$

$$4\pi\epsilon_0 = \frac{1}{9.00 \times 10^9} \left(\frac{C^2}{Nm^2} \right)$$

$$C = ?$$

BASIC EQUATION:

$$C = 4\pi\epsilon_0 r$$

WORKING EQUATION:

same

SUBSTITUTION:

$$C = \left(\frac{1}{9.00 \times 10^9} \frac{C^2}{Nm^2} \right) (6.4 \times 10^6 \text{ m})$$

$$\frac{C^2}{Nm^2} m = \frac{C^2}{Nm} = \frac{C^2}{J} = \frac{C}{\dfrac{J}{C}} = \frac{C}{V} = F$$

$$= 7.1 \times 10^{-4} \text{ F}$$

It should be noted that a capacitance of 1 F is a large unit. In practice, it is common to express the capacitance of an object in microfarads (1 $\mu F = 10^{-6}$ F) or in picofarads (1 pF = 10^{-12} F).

(b) DATA: $\qquad\qquad\qquad$ $C = 7.1 \times 10^{-4}$ F $\qquad$ [from part (a)]

$\qquad\qquad\qquad\qquad\qquad\qquad$ $q = 1.0$ C

$\qquad\qquad\qquad\qquad\qquad\qquad$ $V = ?$

BASIC EQUATION: $\qquad\qquad\qquad$ $q = CV$

WORKING EQUATION: $\qquad\qquad$ $V = \dfrac{q}{C}$

SUBSTITUTION: $\qquad\qquad\qquad$ $V = \dfrac{1.0 \text{ C}}{7.1 \times 10^{-4} \text{ F}}$ $\qquad$ $\boxed{\dfrac{\text{C}}{\text{F}} = \text{V}}$

$\qquad\qquad\qquad\qquad\qquad\qquad$ $= 1.4 \times 10^3$ V

We find a potential of 1400 V required to place a 1.0-C charge on a metallic sphere the size of the earth. An isolated static charge of 1.0 C is very large; however, we will see later that a flow of 1.0 C/s is easily achieved.

(c) DATA: $\qquad\qquad\qquad$ $C = 7.1 \times 10^{-4}$ F

$\qquad\qquad\qquad\qquad\qquad\qquad$ $q = 1.0$ C

$\qquad\qquad\qquad\qquad\qquad\qquad$ $U = ?$

BASIC EQUATION: $\qquad$ $U = \dfrac{1}{2}\dfrac{q^2}{C}$

WORKING EQUATION: same

SUBSTITUTION: $\qquad\qquad\qquad$ $U = \dfrac{1}{2}\dfrac{(1.0 \text{ C})^2}{7.1 \times 10^{-4} \text{ F}}$ $\qquad$ $\boxed{\dfrac{\text{C}^2}{\text{F}} = \dfrac{\text{C}^2}{\frac{\text{C}}{V}} = CV = \text{C}\cdot\dfrac{\text{J}}{\text{C}} = \text{J}}$

$\qquad\qquad\qquad\qquad\qquad\qquad$ $= 7.0 \times 10^2$ J

The energy of $7\overline{0}0$ J is small when one considers that an average electric iron dissipates 1000 J of energy each second.

21.8 CAPACITORS IN PARALLEL

When two or more capacitors are connected in such a manner that each capacitor lies between the same two common points, they are said to be *connected in parallel*. The figure on page 415 shows three capacitors of capacitances C_1, C_2, and C_3 connected in parallel; each capacitor lies between the common points A and B. When a potential difference is applied across the points A and B, the same potential difference exists across all three capacitors. The charge on each capacitor is

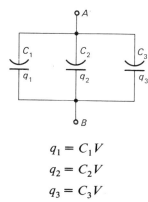

$$q_1 = C_1 V$$
$$q_2 = C_2 V$$
$$q_3 = C_3 V$$

where q_1, q_2, and q_3 are the charges on the respective capacitors,
$\quad$ C_1, C_2, and C_3 are the capacitances of the respective capacitors, and
$\qquad\qquad$ V is the voltage common to all the capacitors.

$\quad$ This parallel assembly of capacitors may be considered as a single capacitor with an equivalent capacitance C. Thus,

$$q = CV$$

where q is the total charge on the parallel assembly of capacitors,
$\quad$ C is the equivalent capacitance of the parallel assembly of capacitors, and
$\qquad$ V is the potential difference across this equivalent capacitance.

Also, the total charge on the assembly of capacitors is the sum of the individual charges.

$$q = q_1 + q_2 + q_3$$

where $\qquad$ q is the total charge on the equivalent capacitor and
q_1, q_2, and q_3 are the charges on the respective individual capacitors.
Combining these equations gives

$$CV = C_1 V + C_2 V + C_3 V$$

or

$$C = C_1 + C_2 + C_3$$

where $\qquad$ C is the equivalent capacitance of the parallel assembly of capacitors,
$\qquad\qquad$ and
C_1, C_2, and C_3 are the capacitances of the respective individual capacitors.
For a number of capacitors placed together in parallel, the generalization is

$$\boxed{\begin{aligned} C &= C_1 + C_2 + C_3 + \cdots \\ q &= q_1 + q_2 + q_3 + \cdots \\ V &= V_1 = V_2 = V_3 = \cdots \end{aligned}}$$

example $\quad$ Three capacitors of 2.0 μF, 3.0 μF, and 4.0 μF are connected in parallel. A potential difference of 24 V is applied to this parallel assembly.

(a) What is the equivalent capacitance of the three capacitors connected together?
(b) What is the charge on each capacitor?

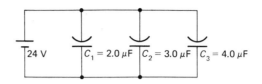

(a) DATA:

$$C_1 = 2.0 \times 10^{-6} \text{ F}$$

$$C_2 = 3.0 \times 10^{-6} \text{ F}$$

$$C_3 = 4.0 \times 10^{-6} \text{ F}$$

$$C = ?$$

BASIC EQUATION:

$$C = C_1 + C_2 + C_3$$

WORKING EQUATION: same

SUBSTITUTION:

$$C = 2.0 \times 10^{-6} \text{ F} + 3.0 \times 10^{-6} \text{ F} + 4.0 \times 10^{-6} \text{ F}$$

$$= 9.0 \times 10^{-6} \text{ F or } 9.0 \text{ } \mu\text{F}$$

(b) DATA:

$$C_1 = 2.0 \times 10^{-6} \text{ F}$$

$$C_2 = 3.0 \times 10^{-6} \text{ F}$$

$$C_3 = 4.0 \times 10^{-6} \text{ F}$$

$$V = 24 \text{ V}$$

$$q_1 = ?$$

$$q_2 = ?$$

$$q_3 = ?$$

BASIC EQUATION:

$$q = CV$$

WORKING EQUATION: same

SUBSTITUTIONS:

$$q_1 = (2.0 \times 10^{-6} \text{ F})(24 \text{ V}) = 4.8 \times 10^{-5} \text{ C}$$

$$q_2 = (3.0 \times 10^{-6} \text{ F})(24 \text{ V}) = 7.2 \times 10^{-5} \text{ C}$$

$$q_3 = (4.0 \times 10^{-6} \text{ F})(24 \text{ V}) = 9.6 \times 10^{-5} \text{ C}$$

21.9 CAPACITORS IN SERIES

When two or more capacitors are connected in such a manner that the capacitors form a single continuous path between two points, they are said to be *connected in series*. The figure below shows three capacitors of capacitances C_1, C_2, and C_3 connected in series; the three capacitors form a single continuous path from A to B.

When a potential difference exists across the points A and B, the capacitors are charged. No charge is created, but rather charge is separated. Referring to the figure below, a dashed box is shown which encloses both a plate from C_1 and a plate from C_2. When there is no potential difference between A and B, the capacitors are not charged and all points remain electrically neutral. When a

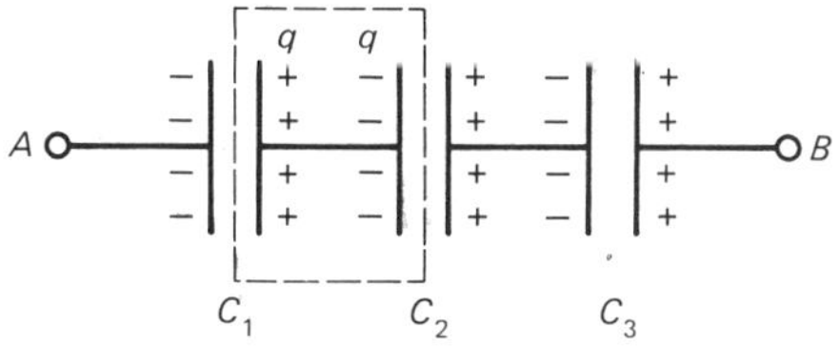

potential difference is maintained between A and B, the charge separates with $+q$ on the plate of C_1 and $-q$ on the plate of C_2. The values of the charges on the capacitors connected in series is the same for all capacitors and the potential difference across each is

$$V_1 = \frac{q}{C_1}$$

$$V_2 = \frac{q}{C_2}$$

$$V_3 = \frac{q}{C_3}$$

where q is the charge common to all the capacitors,
C_1, C_2, and C_3 are the capacitances of the respective capacitors, and
V_1, V_2, and V_3 are the potential difference across the respective capacitors.
The value of the potential difference across the assembly of series capacitors is

$$V = V_1 + V_2 + V_3$$

$$\frac{q}{C} = \frac{q}{C_1} + \frac{q}{C_2} + \frac{q}{C_3}$$

or

$$\frac{1}{C} = \frac{1}{C_1} + \frac{1}{C_2} + \frac{1}{C_3}$$

where C is the equivalent capacitance of the assembly of series capacitors.
For a number of capacitors placed together in series, the generalization is

$$\frac{1}{C} = \frac{1}{C_1} + \frac{1}{C_2} + \frac{1}{C_3} + \cdots$$

$$q = q_1 = q_2 = q_3 = \cdots$$

$$V = V_1 + V_2 + V_3 + \cdots$$

example   Three capacitors of 2.0 μF, 3.0 μF, and 4.0 μF are connected in series. A potential of 24 V is applied to this series assembly.

(a) What is the equivalent capacitance of the three capacitors connected together?
(b) What is the charge on each capacitor?
(c) What is the potential difference across each capacitor?

(a) DATA:

$$C_1 = 2.0 \times 10^{-6} \text{ F}$$

$$C_2 = 3.0 \times 10^{-6} \text{ F}$$

$$C_3 = 4.0 \times 10^{-6} \text{ F}$$

$$C = ?$$

BASIC EQUATION:

$$\frac{1}{C} = \frac{1}{C_1} + \frac{1}{C_2} + \frac{1}{C_3}$$

WORKING EQUATION   same

SUBSTITUTION:

$$\frac{1}{C} = \frac{1}{2.0 \times 10^{-6} \text{ F}} + \frac{1}{3.0 \times 10^{-6} \text{ F}} + \frac{1}{4.0 \times 10^{-6} \text{ F}}$$

Solve for C:

$$\frac{1}{C} = \frac{6}{12 \times 10^{-6} \text{ F}} + \frac{4}{12 \times 10^{-6} \text{ F}} + \frac{3}{12 \times 10^{-6} \text{ F}} = \frac{13}{12 \times 10^{-6} \text{ F}}$$

$$C = \frac{12 \times 10^{-6} \text{ F}}{13}$$

$$= 0.92 \ \mu\text{F}$$

(b) DATA:

$$C = 0.92 \times 10^{-6} \text{ F} \qquad \text{[from part (a)]}$$

$$V = 24 \text{ V}$$

$$q = q_1 = q_2 = q_3$$

BASIC EQUATION: $q = CV$

WORKING EQUATION: same

SUBSTITUTION:

$$q = (0.92 \times 10^{-6} \text{ F})(24 \text{ V})$$

$$q = 2.2 \times 10^{-5} \text{ C}$$

$$q_1 = q_2 = q_3 = 2.2 \times 10^{-5} \text{ C}$$

(c) DATA:

$$C_1 = 2.0 \times 10^{-6} \text{ F}$$

$$C_2 = 3.0 \times 10^{-6} \text{ F}$$

$$C_3 = 4.0 \times 10^{-6} \text{ F}$$

$$q = 2.2 \times 10^{-5} \text{ C} \qquad \text{[from part (b)]}$$

$$V_1 = ?, \ V_2 = ?, \ V_3 = ?$$

BASIC EQUATION: $q = CV$

WORKING EQUATION: $\quad V = \dfrac{q}{C}$

SUBSTITUTIONS: $\quad V_1 = \dfrac{2.2 \times 10^{-5}\,\text{C}}{2.0 \times 10^{-6}\,\text{F}} = 11\,\text{V}$

$$V_2 = \dfrac{2.2 \times 10^{-5}\,\text{C}}{3.0 \times 10^{-6}\,\text{F}} = 7.3\,\text{V}$$

$$V_3 = \dfrac{2.2 \times 10^{-5}\,\text{C}}{4.0 \times 10^{-6}\,\text{F}} = 5.5\,\text{V}$$

EXERCISES

Sketch	Data	Basic Equation	Working Equation	Substitution
12 cm² $h = ?$ $b = 6.0$ cm	$A = 12$ cm² $b = 6.0$ cm $h = ?$	$A = bh$	$h = \dfrac{A}{b}$	$h = \dfrac{12 \text{ cm}^2}{6.0 \text{ cm}}$ $= 2.0$ cm

1. How many electrons are present in 1.00 C of negative charge?

2. What is the net charge on an atom of copper that has two electrons removed from the neutral atom?

3. What is the charge on a negative chlorine ion?

4. One gram of copper contains 9.4×10^{21} atoms. Each atom has 29 electrons and 29 protons.

   (a) What is the total positive charge contained in the 1.0 g of copper?
   (b) What is the total negative charge contained in the 1.0 g of copper?
   (c) What is the net charge on the 1.0 g of copper?

5. Two point charges of 3.0×10^{-7} C and 5.0×10^{-7} C are placed 25 cm apart. What is the force on each charge?

6. (a) What is the force on a -6.0×10^{-8} C point charge placed 0.50 m from a 1.0×10^{-7} C charge?

   (b) What is the force on the 1.0×10^{-7} C charge?

7. Point charges of 2.0×10^{-6} C are placed on each of the vertices of an equilateral triangle 0.20 m on a side. What is the force on each charge?

8. A charge of -3.0×10^{-8} C is placed midway between two charges of 2.0×10^{-8} C and 4.0×10^{-8} C, which are separated by a distance of 1.00 m.

   (a) What is the force on the -3.0×10^{-8} C charge?
   (b) What is the force on the 2.0×10^{-8} C charge?

9. Two charges, $q_1 = 3.0 \times 10^{-8}$ C and $q_2 = -4.0 \times 10^{-8}$ C, are separated by a distance of 0.25 m.

   (a) Find the field at q_2 due to q_1.
   (b) Find the field at q_1 due to q_2.
   (c) Find the forces on each charge.

10. Charges of 2.0×10^{-9} C are placed on two vertices of an equilateral triangle that is 12 cm on a side.

(a) What is the electric field at the third vertex?

(b) What force would be exerted on an electron placed at the third vertex?

11. Charges of 1.0×10^{-8} C, -1.0×10^{-8} C, -2.0×10^{-8} C, and 2.0×10^{-8} C are placed at respective corners of a square 0.10 m on a side.

(a) What is the electric field at the center of the square?

(b) What force would be exerted on a -3.0×10^{-8} C charge placed at the center of the square?

12. Two charges of values, 7.0×10^{-8} C and 9.0×10^{-8} C, are separated by a distance of 1.0 m. At what point on a line adjoining the two is the electric field zero?

13. Two points, A and B, are 0.25 m and 0.35 m, respectively, from a point charge of 4.0×10^{-6} C.

(a) What is the potential difference between points A and B?

(b) Is it necessary for the position of the charge, point A, and point B to all lie in a straight line?

14. A rectangle $ABCD$ has sides $AB = 0.25$ m and $BC = 0.10$ m. Point charges are at corners A and B; $q_A = 2.0 \times 10^{-7}$ C and $q_B = 3.0 \times 10^{-7}$ C.

(a) What is the electric field at C?

(b) What is the electric field at D?

15. Refer to Exercise 14.

(a) What is the potential difference between points C and D?

(b) How much work would be required to move a charge of -2.0×10^{-7} C from point C to point D?

16. Charges of $+5.0 \times 10^{-9}$ C are placed at the corners of a square that is 0.25 m on a side.

(a) What is the value of the electric field at the center of the square?

(b) What is the potential at the center of the square?

(c) How much work would be required to place a 2.0×10^{-8} C charge at the center of the square?

17. An equilateral triangle, 0.35 m on a side, has charges of 1.0×10^{-7} C, -2.0×10^{-7} C, and 3.0×10^{-7} C at each vertex.

(a) What is the potential of each charge at the center of the triangle?

(b) What is the total potential at the center of the triangle?

18. How much work was required to assemble the charges to the charge configuration of Exercise 17?

19. (a) What is the capacitance of a metal ball of 0.10-m radius?

(b) What potential is required to place a charge of 1.0 C on this ball?

20. What charge would be placed on a 0.20-m-diameter metal sphere by an applied potential of $1\bar{0}0$ V?

21. A parallel-plate capacitor has two plates separated by 0.0010 m and a capacitance of 1.0 μF.

(a) What is the area of the plates?

(b) What charge must the capacitor have to have a potential of $5\bar{0}$ V?

(c) What is the energy of the charged capacitor?

22. Given three capacitors of 2.0 μF, 3.0 μF, and 4.0 μF, find the value of capacitors that can be constructed using these capacitors two at a time.

23. Given three capacitors of 2.0 μF, 3.0 μF, and 4.0 μF, find the value of capacitors that can be constructed using these capacitors three at a time.

24. Two capacitors of 5.0 μF and 10.0 μF are each charged to a potential difference of $5\bar{0}$ V.

   (a) What are the values of the charges on the individual capacitors? The two capacitors are then connected with their positive plates together and their negative plates together.

   (b) What is the equivalent capacitance of this assembly?

   (c) What is the charge on each capacitor?

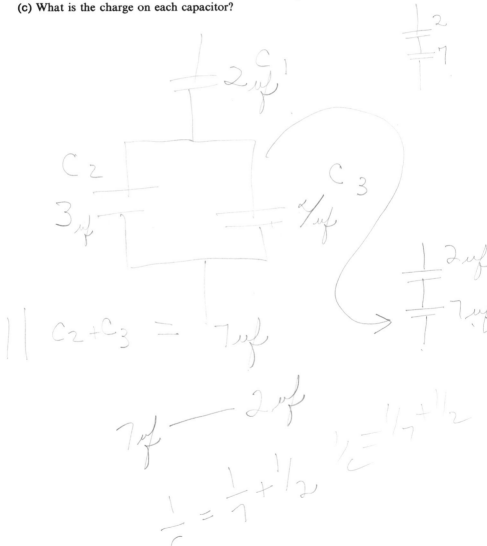

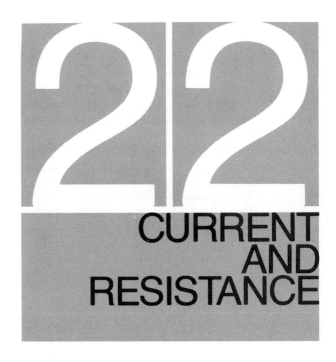

CURRENT AND RESISTANCE

22.1 CONDUCTORS

A conductor or metal is a material in which some electrical charge is free to move. The conductor is composed of an orderly array or lattice of atomic nuclei. Associated with each nucleus is a number of electrons which are bound to it and must remain in that spatial area. In addition, there are electrons that are free to move within the conductor and are not bound to any site. These electrons are called *conduction electrons* and have an average concentration of about one conduction electron per atom or less.

The conduction electrons are sometimes described as behaving like a gas and the conductor as acting like a container. The conduction electrons travel at very high speeds, of the order of 10^5 or 10^6 m/s, because of their thermal energies. In spite of the high speeds of these electrons, there is no net transfer of electrons within the conductor in the absence of any electrical fields. Their motion is random and they make collisions with other electrons and nuclei within the conductor, with the result that their paths are continually changing directions.

If a potential difference is placed between the ends of the conductor, an electric field is introduced within the conductor, which gives a directional acceleration to the electrons. This directional acceleration of the electrons causes a movement of charge in that direction. The movement of any electron is continually interrupted by collisions and its motion may well reverse from time to time, but the net result is a movement of charge at an average rate called the *drift velocity*. A mechanical analogy is a ball bouncing down a long flight of stairs. Its instantaneous velocity is continually changing, but over a period of time it has an average velocity toward the bottom of the stairs. The drift velocity of the electrons in a conductor is of the order of 10^{-4} m/s. This velocity is very slow compared with the thermal speeds of the electrons; yet because this drift velocity is directed, it gives a net transfer of charge. At this low speed, it is difficult to see how any large quantity of charge can flow until one realizes that the density of conduction electrons is of the order of 10^{28} electrons/m³ and that on the average each of these electrons moves with a drift velocity produced by the applied electric field.

22.2 CURRENT

A movement of electric charge is called a *current*. In a steady state the amount of charge, q, per unit of time, t, which flows through a cross section of the conductor is the current, I:

$$I = \frac{q}{t}$$

The unit of current is the ampere, A, and, of course, is 1 coulomb per second.

In discussing the motion of the charges, the electron flow has been continually used. Historically, before the microscopic nature of charges was understood, it was assumed that the charge carrier was positive and, of course, flowed from a high potential to a lower potential. This has become known as *conventional current flow*. As long as only microscopic properties are considered, the same result is obtained by a positive charge flowing from high potential to a lower potential as is obtained by a negative charge flowing from a low potential to a higher potential. To be consistent with the majority of related courses, conventional current flow will be assumed and used unless specifically stated as electron flow.

If the amount of charge per unit time through a cross section of the conductor is a function of time, then the current, I, is the average rate of current flow and has the units of amperes

$$I = \frac{\Delta q}{\Delta t} \qquad \boxed{A = \frac{C}{s}}$$

or coulombs per second.

22.3   RESISTANCE

When a current flows through a conductor, the charges lose energy or a potential difference V exists between the ends of the conductor. The *resistance*, R, of the conductor is defined by the relation

$$R = \frac{V}{I}$$

$$\Omega = \frac{V}{A}$$

where V is the potential drop in volts, and
$\qquad I$ is the current in amperes flowing through the conductor.
The unit of resistance is ohms and is abbreviated by the symbol Ω (Greek capital letter omega). If the object is a conductor, R is a constant and the object is said to obey Ohm's law. If the object is some electronic component such as a diode, the value of R for that diode may be different for different applied potentials. Resistance is a property of an object whether that object is in a circuit or on the shelf. A schematic representation of a resistor is shown below.

Consider a wire of conducting material. The resistance of the wire is given by the relation

$$R = \rho \frac{l}{A}$$

where R is the resistance of the wire, in ohms,
$\qquad l$ is the length of the wire,
$\qquad A$ is the cross-sectional area of the wire, and
$\qquad \rho$ (Greek lower case letter rho) is the resistivity in Ω m.
The term ρ is a property of the material of which the wire is constructed. The table

Resistivities of Some Common Materials

Material	ρ (Ω m)	α (per $C°$)
Aluminum	2.8×10^{-8}	3.9×10^{-3}
Constantan	49×10^{-8}	0.01×10^{-3}
Copper	1.72×10^{-8}	3.9×10^{-3}
Gold	2.44×10^{-8}	3.4×10^{-3}
Mercury	95.8×10^{-8}	0.89×10^{-3}
Nickel	3.8×10^{-8}	6.0×10^{-3}
Nichrome	100×10^{-8}	0.4×10^{-3}
Steel	$10 - 70 \times 10^{-8}$	—
Tungsten	5.6×10^{-8}	4.5×10^{-3}

shows the resistivity of several common metals and alloys. It should be noted that the resistance is a property of the object which depends upon the material of which the object is constructed and the geometry of the object.

The table also presents the temperature coefficient of resistivity:

$$\alpha = \frac{\Delta\rho}{\rho \, \Delta T}$$

where $\Delta\rho$ gives the relative change in resistivity per Celsius degree of temperature change.

example

(a) What is the resistance of 1.00 km of No. 12 copper wire at $2\bar{0}°C$?
(b) What is the resistance of this wire at $8\bar{0}°C$?
  The diameter of No. 12 wire is 2.05 mm.

(a) DATA:

$$l = 10\bar{0}0 \text{ m}$$

$$A = \frac{\pi d^2}{4} = \frac{\pi}{4}(2.05 \times 10^{-3} \text{ m})^2$$

$$= 3.30 \times 10^{-6} \text{ m}^2$$

$$\rho = 1.72 \times 10^{-8} \ \Omega \text{ m} \qquad \text{(from resistivity table)}$$

$$R = ?$$

BASIC EQUATION:

$$R = \rho \frac{l}{A}$$

WORKING EQUATION:

same

SUBSTITUTION:

$$R = (1.72 \times 10^{-8} \ \Omega \text{ m})\left(\frac{10\bar{0}0 \text{ m}}{3.30 \times 10^{-6} \text{ m}^2}\right)$$

$$= 5.21 \ \Omega$$

(b) DATA:

$$\rho = 1.72 \times 10^{-8} \ \Omega \text{ m}$$

$$\Delta T = 8\bar{0}°C - 2\bar{0}°C = 6\bar{0} \ C°$$

$$\alpha = 3.9 \times 10^{-3}/C°$$

$$\Delta\rho = ?$$

BASIC EQUATION:

$$\alpha = \frac{\Delta\rho}{\rho \, \Delta T}$$

WORKING EQUATION:

$$\Delta\rho = \alpha\rho \, \Delta T$$

SUBSTITUTION:

$$\Delta\rho = (3.9 \times 10^{-3}/C°)(1.72 \times 10^{-8} \ \Omega \text{ m})(6\bar{0} C°)$$

$$= 4.0 \times 10^{-9} \ \Omega \text{ m}$$

Then

$$\rho_T = 1.72 \times 10^{-8} \ \Omega \text{ m} + 4.0 \times 10^{-9} \ \Omega \text{ m}$$

$$= 2.12 \times 10^{-8} \ \Omega \text{ m}$$

The thermal changes in the dimensions of the wire are small compared to the change in resistivity and may be ignored in this exercise. Then the resistance varies directly as the resistivity.

DATA:

$$R = 5.21 \ \Omega \qquad \text{[from part (a)]}$$

$$\rho = 1.72 \times 10^{-8} \ \Omega \ \text{m}$$

$$\rho_T = 2.12 \times 10^{-8} \ \Omega \ \text{m}$$

$$R_T = ?$$

BASIC EQUATION:

$$\frac{R_T}{R} = \frac{\rho_T}{\rho}$$

WORKING EQUATION:

$$R_T = \frac{\rho_T}{\rho} R$$

SUBSTITUTION:

$$R_T = \left(\frac{2.12 \times 10^{-8} \ \Omega \ \text{m}}{1.72 \times 10^{-8} \ \Omega \ \text{m}} \right) (5.21 \ \Omega)$$

$$= 6.42 \ \Omega$$

22.4   POWER

In Chapter 21, the potential difference between two points was defined as the energy in joules required to move 1 C of charge from one point to the other. In current electricity, there must be a concern for the rate at which this energy loss occurs. Instead of work, consider the power required to produce a charge flow rate or current:

$$V = \frac{\Delta W}{\Delta q} = \frac{\Delta W / \Delta t}{\Delta q / \Delta t} = \frac{P}{I}$$

or

$$\boxed{P = IV}$$

where P is the power in watts,
   I is the current in amperes, and
   V is the voltage in volts.
Using the definition of resistance ($R = V/I$), we have

$$P = IV = I^2 R = \left(\frac{V}{R} \right)^2 R = \frac{V^2}{R}$$

example   An electric iron is rated at $15\overline{0}0$ W when used on a $12\overline{0}$-V line. Household lines are alternating current, but for this purpose may be considered to have a constant voltage. What is the resistance of the heating element of the iron?

DATA:

$$P = 15\overline{0}0 \text{ W}$$

$$V = 12\overline{0} \text{ V}$$

$$R = ?$$

BASIC EQUATION:

$$P = \frac{V^2}{R}$$

WORKING EQUATION:

$$R = \frac{V^2}{P}$$

SUBSTITUTION:

$$R = \frac{(12\overline{0} \text{ V})^2}{15\overline{0}0 \text{ W}}$$

$$= 9.60 \; \Omega$$

EXERCISES

Sketch	Data	Basic Equation	Working Equation	Substitution
$A = 12 \text{ cm}^2$, $h = ?$, $b = 6.0 \text{ cm}$	$A = 12 \text{ cm}^2$ $b = 6.0 \text{ cm}$ $h = ?$	$A = bh$	$h = \dfrac{A}{b}$	$h = \dfrac{12 \text{ cm}^2}{6.0 \text{ cm}}$ $= 2.0 \text{ cm}$

1. What is the resistance of an aluminum wire 2.59 mm in diameter and 375 m in length?

2. What is the resistance of 632 m of No. 12 copper wire?

3. A copper bar has a cross-sectional area of $1.00 \times 10^{-4} \text{ m}^2$. What is the resistance per unit length of the bar?

4. What length of aluminum wire has the same resistance as $10\overline{0}0$ m of copper wire of the same diameter?

5. At what temperature would the resistance of a copper wire be 50 percent greater than its resistance at $2\overline{0}°C$?

6. A copper wire has a resistance of 7.6 Ω at $2\overline{0}°C$. What is the resistance of this wire at $87°C$?

7. A heating element is designed to provide $10\overline{0}$ W on a $12\overline{0}$-V line. What is the resistance of the element?

8. What current must flow through a $1\overline{0}$-Ω resistor to produce $10\overline{0}$ W?

9. What current flows through a $4\overline{0}$-W light bulb in a 120-V circuit?

10. A wire coil of tungsten requires $5\overline{0}$ W when operated from a $5\overline{0}$-V source at $2\overline{0}°C$. What power will this coil require when operated from the same source at a temperature of $27\overline{0}°C$?

23

DC CURRENT

23.1 SOURCES OF ELECTROMOTIVE FORCE

There exists a class of objects: cells, batteries, generators, thermopiles, alternators, and so on, which supply energy to electric charges. These objects are called sources of *electromotive force* (emf). A current passing through a source of emf receives an increase of energy just as a current flowing through a resistor loses energy. This gain in energy per unit charge given by a source of emf is expressed in volts and is called a *gain of potential*. The loss in energy per unit of charge dissipated by a resistor is expressed in volts and is called *loss of potential* or *voltage drop*.

The circuit diagram is the most common and useful way to show a circuit. Note how each component (part) of the figure is represented by its symbol in the circuit diagram in its relative position.

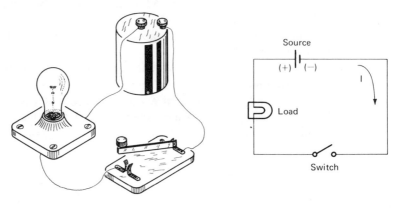

Picture diagram   Circuit diagram

The light bulb may be represented as a resistance. The circuit diagram would then appear:

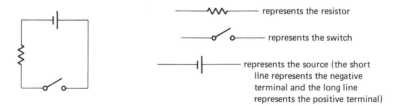

An analogy may be drawn between this simple electric circuit and a closed water system. This circuit consists of a source of emf (dry cell) and a resistance or load (light bulb). Let us compare this electrical circuit with a garden water display shown below.

The water display has a pump and an ornamental waterwheel. The water is raised by the pump to a height from which it falls down a cascade and turns the wheel. The water pump and the dry cell play analogous roles. The pump provides an increase in pressure to the water, giving it an increased energy per unit volume. The dry cell provides an emf to the electric charge, giving it an increased energy per unit charge. The ornamental garden display has the water flowing down a cascade, turning a waterwheel. This takes energy from the water and results in a drop in the pressure of the water. The light bulb heats to incandescence, taking energy from the electric charge, which results in a voltage drop. The water pump did not produce water, but rather supplied energy to the water—the dry cell did not produce charge, but rather supplied energy to the charge. The waterwheel did not consume water, but rather took energy from it. The light bulb did not consume electric charge but rather took energy from it. In both cases, no matter was created or destroyed; rather, energy was taken from a source and dissipated through a load.

We will adopt the convention throughout this book that electric current flows from the higher potential to the lower potential just as water flows from a high pressure to a lower pressure.

23.2   EQUILIBRIUM CURRENT FLOW

When the current is flowing through the light circuit, an equilibrium condition exists. The dry cell is furnishing energy at the same rate that the bulb is dissipating energy. The power (Section 22.5) supplied by the battery is $\mathscr{E}I$ and the power dissipated by the bulb is I^2R, so the condition for equilibrium is

$$\mathscr{E}I = I^2R$$

or

$$\boxed{\mathscr{E} = IR}$$

where $\mathscr{E}$ is the emf of the source,
   R is the total resistance of the circuit, and
   I is the equilibrium value of the current in that circuit.

This relationship says that the emf supplied by the source is just equal to the potential drop across the load and is a condition for a steady current flow.

A source of emf in addition to the energy it supplies to the current also has a resistance. This resistance is a property of the source itself and, in general, is small. In the circuit diagram below

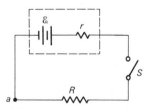

r is the internal resistance of the cell. Starting at a and going clockwise around the circuit and recording each potential difference as we pass through, calling it positive if it is a source, negative if it is a drop, gives

$$\mathscr{E} - Ir - IR = 0$$

The sum of the potential differences must be zero because we have returned to the same point in the circuit.

$$\mathscr{E} - Ir = IR$$

where $\mathscr{E}$ is the emf of the source,

r is the internal resistance of the source,
R is the resistance of the load, and
I is the total current.

The quantity IR is the voltage drop across the load, and it is numerically equal to $\mathcal{E} - Ir$. This last quantity is the applied voltage, whereas our source had an emf of $\mathcal{E}$, a potential drop of Ir occurs inside the source itself.

$$V = \mathcal{E} - Ir$$

23.3 CIRCUIT DIAGRAMS

In order to communicate about problems in electricity, technicians have developed a "language" of their own. It is a picture language using symbols and diagrams. Some of the most often used symbols appear below and on page 432.

ELECTRICAL SYMBOLS

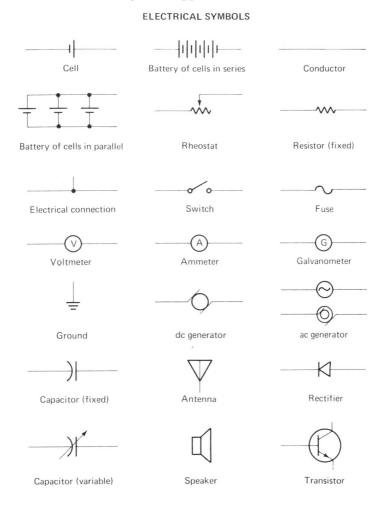

Cell	Battery of cells in series	Conductor
Battery of cells in parallel	Rheostat	Resistor (fixed)
Electrical connection	Switch	Fuse
Voltmeter	Ammeter	Galvanometer
Ground	dc generator	ac generator
Capacitor (fixed)	Antenna	Rectifier
Capacitor (variable)	Speaker	Transistor

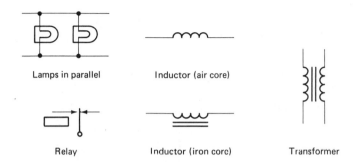

Lamps in parallel   Inductor (air core)

Relay   Inductor (iron core)   Transformer

example   A 6.5-V battery has an internal resistance of 1.0 Ω and is connected to a 5.5-Ω load.

(a) What is the current through the circuit?
(b) What is the voltage drop across the load?
(c) What is the applied voltage?

(a) SKETCH:

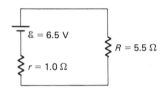

DATA:

$$\mathcal{E} = 6.5 \text{ V}$$
$$r = 1.0 \text{ } \Omega$$
$$R = 5.5 \text{ } \Omega$$
$$I = ?$$

BASIC EQUATION: $\mathcal{E} - Ir = IR$

WORKING EQUATION: $I = \dfrac{\mathcal{E}}{R + r}$

SUBSTITUTION: $I = \dfrac{6.5 \text{ V}}{5.5 \text{ } \Omega + 1.0 \text{ } \Omega}$

$$= 1.0 \text{ A} \qquad \boxed{\dfrac{\text{V}}{\Omega} = \text{A}}$$

(b) DATA: $I = 1.0 \text{ A}$ [from part (a)]

$$R = 5.5 \text{ } \Omega$$
$$V = ?$$

BASIC EQUATION: $V = IR$

WORKING EQUATION: same

SUBSTITUTION: $V = (1.0 \text{ A})(5.5 \ \Omega)$

$= 5.5 \text{ V}$

(c) DATA: $\mathcal{E} = 6.5 \text{ V}$

$I = 1.0 \text{ A}$

$r = 1.0 \ \Omega$

BASIC EQUATION: $V = \mathcal{E} - Ir$

WORKING EQUATION: same

SUBSTITUTION: $V = 6.5 \text{ V} - (1.0 \text{ A})(1.0 \ \Omega)$

$= 5.5 \text{ V}$

The applied voltage and the voltage drop across the load are the same and would be measured between the same two points of the circuit.

23.4   CIRCUITS CONTAINING RESISTANCE

A circuit may have several resistors connected together in series to form the load. When resistors are connected in series, the current has only one path to flow through, and the same current must flow through each of the resistors.

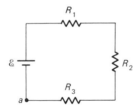

$$I = I_1 = I_2 = I_3 = \cdots$$

where I is the total current,

I_1 is the current through R_1,

I_2 is the current through R_2, and

I_3 is the current through R_3.

Starting at point a in the figure above and listing our potential differences—positive for a source, negative for a load—give

$$\mathcal{E} - V_1 - V_2 - V_3 = 0$$
$$\mathcal{E} - IR_1 - IR_2 - IR_3 = 0$$
$$\mathcal{E} = I(R_1 + R_2 + R_3)$$

where $\mathcal{E}$ is the emf of the source,
   V_1 is the voltage drop sum R_1,
   V_2 is the voltage drop sum R_2,
   V_3 is the voltage drop sum R_3, and
   I is the current through the circuit.

From these expressions it is seen that the total resistance of the circuit is the sum of the resistances of the individual resistors:

$$R = R_1 + R_2 + R_3 + \cdots$$

The resistors connected in series R_1, R_2, R_3, ... can be replaced in the circuit by a single equivalent resistance R. The equivalent resistance is greater than the largest resistance of the component resistors. The current through each component resistor is the same, and the potential drop across the equivalent resistor is the sum of the voltage drops across the individual resistors.

Characteristics of series circuits

Voltage	$\mathcal{E} = V_1 + V_2 + V_3 + \cdots$
Current	$I = I_1 = I_2 = I_3 = \cdots$
Resistance	$R = R_1 + R_2 + R_3 + \cdots$

---

example 1   Resistors of $3\bar{0}\ \Omega$, $4\bar{0}\ \Omega$, and $5\bar{0}\ \Omega$ are connected in series to a 12-V battery with negligible internal resistance.

(a) What is the resistance of the circuit?
(b) What is the current through each resistor?
(c) What is the potential drop across each resistor?

(a) SKETCH:

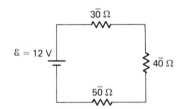

DATA:
$$R_1 = 3\bar{0}\ \Omega$$
$$R_2 = 4\bar{0}\ \Omega$$
$$R_3 = 5\bar{0}\ \Omega$$
$$R = ?$$

BASIC EQUATION: $R = R_1 + R_2 + R_3$

WORKING EQUATION: same

SUBSTITUTION:	$R = 3\bar{0}\ \Omega + 4\bar{0}\ \Omega + 5\bar{0}\ \Omega$
	$R = 12\bar{0}\ \Omega$
(b) DATA:	$R = 12\bar{0}\ \Omega$ [from part (a)]
	$\mathcal{E} = 12\ \text{V}$
	$I_1 = ?$
	$I_2 = ?$
	$I_3 = ?$
BASIC EQUATIONS:	$\mathcal{E} = IR$
	$I = I_1 = I_2 = I_3$
WORKING EQUATION:	$I_1 = I_2 = I_3 = \dfrac{\mathcal{E}}{R}$
SUBSTITUTION:	$I_1 = I_2 = I_3 = \dfrac{12\ \text{V}}{12\bar{0}\ \Omega}$
	$= 0.10\ \text{A}$
(c) DATA:	$R_1 = 3\bar{0}\ \Omega$
	$R_2 = 4\bar{0}\ \Omega$
	$R_3 = 5\bar{0}\ \Omega$
	$I_1 = I_2 = I_3 = 0.10\ \text{A}$ [from part (b)]
	$V_1 = ?$
	$V_2 = ?$
	$V_3 = ?$
BASIC EQUATION:	$V = IR$
WORKING EQUATION:	same
SUBSTITUTION:	$V_1 = (0.10\ \text{A})(3\bar{0}\ \Omega) = 3.0\ \text{V}$
	$V_2 = (0.10\ \text{A})(4\bar{0}\ \Omega) = 4.0\ \text{V}$
	$V_3 = (0.10\ \text{A})(5\bar{0}\ \Omega) = 5.0\ \text{V}$

An electrical circuit with more than one path for the current to flow is a parallel circuit.

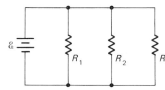

The current in a parallel circuit is divided among the branches of the circuit. The current from the source is equal to the sum of the currents through the branches.

$$I = I_1 + I_2 + I_3 + \cdots$$

where I is the total current in the circuit,
   I_1 is the current through R_1,
   I_2 is the current through R_2, and
   I_3 is the current through R_3.

Regardless of which branch is considered, the emf of the source is just equal to the voltage drop of that branch. It follows that the voltage drops across the branches are equal.

$$\mathcal{E} = V_1 = V_2 = V_3 = \cdots$$

where $\mathcal{E}$ is the emf of the source,
   V_1 is the voltage drop across R_1,
   V_2 is the voltage drop across R_2, and
   V_3 is the voltage drop across R_3.

These two conditions combine to give us the relationship among resistors connected in parallel.

$$\frac{1}{R} = \frac{1}{R_1} + \frac{1}{R_2} + \frac{1}{R_3} + \cdots$$

where R is the equivalent resistance,
   R_1 is the resistance of R_1,
   R_2 is the resistance of R_2, and
   R_3 is the resistance of R_3.

It should be noted that where resistors are connected in parallel, the equivalent resistance of the combination is less than the resistance of the least of the component resistors. The resistor with the lowest resistance will have the highest current flow through it, but the voltage drop across each resistor is the same.

Characteristics of parallel circuits	
Voltage	$\mathcal{E} = V_1 = V_2 = V_3 = \cdots$
Current	$I = I_1 + I_2 + I_3 + \cdots$
Resistance	$\frac{1}{R} = \frac{1}{R_1} + \frac{1}{R_2} + \frac{1}{R_3} + \cdots$

example 2   Resistors of $3\bar{0}$ Ω, $4\bar{0}$ Ω, and $5\bar{0}$ Ω are connected in parallel to a 12-V battery with negligible internal resistance.

(a) What is the resistance of the circuit?

(b) What is the current through each resistor?

(c) What is the potential drop across each resistor?

(a) SKETCH:

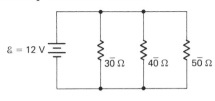

DATA:
$$R_1 = 3\bar{0} \ \Omega$$
$$R_2 = 4\bar{0} \ \Omega$$
$$R_3 = 5\bar{0} \ \Omega$$
$$R = ?$$

BASIC EQUATION:
$$\frac{1}{R} = \frac{1}{R_1} + \frac{1}{R_2} + \frac{1}{R_3}$$

WORKING EQUATION: same

SUBSTITUTION:
$$\frac{1}{R} = \frac{1}{3\bar{0} \ \Omega} + \frac{1}{4\bar{0} \ \Omega} + \frac{1}{5\bar{0} \ \Omega}$$
$$R = 13 \ \Omega$$

(b) DATA:
$$\mathcal{E} = 12 \ \text{V}$$
$$R_1 = 3\bar{0} \ \Omega$$
$$R_2 = 4\bar{0} \ \Omega$$
$$R_3 = 5\bar{0} \ \Omega$$
$$I_1 = ?$$
$$I_2 = ?$$
$$I_3 = ?$$

BASIC EQUATION:
$$I = \frac{\mathcal{E}}{R}$$

WORKING EQUATION: same

SUBSTITUTIONS:
$$I_1 = \frac{12 \ \text{V}}{3\bar{0} \ \Omega} = 0.40 \ \text{A}$$
$$I_2 = \frac{12 \ \text{V}}{4\bar{0} \ \Omega} = 0.30 \ \text{A}$$
$$I_3 = \frac{12 \ \text{V}}{5\bar{0} \ \Omega} = 0.24 \ \text{A}$$

(c) DATA:

$$\mathcal{E} = 12 \text{ V}$$

$$V_1 = ?$$

$$V_2 = ?$$

$$V_3 = ?$$

BASIC EQUATION:

$$\mathcal{E} = V_1 = V_2 = V_3$$

WORKING EQUATION:

$$V_1 = V_2 = V_3 = 12 \text{ V}$$

Many circuits are a combination of series and parallel branches. To solve these circuits, it is necessary to consider parts of the circuit that are all parallel or all series and replace combinations of resistors with their equivalent until we have reduced the circuit to one equivalent resistance.

example 3   In the circuit below, $\mathcal{E} = 24$ V, $R_1 = 1\bar{0}$ Ω, $R_2 = 2\bar{0}$ Ω, $R_3 = 3\bar{0}$ Ω, and $R_4 = 4\bar{0}$ Ω.

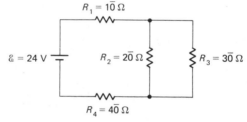

$$R_1 = 1\bar{0} \text{ Ω}$$

$$\mathcal{E} = 24 \text{ V} \qquad R_2 = 2\bar{0} \text{ Ω} \qquad R_3 = 3\bar{0} \text{ Ω}$$

$$R_4 = 4\bar{0} \text{ Ω}$$

(a) What is the equivalent resistance of the circuit?

(b) What is the current through each resistor?

(a) First, let us find the equivalent resistance, R_{23}, of R_2 and R_3, which are in parallel, then the resistance of R_1, R_{23}, R_4, which are all in series.

DATA:

$$R_2 = 2\bar{0} \text{ Ω}$$

$$R_3 = 3\bar{0} \text{ Ω}$$

$$R_{23} = ?$$

BASIC EQUATION:

$$\frac{1}{R} = \frac{1}{R_1} + \frac{1}{R_2}$$

WORKING EQUATION: same

SUBSTITUTION:

$$\frac{1}{R_{23}} = \frac{1}{2\bar{0} \text{ Ω}} + \frac{1}{3\bar{0} \text{ Ω}}$$

$$R_{23} = 12 \text{ Ω}$$

DATA:

$$R_1 = 1\bar{0} \text{ Ω}$$

$$R_{23} = 12 \text{ Ω} \qquad \text{[from part 1 of (a)]}$$

$$R_4 = 4\bar{0} \text{ Ω}$$

$$R = ?$$

BASIC EQUATION:	$R = R_1 + R_2 + R_3$
WORKING EQUATION:	same
SUBSTITUTION:	$R = 1\bar{0}\ \Omega + 12\ \Omega + 4\bar{0}\ \Omega$
	$= 62\ \Omega$

(b) The current through R_1 and R_4 is the same as the current through the system.

DATA:	$\mathcal{E} = 24$ V
	$R = 62\ \Omega$
	$I_1 = ?$
	$I_4 = ?$
BASIC EQUATION:	$\mathcal{E} = IR$
	$I = I_1 = I_4$
WORKING EQUATION:	$I_1 = I_4 = \dfrac{\mathcal{E}}{R}$
SUBSTITUTION:	$I_1 = I_4 = \dfrac{24\text{ V}}{62\ \Omega}$
	$= 0.39$ A

To find the currents through R_2 and R_3, first find the potential difference across these resistors then the currents through each.

DATA:	$R_{23} = 12\ \Omega$ [from part (a)]
	$I_{23} = 0.39$ A
	$V_2 = V_3 = ?$
BASIC EQUATION:	$V = IR$
WORKING EQUATION:	same
SUBSTITUTION:	$V_2 = V_3 = (0.39\text{ A})(12\ \Omega)$
	$= 4.7$ V
DATA:	$R_2 = 2\bar{0}\ \Omega$
	$R_3 = 3\bar{0}\ \Omega$
	$V_2 = V_3 = 4.7$ V
	$I_2 = ?$
	$I_3 = ?$
BASIC EQUATION:	$V = IR$
WORKING EQUATION:	$I = \dfrac{V}{R}$

SUBSTITUTIONS:
$$I_2 = \frac{4.7 \text{ V}}{20 \text{ }\Omega} = 0.24 \text{ A}$$

$$I_3 = \frac{4.7 \text{ V}}{30 \text{ }\Omega} = 0.16 \text{ A}$$

Circuits such as the one shown below cannot be reduced to a simple source and resistance circuit. These circuits are called *multiloop circuits* and can be solved by an extension of the principles of the last section and are called Kirchhoff's laws.

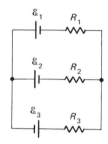

1. The sum of the voltages around any closed continuous loop must be zero.
2. The sum of the currents at any junction must be zero.

If the loop is traversed in the direction of a source of emf, that value of emf is taken as positive. If the loop is traversed in a direction opposite to the direction of the emf, that value of emf is taken to be negative. If a loop is traversed in the direction of an assumed current, the value of the voltage drop across a resistor is $-IR$. If a loop is traversed in the opposite direction of an assumed current, the value of the voltage drop across a resistor is $+IR$.

example 4 Find the current through each resistor in the circuit shown.

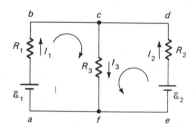

Traversing the loop *abcf* gives the following relation:
$$\mathcal{E}_1 - I_1 R_1 - I_3 R_3 = 0$$
Traversing the loop *edcf* gives the following relation:
$$\mathcal{E}_2 - I_2 R_2 - I_3 R_3 = 0$$
Summing the currents at junction *c* gives
$$I_1 + I_2 - I_3 = 0$$

These three equations may be written as follows:

$$I_1 R_1 + I_3 R_3 = \mathcal{E}_1$$

$$I_2 R_2 + I_3 R_3 = \mathcal{E}_2$$

$$I_1 + I_2 - I_3 = 0$$

This is a set of three independent equations in three unknowns. Their solution is

$$I_1 = \frac{\mathcal{E}_1 R_2 + \mathcal{E}_1 R_3 - \mathcal{E}_2 R_3}{R_1 R_2 + R_1 R_3 + R_2 R_3}$$

$$I_2 = \frac{\mathcal{E}_2 R_1 + \mathcal{E}_2 R_3 - \mathcal{E}_1 R_3}{R_1 R_2 + R_1 R_3 + R_2 R_3}$$

$$I_3 = \frac{R_1 \mathcal{E}_2 + R_2 \mathcal{E}_1}{R_1 R_2 + R_1 R_3 + R_2 R_3}$$

example 5 Find the value of the currents in Example 4 when $\mathcal{E}_1 = 9.00$ V, $\mathcal{E}_2 = 6.00$ V, $R_1 = 5.00 \ \Omega$, $R_2 = 10.0 \ \Omega$, and $R_3 = 15.0 \ \Omega$.

DATA:

$$\mathcal{E}_1 = 9.00 \text{ V}$$

$$\mathcal{E}_2 = 6.00 \text{ V}$$

$$R_1 = 5.00 \ \Omega$$

$$R_2 = 10.0 \ \Omega$$

$$R_3 = 15.0 \ \Omega$$

$$I_1 = ?$$

$$I_2 = ?$$

$$I_3 = ?$$

Derived equation from Example 4:

WORKING EQUATION: $I_1 = \dfrac{\mathcal{E}_1 R_2 + \mathcal{E}_1 R_3 - \mathcal{E}_2 R_3}{R_1 R_2 + R_1 R_3 + R_2 R_3}$

SUBSTITUTION: $I_1 = \dfrac{(9.00 \text{ V})(10.0 \ \Omega) + (9.00 \text{ V})(15.0 \ \Omega) - (6.00 \text{ V})(15.0 \ \Omega)}{(5.00 \ \Omega)(10.0 \ \Omega) + (5.00 \ \Omega)(15.0 \ \Omega) + (10.0 \ \Omega)(15.0 \ \Omega)}$

$$= 0.491 \text{ A}$$

Derived equation from Example 4:

WORKING EQUATION: $I_2 = \dfrac{\mathcal{E}_2 R_1 + \mathcal{E}_2 R_3 - \mathcal{E}_1 R_3}{R_1 R_2 + R_1 R_3 + R_2 R_3}$

SUBSTITUTION: $I_2 = \dfrac{(6.00 \text{ V})(5.00 \ \Omega) + (6.00 \text{ V})(15.0 \ \Omega) - (9.00 \text{ V})(15.0 \ \Omega)}{(5.00 \ \Omega)(10.0 \ \Omega) + (5.00 \ \Omega)(15.0 \ \Omega) + (10.0 \ \Omega)(15.0 \ \Omega)}$

$$= -0.055 \text{ A}$$

The negative sign on the value of the current, I_2, means that the assumption of the direction of the current flow was incorrect for the values assigned to the sources of

emf and resistors in this exercise. The current, I_2, will flow against the direction of $\mathcal{E}_2$.
Derived equation from Example 4:

WORKING EQUATION: $I_3 = \dfrac{R_1\mathcal{E}_2 + R_2\mathcal{E}_1}{R_1R_2 + R_1R_3 + R_2R_3}$

SUBSTITUTION: $I_3 = \dfrac{(5.00\ \Omega)(6.00\ \text{V}) + (10.0\ \Omega)(9.00\ \text{V})}{(5.00\ \Omega)(10.0\ \Omega) + (5.00\ \Omega)(15.0\ \Omega) + (10.0\ \Omega)(15.0\ \Omega)}$

$= 0.436\ \text{A}$

23.5 *RC* CIRCUITS

An *RC* circuit consists of a resistance, *R*, and a capacitor, *C*, in series. The figure below shows an *RC* circuit with a battery and a switch. The capacitor can be charged and discharged. When the switch is closed, a current flows clockwise,

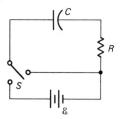

charging the capacitor. During the time that the capacitor is being charged, the potential difference around the circuit is given by

$$\mathcal{E} - IR - \frac{q}{C} = 0$$

where $\mathcal{E}$ is the emf of the battery,
  I is the value of the current flowing in the circuit (this current is a function of time),
  R is the resistance in the circuit,
  C is the capacitance of the circuit, and
  q is the charge collected on the capacitor (the charge, *q*, is a function of time).

The solution of this expression is the result of a differential equation

$$q = C\mathcal{E}(1 - e^{-t/RC})$$
$$I = \frac{\mathcal{E}}{R}e^{-t/RC}$$

where t is the independent variable time,
   $C\mathcal{E}$ is the equilibrium charge on the capacitor as $t \to \infty$, and
   RC is called the time constant of the circuit and is a property of the circuit.

   The figure below shows graphs of both q and I as a function of time. The charge, q, on the capacitor starts at zero and builds up exponentially to $C\mathcal{E}$. The current, I, starts off as if no capacitor were in the circuit, then decays exponentially to zero as the capacitor charges.

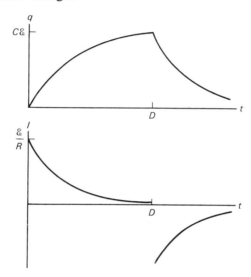

   When the switch is thrown to discharge the capacitor, the sum of potential differences becomes

$$IR + \frac{q}{C} = 0$$

which has a solution

$$q = C\mathcal{E}e^{-t/RC}$$
$$I = \frac{\mathcal{E}}{R}e^{t/RC}$$

where t is the independent variable time,
   C is the value of the capacitor,
   $\mathcal{E}$ is the potential across the capacitor at time $t = D$,
   R is the resistance of the circuit, and
   $\mathcal{E}/R$ represents the value of the current at the time $t = D$.

Referring to the figure, D on the time axis is the instant that the switch is closed for discharge. The charge on the capacitor discharges exponentially while the current starts at a maximum value and decays exponentially to zero. As is expected, the current flow on discharge is in the opposite direction to the current flow during charging. When the capacitor is charging, the voltage developed across it is in opposition to the emf of the source and opposing the current. When the capacitor is discharging, the current flow is reversed and is sustained by the voltage on the capacitor.

example   An *RC* circuit has a capacitor of 0.050 μF and a resistance of 11,000 Ω. The capacitor is charged to 15 V.

(a) What is the value of the time constant?
(b) At what time after the discharge begins will the voltage across the capacitor be 3.0 V?

(a) DATA:

$$R = 11,000 \ \Omega$$

$$C = 0.050 \times 10^{-6} \ \text{F}$$

time constant = ?

BASIC EQUATION:

time constant = *RC*

WORKING EQUATION:

same

SUBSTITUTION:

$$RC = (11,000 \ \Omega)(0.050 \times 10^{-6} \ \text{F})$$

$$\boxed{\Omega\text{F} = \frac{\text{V}}{\text{A}} \times \frac{\text{C}}{\text{V}} = \frac{\text{C}}{\text{A}} = \frac{\text{As}}{\text{A}} = \text{s}}$$

$$= 5.5 \times 10^{-4} \ \text{s}$$

(b) DATA:

$$\mathscr{E} = 15 \ \text{V}$$

$$C = 0.050 \times 10^{-6} \ \text{F}$$

$$RC = 5.5 \times 10^{-4} \ \text{s} \qquad \text{[from part (a)]}$$

$$V = \frac{q}{C} = 3.0 \ \text{V}$$

$$t = ?$$

BASIC EQUATION:

$$q = C\mathscr{E}e^{-t/RC}$$

WORKING EQUATION:

$$\frac{q}{C} = \mathscr{E}e^{-t/RC}$$

SUBSTITUTION:

$$3.0 \ \text{V} = (15 \ \text{V})e^{-t/5.5 \times 10^{-4} \text{s}}$$

Dividing both sides by 15 V, we have

SOLVE:

$$0.20 = e^{-t/(5.5 \times 10^{-4})}$$

Now, take the natural logarithm of both sides:

$$\ln(0.20) = \frac{-t}{5.5 \times 10^{-4}}$$

$$t = (-5.5 \times 10^{-4})[\ln(0.20)]$$

$$= 8.9 \times 10^{-4} \text{ s}$$

EXERCISES

Sketch	Data	Basic Equation	Working Equation	Substitution
12 cm² $h = ?$ $b = 6.0$ cm	$A = 12$ cm² $b = 6.0$ cm $h = ?$	$A = bh$	$h = \dfrac{A}{b}$	$h = \dfrac{12 \text{ cm}^2}{6.0 \text{ cm}}$ $= 2.0$ cm

1. Three resistors of 10.0 Ω, 16.0 Ω, and 22.0 Ω are connected in series with a 24.0-V battery.

    (a) What is the total resistance of the circuit?

    (b) What current flows through each resistor?

    (c) What is the potential drop across each resistor?

2. Four resistors of 18.0 Ω, 22.0 Ω, 26.0 Ω, and 30.0 Ω are connected in series with a 12.0-V battery.

    (a) What is the total resistance of the system?

    (b) What current flows through each resistor?

    (c) What is the potential difference across each resistor?

3. Resistors of 5.00 Ω, 6.00 Ω, and 7.00 Ω are connected in series.

    (a) What emf must be applied to produce a current of 0.750 A?

    (b) What is the resistance of the circuit?

    (c) What is the voltage drop across each resistor?

For each circuit fill in the appropriate numbers in the spaces provided.

4.

	V	I	R
Battery	V	A	24.0 Ω
R_1	V	0.50 A	6.0 Ω
R_2	V	A	8.0 Ω
R_3	V	A	Ω

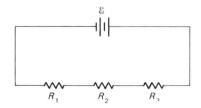

5.

	V	I	R
Battery	12.0 V	A	Ω
R_1	4.0 V	A	Ω
R_2	6.0 V	A	Ω
R_3	V	0.50 A	Ω

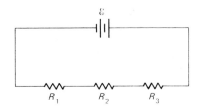

6. Two resistors of $1\bar{0}$ Ω and 15 Ω are connected in parallel to a 5.0-V source of emf.

  (a) What current flows through each?

  (b) What is the potential difference across each?

  (c) What is the total resistance of the circuit?

7. Four resistors are connected in parallel. Their resistances are 10.0 Ω, 15.0 Ω, 20.0 Ω, and 30.0 Ω.

  (a) What is their total resistance?

  (b) What potential must be applied for a 0.750-A current to flow in the 10.0-Ω resistor?

  (c) What current flows in the 15.0-Ω resistor?

8. Three resistors of 4.00 Ω, 6.00 Ω, and 12.0 Ω are connected in parallel. A current of 0.200 A flows through the 12.0-Ω resistor.

  (a) What current flows through the 4.00-Ω resistor?

  (b) What current flows through the 6.00-Ω resistor?

  (c) What is the total resistance of the three resistors?

For each circuit fill in the appropriate numbers in the spaces provided.

9.

	V	I	R
Battery	V	3.00 A	Ω
R_1	V	A	Ω
R_2	V	1.50 A	18.0 Ω
R_3	V	0.50 A	Ω

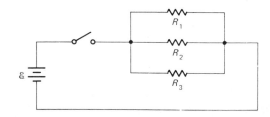

10.

	V	I	R
Battery	10.0 V	1.000 A	Ω
R_1	V	0.400 A	Ω
R_2	5.00 V	A	Ω
R_3	2.00 V	A	Ω
R_4	V	0.600 A	Ω

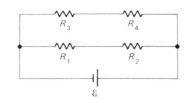

11.

	V	I	R
Battery	24.0 V	A	Ω
R_1	8.00 V	A	Ω
R_2	V	4.00 A	Ω
R_3	V	2.00 A	Ω

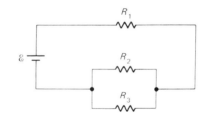

12.

	V	I	R
Battery	V	A	Ω
R_1	12.0 V	A	2.00 Ω
R_2	V	A	4.00 Ω
R_3	24.0 V	A	4.00 Ω
R_4	V	A	8.00 Ω

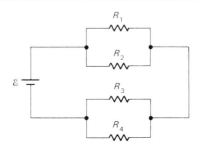

13.

	V	I	R
Battery	30.0 V	A	Ω
R_1	6.00 V	3.00 A	Ω
R_2	V	2.00 A	Ω
R_3	V	A	3.00 Ω
R_4	V	1.00 A	Ω
R_5	8.00 V	A	Ω
R_6	V	A	Ω

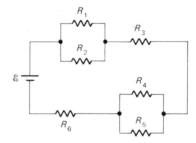

14. A 12.0-V battery has an internal resistance of 1.00 Ω and is connected to a 5.00-Ω load.

(a) What current flows in the circuit?

(b) What is the applied voltage?

15. A 6.00-V battery produces an applied voltage of 5.50 V when connected to an 11.0-Ω load.

(a) What is the current through the battery?

(b) What is the internal resistance of the battery?

16. An amplifier is a source of 4.00 V potential and 8.00 Ω internal resistance. A speaker is connected to this source.

(a) Construct a graph of the current through the speaker as a function of the speaker's resistance. (Do not exceed 20.0-Ω speaker resistance.)

(b) Construct a graph of the voltage drop across the speaker as a function of its resistance.

(c) Multiply the corresponding values from the graphs of parts (a) and (b). This new graph gives a plot of power output of the speaker as a function of its resistance.

(d) How does the value of the resistance of the speaker for maximum value of power output compare with the resistance of the amplifier? Is there an advantage to matching the amplifier and speaker in a system?

17. Find the current through each resistor in the circuit shown. $\mathcal{E}_1 = 8.00$ V, $\mathcal{E}_2 = 6.00$ V, $R_1 = 10.0$ Ω, $R_2 = 20.0$ Ω, and $R_3 = 30.0$ Ω.

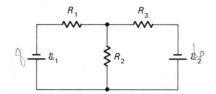

18. Find the current through each resistor in the circuit shown. $\mathcal{E}_1 = 12.0$ V, $\mathcal{E}_2 = 8.0$ V, $\mathcal{E}_3 = 6.0$ V, $R_1 = 5\bar{0}$ Ω, $R_2 = 75$ Ω, and $R_3 = 10\bar{0}$ Ω.

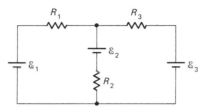

19. (a) What is the potential difference $V_b - V_a$?
   (b) What is the emf of $\mathcal{E}_1$?
   (c) What is the emf of $\mathcal{E}_2$?

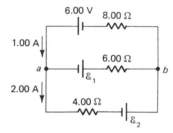

20. (a) What is the time constant of an *RC* circuit where $R = 10,\overline{0}00$ Ω and $C = 3.50 \times 10^{-6}$ F?

   (b) What is the time required for the charge on the capacitor to reach one-half of its maximum value?

21. An *RC* circuit with $C = 0.50$ μF and $R = 3.00 \times 10^5$ Ω is connected to a 45-V battery.

   (a) What time is required for the capacitor to receive half of its maximum charge?
   (b) What time is required for the capacitor to receive half of its maximum energy?

22. In the circuit shown, $C = 2.0 \times 10^{-7}$ F, $R_1 = 5\bar{0}00$ Ω, $R_2 = 5\bar{0}00$ Ω, and $\mathcal{E} = 25$ V.

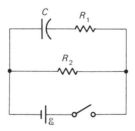

   (a) When the switch is first closed ($t = 0$), what currents flow through R_1 and R_2?
   (b) After the switch has been closed a long time, what currents flow through R_1 and R_2?
   (c) After the switch has been closed for a long time, it is opened. What currents flow through R_1 and R_2 when the switch is just opened?
   (d) What currents flow through R_1 and R_2 a long time after the switch has been opened?

DC SOURCES

24.1 THE LEAD STORAGE CELL

Lead storage batteries are used in automobiles and many other types of vehicles and machinery. A battery is a group of cells connected together. These lead cells are *secondary cells*, which means they are rechargeable. The passing of an electric current through the cell to restore the original chemicals is called recharging. Cells, such as the dry cell, which cannot be efficiently recharged are called *primary cells*.

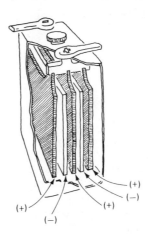

Lead storage cells are made up of two kinds of lead plates (lead and lead oxide) submerged in a solution of distilled water and sulfuric acid. This acid solution is called an *electrolyte*.

Acid solution

The chemical action between the lead plates and the acid solution produces large numbers of free electrons at the negative (−) pole of the battery. These electrons have a large amount of electrical potential energy which is used in the load in the circuit (for instance, to operate headlights or to turn a starter motor).

As the electrical energy is used in the load, the battery must be recharged. This is done by a generator or an alternator. Such devices provide an electric current to reverse the chemical reaction taking place in the battery. The recharging process extends the life of the battery, which would otherwise be very short.

24.2 THE DRY CELL

The *dry cell* is the most widely used primary cell. This is the kind of cell we use in flashlights and portable radios. A dry cell is made of a carbon rod, which is the positive (+) terminal or pole, and a zinc can, which acts as the negative (−) terminal. In between is a paste of chemicals and water which reacts with the terminals to provide energized electrons. These cells are available in a wide range of sizes. Common sizes are usually 1.5 to 9 V.

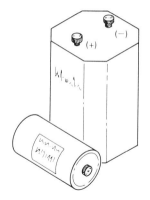

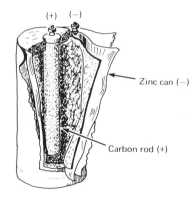

Zinc can (−)

Carbon rod (+)

The dry cell, as well as the lead cell, has resistance within the cell itself which opposes the movement of the electrons from the (+) to (−) poles. This is called the *internal resistance*, *r*, of the cell. Because current flows in the cell, the emf of the cell is reduced by the voltage drop across the internal resistance. The voltage applied to the external circuit is then

$$V = \mathcal{E} - Ir$$

where V is the voltage applied to circuit,
$\mathcal{E}$ is the emf of the cell,
I is the current through cell, and
r is the internal resistance of cell.

Often, the current or voltage available from a single cell is inadequate to do a particular job. Then we usually connect two or more cells in series or parallel.

24.3   CELLS IN SERIES

To connect cells in series, the positive terminal of one is connected to the negative terminal of the next cell. This procedure is continued until the desired number of cells are all connected (see figure below).

The rules for cells connected in series and parallel are similar to those for simple resistances.

CELLS IN SERIES

1. The current in the circuit is the same as in any single cell:

$$I = I_1 = I_2 = I_3 = \cdots$$

2. The internal resistance of the battery is equal to the sum of the individual internal resistances of the cells:

$$r = r_1 + r_2 + r_3 + \cdots$$

3. The emf of the battery is equal to the sum of the emf's of the individual cells:

$$\mathcal{E} = \mathcal{E}_1 + \mathcal{E}_2 + \mathcal{E}_3 + \cdots$$

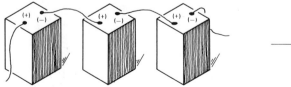

Series connected dry cells   Circuit diagram

example   Two 6.00-V cells with internal resistance of 0.100 Ω each are connected in series to form a battery with a current of 0.750 A in each cell.

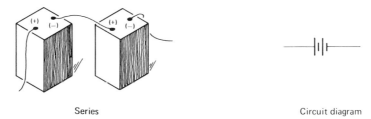

Series

Circuit diagram

(a) What is the emf of the battery?
(b) What is the internal resistance of the battery?
(c) What is the current in the external circuit?

(a) $\mathcal{E} = \mathcal{E}_1 + \mathcal{E}_2 = 6.00 \text{ V} + 6.00 \text{ V} = 12.00 \text{ V}$ (Rule 3)
(b) $r = r_1 + r_2 = 0.100 \ \Omega + 0.100 \ \Omega = 0.200 \ \Omega$ (Rule 2)
(c) 0.750 A   (Rule 1)

24.4   CELLS IN PARALLEL

To connect cells in parallel, the positive terminals of all the cells are connected together and the negative terminals are all connected together. The leads from the external circuit may be connected to any positive and negative terminals. (The external circuit is all the circuit *outside* the battery or cell.)

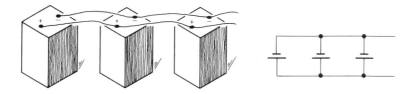

CELLS IN PARALLEL

1.   The total current is the sum of the individual currents in each cell:

$$I = I_1 + I_2 + I_3 + \cdots$$

2.   The internal resistance is equal to the resistance of one cell divided by the number of cells:

$$r = \frac{r \text{ of one cell}}{\text{number of cells}}$$

3.   The emf of the battery is equal to the emf of any single cell:

$$\mathcal{E} = \mathcal{E}_1 = \mathcal{E}_2 = \mathcal{E}_3 = \cdots$$

example Four cells, each 1.50 V and internal resistance of 0.0500 Ω, are connected in parallel to form a battery with a current output of 0.25 A in each cell.

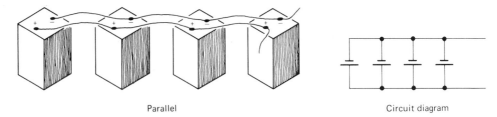

Parallel   Circuit diagram

(a) What is the emf of the battery?
(b) What is the internal resistance of the battery?
(c) What is the current in the external circuit?

(a) 1.50 V   (Rule 3)

(b) $\dfrac{r \text{ of one cell}}{\text{no. of cells}} = \dfrac{0.0500 \ \Omega}{4} = 0.0125 \ \Omega$ (Rule 2)

(c) $I = I_1 + I_2 + I_3 + I_4 = 0.25 \ \text{A} + 0.25 \ \text{A} + 0.25 \ \text{A} + 0.25 \ \text{A} = 1.00 \ \text{A}$ (Rule 1)

EXERCISES

Sketch	Data	Basic Equation	Working Equation	Substitution
$A = 12 \ \text{cm}^2$, $h = ?$, $b = 6.0 \ \text{cm}$	$A = 12 \ \text{cm}^2$ $b = 6.0 \ \text{cm}$ $h = ?$	$A = bh$	$h = \dfrac{A}{b}$	$h = \dfrac{12 \ \text{cm}^2}{6.0 \ \text{cm}}$ $= 2.0 \ \text{cm}$

1. A cell has an emf of 1.50 V and an internal resistance of 0.050 Ω. If there is 0.30 A in the cell, what voltage is applied to the external circuit?

2. The voltage applied to a circuit is 11.8 V when the current through the battery is 0.500 A. If the internal resistance of the battery is 0.300 Ω, what is the emf of the battery?

3. Three 1.50-V cells, each with internal resistance of 0.0500 Ω, are connected in series to form a battery with a current of 0.850 A in each cell.

   (a) What is the current in the external circuit?
   (b) What is the emf of the battery?
   (c) What is the internal resistance of the battery?

4. Five 9.00-V cells with internal resistance of 0.100 Ω each are connected in parallel to form a battery with a current output of 0.750 A in a certain circuit.

   (a) What is the current in the external circuit?
   (b) What is the emf of the battery?
   (c) What is the internal resistance of the battery?

5. What is the current in the circuit below?

$\mathcal{E}_1 = 1.50 \ \text{V}$

$R_1 = 7.40 \ \Omega$

$\mathcal{E}_2 = 1.50 \ \text{V}$

6. Find the current in the circuit below.

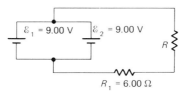

7. If the current in the circuit below is 1.20 A, what is the value of R?

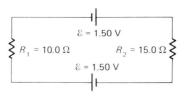

8. Find the current in the circuit below.

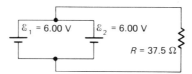

25.1 MAGNETS IN NATURE

Some stones found in nature tend to align their long axes in a north-south direction. It has long been observed that these stones repel and attract each other. They are called *magnets* and were used as navigational aids and sources of amusement and magic by early civilizations.

Electricity has also been known for years, but it was only in the early nineteenth century that Hans Christian Oersted discovered that magnetism and electricity are interwoven. Following Oersted's discovery, the knowledge of electricity and magnetism expanded until by the middle of the nineteenth century the basic laws of electromagnetic behavior were well formulated.

25.2 MAGNETIC POLES

Many different shapes of magnets are available, constructed from iron-based alloys and ceramic materials. Regardless of their shapes, these magnets have one thing in common. Each magnet seems to have regions that act as strong sources of attraction or repulsion for other magnets. These regions are called *poles*. If a magnet in the shape of a bar is suspended by a thread, it will tend to align itself along a geographic north-south line. The pole of the magnet pointing in the geographic north direction is called a north-seeking pole or simply north pole (N), and conversely the pole in the end of the magnet pointing south is called a south pole (S). If several bar magnets have had their N and S poles identified, it is quickly verified that like poles repel each other and unlike poles attract each other.

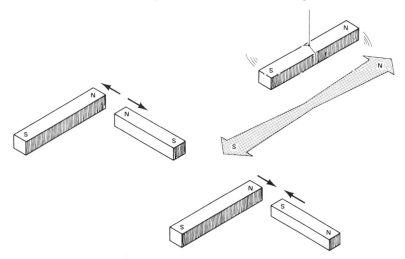

Magnetic poles always come in pairs. There is no way that a single magnetic pole can be isolated. Each time a bar magnet is halved, each half has both a N and S pole. The existence of this N-S pair separated by a distance leads to a description of the magnet in terms of magnetic moment, since a torque will be produced when the magnet is placed near another magnet.

25.3   MAGNETIC FIELD STRENGTH

Oersted's observation that a compass (bar magnet) was affected by passing a current through a nearby wire led to the discovery that a current produced a magnetic field. The field near a long current-carrying wire is circular about the wire and given by *Ampère's law*:

$$B = \frac{\mu_0 I}{2\pi r}$$

where B is the magnetic field, in teslas (T),
   I is the current through the wire,
   r is the perpendicular distance from the center of the wire, and
   $\mu_0 = 4\pi \times 10^{-7}$ Tm/A.

The magnetic field, B, has units of teslas (T) and is defined in terms of electric current by the constant μ_0, the permeability constant (1T = 1N/mA). The value of

μ_0 is not experimentally determined but is an assigned value that explicitly defines magnetic field in terms of electric current. The direction of the magnetic field about a conducting wire can be determined by holding the conductor in your right hand with your thumb extended in the direction of current flow. Your bent fingers point in the direction of the magnetic field.

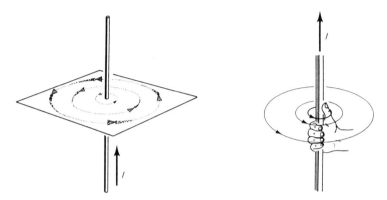

The magnetic field is a vector quantity and must be added by vector methods.

example   A power line carrying $35\bar{0}$ A is 7.50 m above a transit used by a surveying student.

(a) What is the magnetic field because of the power-line current above the transit?
(b) If the earth's horizontal component of magnet field is 5.20×10^{-5} T at that location, what error could be introduced in the student's angular measurement? (Assume the power line to run north-south.)

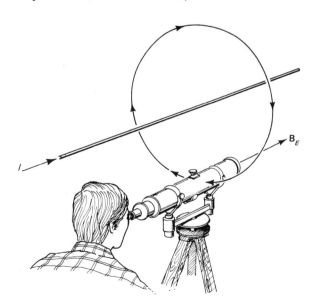

(a) DATA:

$$I = 35\bar{0} \text{ A}$$

$$r = 7.50 \text{ m}$$

$$\mu_0 = 4\pi \times 10^{-7} \text{ Tm/A}$$

$$B = ?$$

BASIC EQUATION:

$$B = \frac{\mu_0 I}{2\pi r}$$

WORKING EQUATION: same

SUBSTITUTION:

$$B = \frac{(4\pi \times 10^{-7} \text{ Tm/A})(35\bar{0} \text{ A})}{2\pi(7.50 \text{ m})}$$

$$= 9.33 \times 10^{-6} \text{ T}$$

With the current from south to north, the rule of thumb shows the direction of **B** to be east to west.

(b) The angle resultant vector—the earth's field plus the wire's field ($\mathbf{B}_E + \mathbf{B}$)—makes with $\mathbf{B}_E$ would be the angular error.

DATA:

$$B_x = B_E = 5.20 \times 10^{-5} \text{ T} \quad \text{Earth's component}$$

$$B_y = B = 9.33 \times 10^{-6} \text{ T} \quad \text{[from part (a)]}$$

$$\theta = ?$$

BASIC EQUATION:

$$\tan \theta = \frac{B_y}{B_x}$$

WORKING EQUATION: same

SUBSTITUTION:

$$\tan \theta = \frac{9.33 \times 10^{-6} \text{ T}}{5.20 \times 10^{-5} \text{ T}} = 0.1794$$

$$\theta = 10.2°$$

The bearing on the surveying student's transit could be in error by 10.2° because of the power line. This example was based on steady-flow direct current and most power lines carry alternating current.

25.4   MAGNETIC DIPOLE

When a wire is formed into a closed loop or circle, the magnetic field at the center of the loop is the vector sum of the magnetic field produced by each increment of the circle circumference. The value of the field at the center of the single circular loop is

$$B = \frac{\mu_0 I}{2r}$$

where B is the magnetic field at the center of the circle,
    I is the current in the loop,
    μ_0 is the permeability constant, and
    r is the radius of the current loop.
The magnetic field in the region of the current loop is shown below.

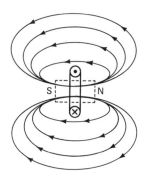

This page bisects the loop so that half of the ring is below the page, half above. The current is shown leaving the page at the top and entering at the bottom. A check by rule of thumb shows that the magnetic field goes from left to right through the loop. Please note that the field pattern is what one would expect from a short bar magnet as shown in phantom (dotted) in the drawing. Indeed, a loop of wire is a magnetic dipole. The lines of force are continuous and do not start or end, so there are no sinks or sources of magnetic field as there were with electric fields. In fact, in a bar magnet the lines of magnetic field continue through the bar magnet from the "S pole" to the "N pole." In the bar magnet we saw that each time it was halved, new "poles" appeared. If we were able to continue halving the size of the bar, we would eventually reach a single magnetic atom whose magnetic moment is produced by unpaired electrons. The bar magnet itself is the vector sum of these atomic magnets and the "poles" seemed to exist because of the concentration of field within the bar magnet. The concept of poles was convenient to use in describing magnetic fields, but poles do not exist as sources or sinks of magnetic fields. The magnetic moment of a coil of wire of N turns is given by

$$\boxed{\mu = NIA}$$

where μ is the magnetic moment of the coil, in Nm/T,
    N is the number of turns of the coil,
    I is the current through any single turns of the coil, in A, and
    A is the area enclosed by the coil, in m^2.
Notice that A is area of coil without regard to shape; that is, it may be circular or rectangular or square or whatever shape is desired. The symbols μ and μ_0 are not to

be confused. μ is a magnetic moment and μ_0 is the permeability constant. The magnetic moment μ is a vector in a direction that is perpendicular to the surface bounded by the loop and in the direction of the magnetic field.

When a magnetic moment is placed in a uniform magnetic field, **B**, it is acted upon by a torque

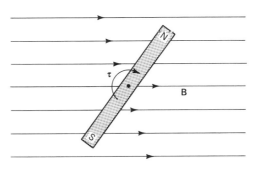

$$\tau = \mu \mathbf{B} \sin \theta$$

where τ is the torque on the magnetic moment,

μ is the magnetic moment of the loop,

B is the magnetic field strength of the uniform magnetic field, and

θ is the angle between the magnetic moment and the uniform magnetic field.

example 1   A large electromagnet is used to produce a uniform magnetic field of 1.50 T. A small bar magnet is suspended in this field with its axis perpendicular to the field. The torque produced on this bar magnet is 2.50×10^{-3} Nm. What is the magnetic moment of this bar magnet?

DATA: $B = 1.50$ T

$\tau = 2.50 \times 10^{-3}$ Nm

$\theta = 90°$

$\mu = ?$

BASIC EQUATION: $\tau = \mu B \sin \theta$

WORKING EQUATION: $\mu = \dfrac{\tau}{B \sin \theta}$

SUBSTITUTION: $\mu = \dfrac{2.50 \times 10^{-3} \, \text{Nm}}{(1.50 \text{ T})(\sin 90°)}$

$= 1.67 \times 10^{-3} \, \text{Nm/T}$

example 2 It is desired to construct a coil with the same magnetic moment as the bar magnet in Example 1 (1.67×10^{-3} Nm/T). The coil would be circular, 0.030 m in diameter, and have 25 turns. What current must flow through the coil?

DATA: $\qquad\qquad\qquad \mu = 1.67 \times 10^{-3}$ Nm/T

$$N = 25 \text{ turns}$$

$$r = \tfrac{1}{2} \text{ diameter} = \tfrac{1}{2}(0.030 \text{ m}) = 0.015 \text{ m}$$

$$A = \pi r^2 = \pi(0.015 \text{ m})^2 = 7.07 \times 10^{-4} \text{ m}^2$$

$$I = ?$$

BASIC EQUATION: $\qquad \mu = NIA$

WORKING EQUATION: $\quad I = \dfrac{\mu}{NA}$

SUBSTITUTION: $\qquad I = \dfrac{1.67 \times 10^{-3} \text{ Nm/T}}{25(7.07 \times 10^{-4} \text{ m}^2)} \qquad \boxed{\dfrac{\dfrac{\text{Nm}}{\text{T}}}{\text{m}^2} = \dfrac{\text{N}}{\text{Tm}} = \dfrac{\text{N}}{\left(\dfrac{\text{N}}{\text{mA}}\right)\text{m}} = \text{A}}$

$$= 0.0945 \text{ A}$$

25.5 SOLENOID

When a number of turns are formed from a wire, a *solenoid* is produced. The figure below shows a loosely packed solenoid. Magnetic field lines, like electric field lines,

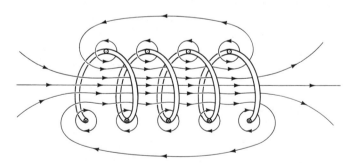

cannot intersect and jam together to form a high density of lines through the solenoid itself. For a long coil that is tightly turned, the field strength at its center is

$$\boxed{B = \mu_0 In}$$

where B is the magnetic field in the region at the center of the solenoid,

  μ_0 is the permeability constant, $4\pi \times 10^{-7}$Tm/A,

  I is the current through the solenoid, and

  n is the turn per unit length of solenoid.

The longer the solenoid is with respect to its radius, the more uniform the magnetic field is inside the solenoid; and for an infinitely long solenoid, the value of B is uniform throughout.

example What is the magnetic field at the center of a solenoid which is 0.425 m long, 0.075 m in diameter, and has three layers of $85\bar{0}$ turns each, when 0.25 A flows throughout?

DATA: $I = 0.25$ A

$$n = \frac{3 \times 85\bar{0} \text{ turns}}{0.425 \text{ m}} = 60\bar{0}0 \text{ turns/m}$$

$$\mu_0 = 4\pi \times 10^{-7} \text{ Tm/A}$$

$$B = ?$$

BASIC EQUATION: $B = \mu_0 In$

WORKING EQUATION: same

SUBSTITUTION: $B = (4\pi \times 10^{-7} \text{ Tm/A})(0.25 \text{ A})(60\bar{0}0 \text{ turns/m})$

$$\boxed{\left(\frac{\text{Tm}}{\text{A}}\right)(\text{A})\left(\frac{\text{turns}}{\text{m}}\right) = T}$$

$$= 1.9 \times 10^{-3} \text{ T}$$

25.6 CURRENT IN A MAGNETIC FIELD

When a charge is set in motion in a magnetic field, a force acting on the charge is perpendicular to both the velocity that the charge is traveling with and also the direction of the magnetic field. Notice that it is necessary for the charge to be in motion in a direction different from the direction of the magnetic field in order for a magnetic force to act on the charge.

Consider a current flowing through a length of wire in a magnetic field. The force on the wire is given by

$$\boxed{F = IlB \sin \theta}$$

where F is the force on the wire,
  I is the current through the wire,
  B is the magnetic field strength,
  l is the length of the wire, and
  θ is the angle between the current and the magnetic field.
The relationship of I, B, and F is such that a right-hand screw rotation through the acute angle from I to B would advance in the direction of F.

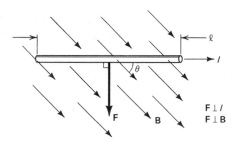

example   A straight wire with a current of 7.50 A passes through the center of a uniform magnetic field of 1.60 T. The magnetic field is circular with 0.125 m diameter and perpendicular to the wire. What force acts on the wire?

DATA: $I = 7.50$ A

                               $l = 0.125$ m   (l is the length of wire which
                                                      is in the magnetic field.)

                               $B = 1.60$ T

                               $\theta = 90°$

                               $F = ?$

BASIC EQUATION: $F = IlB \sin \theta$

WORKING EQUATION: same

SUBSTITUTION: $F = (7.50 \text{ A})(0.125 \text{ m})(1.60 \text{ T})(\sin 90°)$

$$\boxed{\text{AmT} = \text{Am}\frac{\text{N}}{\text{mA}} = \text{N}}$$

                               $= 1.50$ N

25.7   MAGNETIC FLUX AND INDUCTION

Magnetic field strength is often called *magnetic flux density*. We define flux to be the total field over an area perpendicular to the direction of the magnetic field.

$$\phi_B = BA \cos \theta$$

where ϕ_B is the magnetic flux through area A,

  B is the magnetic field strength,

  A is the area of the region of flux, and

  θ is the angle between the field direction and the normal to surface A.

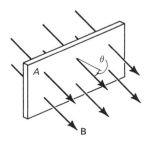

The term weber (Wb) needs to be introduced. It is the unit of magnetic flux and is equivalent to a Tm^2.

example 1 A coil of 75 turns and 0.050 m diameter is placed at the center of the solenoid of the example in Section 25.5 with their axes parallel. What is the total flux through this coil?

DATA: $B = 1.9 \times 10^{-3} \text{ T}$

$r = \frac{1}{2} \text{ diameter} = \frac{1}{2}(0.050 \text{ m}) = 0.025 \text{ m}$

$A = \pi r^2 = \pi (0.025 \text{ m})^2 = 2.0 \times 10^{-3} \text{ m}^2$

$N = 75 \text{ turns}$

$\theta = 0.0°$

$N\phi_B = ?$

BASIC EQUATION: $\phi_B = BA \cos \theta$

WORKING EQUATION: $N\phi_B = NBA \cos \theta$

SUBSTITUTION: $N\phi_B = (75 \text{ turns})(1.9 \times 10^{-3} \text{ T})(2.0 \times 10^{-3} \text{ m}^2)(\cos 0.0°)$

$= 2.9 \times 10^{-4} \text{ Wb}$

If the flux through an area inside a closed path is increased or decreased, an emf is produced around that path. *Faraday's law* states that the value of the emf depends upon the rate that the flux changes:

$$\mathcal{E} = -\frac{\Delta \phi_B}{\Delta t}$$

where $\mathcal{E}$ is the induced emf in volts and

$\dfrac{\Delta\phi_B}{\Delta t}$ is the time rate of change of flux.

The negative sign ensures that the induced emf tends to oppose the change in flux. If the emf could be in a direction to enhance the change in flux, the result would be a runaway with the flux and emf ever enhancing each other. The negative sign ensures that energy will be conserved. Notice that the flux passing through a given area can be changed in two ways: change B or change the angle θ. Also, notice that the emf is present whether the path surrounding the area is a conductor or not; but when the path is a conductor, current will flow and energy will be required to produce the emf. If the area is defined by a number of loops of conductor, the induced emf depends upon the time rate of change of the flux linkage:

$$\mathcal{E} = -\frac{N\Delta\phi_B}{\Delta t}$$

where N is the number of turns of the loop.

example 2   The current through the solenoids of the example in Section 25.5 and Example 1 in this section is reversed $6\bar{0}$ times per second. (It requires $\frac{1}{120}$ s to go from maximum in one direction to maximum in the other direction. Hence, note the factor 2 in the data.) What is the emf induced in the coil of Example 1?

DATA:

$$N\Delta\phi_B = 2(2.9 \times 10^{-4}\ \text{Wb})$$

$$\Delta t = \tfrac{1}{120}\ \text{s}$$

$$\mathcal{E} = ?$$

BASIC EQUATION:

$$\mathcal{E} = -\frac{N\Delta\phi_B}{\Delta t}$$

WORKING EQUATION: same

SUBSTITUTION:

$$\mathcal{E} = -\frac{2(2.9 \times 10^{-4}\ \text{Wb})}{\frac{1}{120}\ \text{s}} \qquad \boxed{\frac{\text{Wb}}{\text{s}} = \text{V}}$$

$$= -0.070\ \text{V}$$

The negative sign reminds us that the induced emf always acts to oppose the change.

25.8   INDUCTANCE

A solenoid or coil with multiple turns has flux linkage between the different turns of the coil. The flux linkage is proportional to the current through the solenoid. The constant of proportionality is called *inductance* and has the unit henry (H). This constant is a property of the electrical element called an inductor and is not dependent upon whether the element is in a circuit or on the shelf.

$$N\Phi_B = LI$$

where $N\phi_B$ is the flux linkage,
      L is the inductance, and
      I is the current.

It follows from Faraday's law that an emf is induced by a changing current:

$$\mathcal{E} = -L\frac{\Delta I}{\Delta t}$$

where $\mathcal{E}$ is the induced emf,
      L is the inductance, and
      $\dfrac{\Delta I}{\Delta t}$ is the time rate of change of current.

The induced emf is in such a direction as to oppose the change of current. Consider a circuit composed of a battery, an inductance coil, and a resistance as shown below.

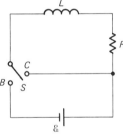

When the switch, S, is closed, the potential differences about the circuit are

$$\mathcal{E}_0 - L\frac{\Delta I}{\Delta t} - IR = 0$$

where $\mathcal{E}_0$ is the emf of the battery,
      $L\dfrac{\Delta I}{\Delta t}$ is the induced emf of inductor, and
      IR is the voltage drop across resistor.

The equation above is a differential equation and its solution will be simply stated here as

$$I = \frac{\mathcal{E}_0}{R}(1 - e^{-Rt/L})$$

$$\frac{\Delta I}{\Delta t} = \frac{\mathcal{E}_0}{L}e^{-Rt/L}$$

Values of IR and $L(\Delta I/\Delta t)$ are plotted below as a function of time. The term IR is the voltage drop across the resistor, while the value $L(\Delta I/\Delta t)$ is the induced emf of the inductor.

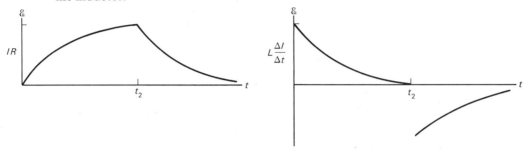

At the instant when the switch is first closed, no current flows—there is no voltage drop across the resistor. All of the emf of the battery is opposed by the induced emf of the inductor. As time passes, the current builds up to a steady value; the induced emf is zero and all the energy supplied by the battery is used in heating the resistor. However, energy was used in establishing a magnetic field in the inductor which is present during the steady-state current flow. The magnetic field has energy in it. When the switch in the circuit is moved from B to C in the figure on page 467, the battery is excluded from the circuit and only the resistor and inductor are left in series, and the voltage around the circuit is

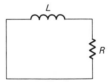

$$L\frac{\Delta I}{\Delta t} + IR = 0$$

where $L\dfrac{\Delta I}{\Delta t}$ is the induced emf of the inductor and
IR is the voltage drop across the resistor.

The equation above is a differential equation and its solution will simply be stated here:

$$I = \frac{\mathcal{E}_0}{R}e^{-Rt/L}$$

$$\frac{\Delta I}{\Delta t} = -\frac{\mathcal{E}_0}{L}e^{-Rt/L}$$

Values of IR and $L(\Delta I/\Delta t)$ are shown on the graphs above, starting at the time t_2, when the switch was moved to exclude the battery. Notice that the direction of the induced emf is now changed. The induced emf always opposes change—it opposed the current buildup—now it opposes the current decay. When the current was building up, the inductor took energy to establish its magnetic field; when the current is decaying, the inductor exhausts its magnetic field trying to maintain the current. The current through the resistor falls off exponentially. The rate at which the current builds up and decays is determined by the value of L/R. The term L/R is called the time constant of the circuit.

example 1 An *LR* circuit consists of a 45.0-V battery, a 1.50-mH inductor, and a 475-Ω resistor.

(a) What will be the equilibrium current in this circuit?
(b) How much time is required for the current to reach one-half of its maximum value?

(a) DATA:

$$\mathcal{E}_0 = 45.0 \text{ V}$$

$$L = 1.50 \times 10^{-3} \text{ H}$$

$$R = 475 \ \Omega$$

$$I = ? \quad \text{when } t \rightarrow \infty$$

BASIC EQUATION:

$$I = \frac{\mathcal{E}_0}{R}(1 - e^{-Rt/L})$$

WORKING EQUATION:

$$\text{When } t \rightarrow \infty, \ I = \frac{\mathcal{E}_0}{R}$$

SUBSTITUTION:

$$I = \frac{45.0 \text{ V}}{475 \ \Omega}$$

$$= 0.0947 \text{ A}$$

(b) DATA:

$$\mathcal{E}_0 = 45.0 \text{ V}$$

$$L = 1.50 \times 10^{-3} \text{ H}$$

$$R = 475 \ \Omega$$

$$I = \frac{1}{2}\frac{\mathcal{E}_0}{R} \quad [\text{see part (a)}]$$

$$t_{1/2} = ?$$

BASIC EQUATION:

$$I = \frac{\mathcal{E}_0}{R}(1 - e^{-Rt/L})$$

WORKING EQUATION:

same

SUBSTITUTION:

$$\frac{1}{2}\frac{\mathcal{E}_0}{R} = \frac{\mathcal{E}_0}{R}\left[1 - e\left(-\frac{475 \ \Omega}{1.50 \times 10^{-3} \text{ H}}\right)t\right]$$

Multiply both sides by $R/\mathcal{E}_0$:

$$\frac{1}{2} = 1 - e\left(-\frac{475 \ \Omega}{1.50 \times 10^{-3} \text{ H}}\right)t$$

(Note that the exponent has units of $\dfrac{\Omega s}{H}$ which is equivalent to unity.)

$$\frac{1}{2} = e^{-3.17 \times 10^5 t}$$

Take the natural logarithm of both sides:

$$-\ln 2 = -3.17 \times 10^5 t$$

$$t = \frac{\ln 2}{3.17 \times 10^5}$$

$$= 2.19 \times 10^{-6} \text{ s}$$

It is common to refer to a value like this as 2.19 μs.

example 2   In the figure shown below, $R_1 = 10\bar{0}\ \Omega$, $R_2 = 15\bar{0}\ \Omega$, $L = 15.0 \times 10^{-3}$ H, and $\mathcal{E} = 22.5$ V.

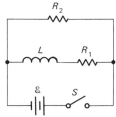

At time $t = 0$, the switch, S, is first closed.

(a) What is the current through R_1?
(b) What is the current through R_2?
(c) What is the potential difference across L?
   After the switch S has been closed for a long time, a steady state is reached.
(d) What is the current through R_1?
(e) What is the current through R_2?
(f) What is the potential difference across L?
   The switch S is now opened.
(g) What is the current through R_1?
(h) What is the current through R_2?
(i) What is the potential difference across L?
   The switch S has now been open for a long time.
(j) What is the current through R_1?
(k) What is the current through R_2?
(l) What is the potential difference across L?

(a) DATA: $t = 0$ s

$$I_1 = ?$$

BASIC EQUATION: $I_1 = \dfrac{\mathcal{E}_0}{R_1}(1 - e^{-Rt/L})$

WORKING EQUATION: same

SUBSTITUTION:
$$I_1 = \frac{\mathcal{E}_0}{R_1}[1 - e^{(-R/L)(0)}] = \frac{\mathcal{E}_0}{R_1}(1 - 1) = 0$$

$$I_1 = 0 \text{ A}$$

(b) DATA:
$$R_2 = 15\overline{0} \ \Omega$$
$$\mathcal{E}_0 = 22.5 \text{ V}$$
$$I_2 = ?$$

BASIC EQUATION:
$$\mathcal{E}_0 = I_2 R$$

WORKING EQUATION:
$$I_2 = \frac{\mathcal{E}_0}{R_2}$$

SUBSTITUTION:
$$I_2 = \frac{22.5 \text{ V}}{15\overline{0} \ \Omega}$$

$$= 0.150 \text{ A}$$

(c) DATA:
$$\mathcal{E}_0 = 22.5 \text{ V}$$
$$t = 0 \text{ s}$$
$$\mathcal{E}_L = ?$$

BASIC EQUATIONS:
$$\frac{\Delta I}{\Delta t} = \frac{\mathcal{E}_0}{L} e^{-Rt/L}$$

$$\mathcal{E}_L = L\frac{\Delta I}{\Delta t}$$

WORKING EQUATIONS: same

SUBSTITUTIONS: For $t = 0$ s
$$\frac{\Delta I}{\Delta t} = \frac{\mathcal{E}_0}{L} e^{(-R/L)(0)} = \frac{\mathcal{E}_0}{L}$$

and
$$\mathcal{E}_L = L\left(\frac{\mathcal{E}_0}{L}\right) = \mathcal{E}_0$$

SUBSTITUTIONS:
$$\mathcal{E}_L = 22.5 \text{ V}$$

(d) DATA:
$$R_1 = 10\overline{0} \ \Omega$$
$$\mathcal{E}_0 = 22.5 \text{ V}$$
$$I_1 = ? \text{ as } t \to \infty$$

BASIC EQUATION:
$$I_1 = \frac{\mathcal{E}_0}{R_1}(1 - e^{-Rt/L})$$

WORKING EQUATION: same

SUBSTITUTION:	$I_1 = \frac{\mathscr{E}_0}{R_1}[1 - e^{(-R/L)(\infty)}] = \frac{\mathscr{E}_0}{R_1}(1 - 0) = \frac{\mathscr{E}_0}{R_1}$
	$I_1 = \frac{22.5 \text{ V}}{10\overline{0} \ \Omega}$
	$= 0.225 \text{ A}$
(e) DATA:	$\mathscr{E}_0 = 22.5 \text{ V}$
	$R_2 = 15\overline{0} \ \Omega$
	$I_2 = ? \text{ as } t \to \infty$
BASIC EQUATION:	$\mathscr{E}_0 = I_2 R_2$
WORKING EQUATION:	$I_2 = \mathscr{E}_0 / R_2$
SUBSTITUTION:	$I_2 = \frac{22.5 \text{ V}}{15\overline{0} \ \Omega}$
	$= 0.150 \text{ A}$
(f) DATA:	$\mathscr{E}_0 = 22.5 \text{ V}$
	$\mathscr{E}_L = ? \text{ as } t \to \infty$
BASIC EQUATIONS:	$\frac{\Delta I}{\Delta t} = \frac{\mathscr{E}_0}{L} e^{(-R/L)t}$
	$\mathscr{E}_L = L \frac{\Delta I}{\Delta t}$
WORKING EQUATIONS:	same
SUBSTITUTION:	$\frac{\Delta I}{\Delta t} = \frac{\mathscr{E}_0}{L} e^{(-R/L)(\infty)} = \left(\frac{\mathscr{E}_0}{L}\right)(0) = 0$
	$\mathscr{E}_L = L\left(\frac{\Delta I}{\Delta t}\right) = L(0) = 0$
	$\mathscr{E}_L = 0 \text{ V}$
(g) DATA:	$\mathscr{E}_0 = 22.5 \text{ V}$
	$R_1 = 10\overline{0} \ \Omega$
	$t = 0 \text{ s}$
	$I_1 = ?$
BASIC EQUATION:	$I_1 = \frac{\mathscr{E}_0}{R_1} e^{(-R/L)t}$

Note: The quantity $\mathscr{E}_0/R_1$ gives the current through the inductor at the time the switch is to be opened (part (d)). The value of R in the exponent is $R_1 + R_2$ since the components of the circuit are in series when the switch is open.

WORKING EQUATION: same

SUBSTITUTION:

$$I_1 = \frac{\mathcal{E}_0}{R_1} e^{(-R/L)(0)} = \frac{\mathcal{E}_0}{R_1}$$

$$I_1 = \frac{22.5\ \text{V}}{10\bar{0}\ \Omega}$$

$$= 0.225\ \text{A}$$

(h) DATA:

$I_1 = 0.225\ \text{A}$ [part (g)]

$I_2 = ?$

WORKING EQUATION: $I = I_1 = I_2$

SUBSTITUTION: $I_2 = I_1 = 0.225\ \text{A}$

(i) DATA:

$\mathcal{E}_0 = 22.5\ \text{V}$

$t = 0\ \text{s}$

$\mathcal{E}_L = ?$

BASIC EQUATIONS:

$$\frac{\Delta I}{\Delta t} = -\frac{\mathcal{E}_0}{L} e^{(-R/L)t}$$

$$\mathcal{E}_L = L \frac{\Delta I}{\Delta t}$$

WORKING EQUATIONS: same

SUBSTITUTION: For $t = 0$ s,

$$\frac{\Delta I}{\Delta t} = -\frac{\mathcal{E}_0}{L} e^{(-R/L)t} = -\frac{\mathcal{E}_0}{L}$$

$$\mathcal{E}_L = L\left(\frac{\Delta I}{\Delta t}\right) = L\left(\frac{-\mathcal{E}_0}{L}\right) = -\mathcal{E}_0$$

$$\mathcal{E}_L = -22.5\ \text{V}$$

(j) DATA.

$\mathcal{E}_0 = 22.5\ \text{V}$

$R_1 = 10\bar{0}\ \Omega$

$I_1 = ?$ as $t \to \infty$

BASIC EQUATION:

$$I_1 = \frac{\mathcal{E}_0}{R_1} e^{(-R/L)t}$$

WORKING EQUATION:

$$I_1 = \frac{\mathcal{E}_0}{R_1} e^{(-R/L)(\infty)} = \left(\frac{\mathcal{E}_0}{R_1}\right)(0) = 0$$

SUBSTITUTION: $I_1 = 0\ \text{A}$

(k) DATA: $\mathscr{E}_0 = 0\text{V}$

$I_2 = ?$

BASIC EQUATION: $\mathscr{E}_0 = I_2 R$

WORKING EQUATION: $I_2 = \dfrac{\mathscr{E}_0}{R}$

SUBSTITUTION: $I_2 = \dfrac{0\text{V}}{R} = 0\text{ A}$

(l) DATA: $\mathscr{E}_0 = 22.5\text{ V}$

$\mathscr{E}_L = ?$ as $t \rightarrow \infty$

BASIC EQUATIONS: $\dfrac{\Delta I}{\Delta t} = -\dfrac{\mathscr{E}_0}{R} e^{(-R/L)t}$

$\mathscr{E}_L = L\left(\dfrac{\Delta I}{\Delta t}\right)$

WORKING EQUATIONS: same

SUBSTITUTION: $\dfrac{\Delta I}{\Delta t} = \dfrac{\mathscr{E}_0}{R} e^{(-R/L)(\infty)} = 0$

$\mathscr{E}_L = L\left(\dfrac{\Delta I}{\Delta t}\right) = L(0) = 0\text{V}$

25.9 ENERGY IN A MAGNETIC FIELD

When a current is flowing through a solenoid, a magnetic field is produced inside the solenoid. As shown in the last section, energy is used to create the magnetic field; energy is present in the field when the current through the solenoid is steady; and energy from the magnetic field is dissipated to try to maintain a decaying current. The energy in a magnetic field in a solenoid is given by

$$U = \tfrac{1}{2} L I^2$$

where U is the energy of the magnetic field,
      L is the inductance of the solenoid, and
      I is the current through the solenoid.
In a capacitor, energy is stored in the electric field and can remain there even when the capacitor is disconnected from any circuit; however, the magnetic field of an inductor does not exist unless a current flows through it. An inductor cannot possess magnetic field energy outside a circuit.

example A solenoid with an inductance of 3.50 mH has a steady current of 5.00 A flowing through it. What energy is stored in the magnetic field?

DATA:

$$L = 3.50 \, \text{mH} = 3.50 \times 10^{-3} \, \text{H}$$

$$I = 5.00 \, \text{A}$$

$$U = ?$$

BASIC EQUATION:

$$U = \tfrac{1}{2} L I^2$$

WORKING EQUATION:

same

SUBSTITUTION:

$$U = \tfrac{1}{2}(3.50 \times 10^{-3} \, \text{H})(5.00 \, \text{A})^2$$

$$\boxed{\text{HA}^2 = \frac{\text{Vs}}{\text{A}} \text{A}^2 = \text{VAs} = \text{Ws} = \text{J}}$$

$$= 43.8 \, \text{mJ}$$

EXERCISES

Sketch	Data	Basic Equation	Working Equation	Substitution
$h = ?$	$A = 12 \, \text{cm}^2$ $b = 6.0 \, \text{cm}$ $h = ?$	$A = bh$	$h = \dfrac{A}{b}$	$h = \dfrac{12 \, \text{cm}^2}{6.0 \, \text{cm}}$ $= 2.0 \, \text{cm}$

1. What is the magnetic field at 0.10 m from a long wire carrying a current of 15.0 A?

2. A wire carrying a current of 50.0 A runs perpendicular to a uniform magnetic field of 1.50×10^{-3} T. Where will the resultant magnetic field be zero?

3. A small bar magnet is placed in a uniform field of 2.25 T. The maximum torque produced on a magnet is 1.79×10^{-4} N m. What is the magnetic moment of the magnet?

4. What is the maximum torque that can be produced by a uniform magnetic field of 1.37 T acting on a magnet whose moment is 1.93×10^{-3} N m/T?

5. A coil of wire has $10\bar{0}$ turns of wire, is 0.32 m by 0.16 m, and carries a current of 5.0 A. What is the magnetic moment of this coil?

6. A solenoid has $300\bar{0}$ turns of wire and is 0.35 m long. What current is required to produce a flux density of 0.10 T at the center of this solenoid?

7. A small electromagnet is 0.15 m in length, 0.015 m in diameter, and has $60\bar{0}$ turns of wire. What current is required to produce a magnetic field of 1.25×10^{-3} T at the center of the solenoid?

8. Two wires of 0.25-m length are parallel to each other and separated by a distance of 0.055 m. Each wire is carrying a current of 10.0 A.

(a) What is the force between the two wires when the currents are in the same direction?

(b) What is the force between the two wires when the currents are in opposite directions?

9. A uniform magnetic field of 5.50×10^{-5} T passes through an area that is 2.50 m square. The normal to the area makes an angle of 25° with the magnetic field. What is the flux through the area?

10. An area of 0.125 m² intercepts a uniform field of 0.162 T. What is the flux through the area?

(a) When is the area parallel to the magnetic field?

(b) When is the area perpendicular to the magnetic field?

(c) When will the area make an angle of 45° with the magnetic field?

11. A circular coil of $10\bar{0}$ turns has a radius of 0.115 m. The flux density through the coil is changing at the rate of 0.140 T/s. The coil is connected in series to a $10\bar{0}$-Ω resistor.

(a) What is the value of the induced emf of the circuit?

(b) What current flows in this circuit? (The resistance of the coil is negligible.)

12. A magnetic field of 0.75 T is changed to 1.50 T in a period of 0.0012 s. This uniform field passes through a circular coil of 0.125-m radius and $20\bar{0}$ turns. What emf is induced in the coil?

13. An *LR* circuit consists of a 7.50-V battery, a 1.50-H inductor, and a 57.5-Ω resistor.

(a) What will be the equilibrium current in this circuit?

(b) How much time is required for the current to reach one-half of its maximum value?

(c) When the current is at one-half of its maximum value, what is the voltage across the inductor?

14. A solenoid with an inductance of 1.05×10^{-3} H has a steady current of 1.25 A flowing through it. What energy is stored in the magnetic field?

15. An *LR* circuit has a steady current flowing through it. The resistance is 75.0 Ω, the inductance is 0.137 H, and the applied emf is 25.0 V.

(a) At what rate is energy being dissipated by the resistance?

(b) At what rate is energy being used to form the magnetic field?

(c) How much energy is present in the resistor?

(d) How much energy is present in the magnetic field?

16. (a) What is the time constant of Exercise 15?

(b) If the switch were suddenly opened, how long would it require for the current to drop to one-half of its value?

17. In the figure below, $R_1 = 25.0$ Ω, $R_2 = 75.0$ Ω, $L = 25.0$ H, and $\mathcal{E} = 50.0$ V.

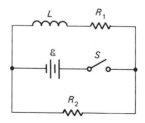

At the time $t = 0$, when the switch, S, is first closed,

(a) What is the current through R_1?
(b) What is the current through R_2?
(c) What is the potential difference across L?

After the switch has been closed for a long time, a steady state is reached.

(d) What is the current through R_1?
(e) What is the current through R_2?
(f) What is the potential difference across L?

The switch is now opened.

(g) What is the current through R_1?
(h) What is the current through R_2?
(i) What is the potential difference across L?

The switch has now been open for a long time.

(j) What is the current through R_1?
(k) What is the current through R_2?
(l) What is the potential difference across L?

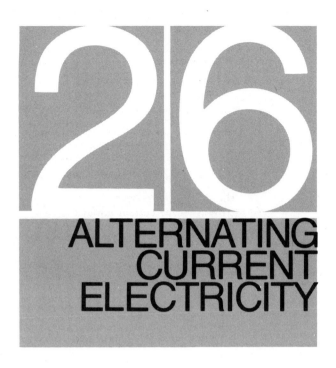

ALTERNATING CURRENT ELECTRICITY

26.1 ALTERNATING CURRENT

The electric power that we have in our homes and industries is called alternating current electricity (ac). We have studied only direct current electricity (dc). Our daily uses of dc electricity are primarily battery-operated devices such as portable radios, recorders, and automobile electrical systems. Our daily uses of ac electricity include home lighting, appliances, air conditioning, and power equipment. Dc electricity is convenient because the sources of emf are portable, without moving parts, and can be located in the vicinity of their loads. Ac electricity is usually supplied by a power company and is transmitted and distributed through a network of cables to many industrial, commercial, and residential areas. The power companies supply ac electricity rather than dc electricity because the production and distribution of ac power is more efficient than dc power.

Ac electricity, as the word "alternating" implies, goes through some sort of repeating cycle. Take an ac source of emf and assign the value of zero potential to one terminal. Then the potential of the other terminal will alternate between a positive and a negative potential. In the United States, commercially available ac power makes 60 complete alternations in every second and hence is called a 60-cycle power source. A plot of the potential difference across an ac source of emf is a sine or cosine function of time. If the load is a purely resistive load, such as a heating element, the current that flows through the circuit is also a sine or cosine function of time, is of the same frequency, and is in phase with the emf of the source.

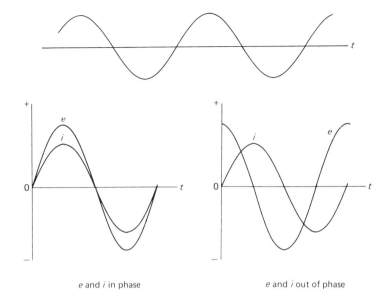

e and i in phase   e and i out of phase

When inductors and capacitors are added to the load, the current may be as much as 90° out of phase with the emf. This phase difference, of course, makes a difference in the rate that energy is dissipated by the resistive load. This power factor will be discussed in Chapter 27, which deals with ac circuits.

An ac source furnishes a source of emf which can be written as

$$e = \mathcal{E}_{max} \sin 2\pi ft$$

where e is the instantaneous value of the emf,

$\mathcal{E}_{max}$ is the maximum value of the emf,

f is the frequency of the ac source of emf, and

t is the time.

An ac source has a maximum emf, $\mathcal{E}_{max}$, which might be used to describe the source; however, a dc source described in the same manner would provide more power to a given load, since its emf is constant. For this reason, we refer to the effective values of the emf of ac sources. The effective value is a root-mean-square value of the emf and gives the same power as a dc source of the same value.

$$\mathcal{E} = 0.707\mathcal{E}_{max}$$

where $\mathcal{E}$ is the effective value of voltage and

$\mathcal{E}_{max}$ is the maximum instantaneous voltage.

The same argument can be made for ac current with the result

$$I = 0.707 I_{max}$$

where I is the effective value of current and
  I_{max} is the maximum instantaneous current.

> **example**   The power lines supplied to a house are rated at 120 V. What is the maximum value of the voltage supplied?
>
> DATA: $\mathscr{E} = 120$ V
>
> $\mathscr{E}_{max} = ?$
>
> BASIC EQUATION: $\mathscr{E} = 0.707 \mathscr{E}_{max}$
>
> WORKING EQUATION: $\mathscr{E}_{max} = \dfrac{\mathscr{E}}{0.707}$
>
> SUBSTITUTION: $\mathscr{E}_{max} = \dfrac{120 \text{ V}}{0.707}$
>
> $= 170$ V

26.2   GENERATORS

A generator is a mechanical–electrical device for converting mechanical energy to electrical energy. The principle involved is Faraday's law of induction (Section 25.7):

$$\mathscr{E} = \frac{-N\Delta\phi_B}{\Delta t}$$

where $\mathscr{E}$ is the induced emf,
  N is the number of turns in the coil, and
  $\dfrac{\Delta\phi_B}{\Delta t}$ is the time rate of change of the magnetic flux through the coil.
A simple generator is shown below. It consists of a coil of wire and a permanent magnet.

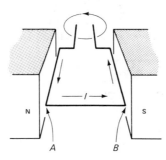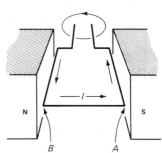

The magnet provides a magnetic field. The flux through the coil is changed by rotating the coil. Note that a greater induced emf is produced by an increased number of turns on the coil, a stronger magnet, and a higher rotational speed of the coil. The emf produced in the wire coil by its rotating in the magnetic field changes directions and is referred to as an ac source. If the coil is rotated at a constant angular speed, the frequency of the ac source is the number of rotations of the coil in one second. As side A of the current loop passes downward by the north pole, the induced emf is in one direction. As side A (same side of the rotating loop) passes upward by the south pole, the induced emf is in the opposite direction. As side B (the other side of the rotating loop) passes upward and downward, the induced emf also alternates. Refer to the figure below.

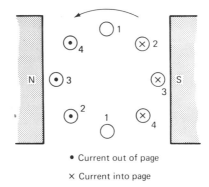

• Current out of page

× Current into page

The pair of wires corresponding to different positions of the loop are labeled 1 to 4. Note that in position 1 the magnetic flux through the coil is a maximum. In position 3 the flux through the coil is zero.

In position 2, the magnetic flux through the coil is increasing until the flux through the coil reaches a maximum at position 1. As the coil passes on to position 4, the flux through the coil is now decreasing. The direction of the rate of change of flux reversed and the current changed directions at position 1. The coil moves on to position 3, where the flux through the coil is zero. In moving from position 1 to position 3, the flux has changed from a positive value to zero or the change of flux has been negative. As the coil moves from position 3 through position 2, to position 1, the value of flux goes from a zero value to a negative value; or the change of flux has been negative, the same as it was from position 1 through position 4 to position 3. The change of flux has been in the same direction, so the current flow has been in the same direction. As a matter of fact, it was a maximum when the coil passed through position 3. The student should use the rule of thumb to verify the directions of current flow in these illustrations.

A graph of the induced current is shown on page 482. One cycle is produced by one revolution of the wire. The time required for one cycle depends on the rotational speed of the coil. If the coil rotates 60 times each second, an alternating current of frequency 60 Hz is produced.

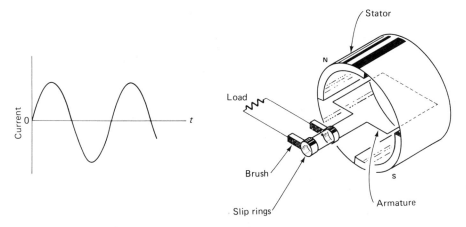

The current produced in the coil is conducted by brushes on slip rings to the external circuit, as shown above. The rotating coil is called the *rotor* or *armature*, and the field magnets are called the *stator*.

The generator does not actually create electrical energy; it changes the mechanical energy of rotation into electrical energy. The energy to turn the rotor may be supplied by water falling down a waterfall, a diesel engine, or a steam turbine.

Power companies use large commercial ac generators to produce the current they need to supply to their customers. These generators work in the same manner as the generator discussed here; but they have many coils, and electromagnets are used instead of permanent magnets.

The large generators used by electrical power companies can produce voltages as large as 13,000 V and currents up to 10 A. The alternator used in automobiles is an ac generator that produces about 14 V and up to 40 A.

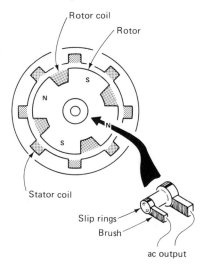

26.3 DC GENERATORS

By the use of a special device called a commutator, the ac generator can be used to produce direct current. The commutator is a split ring which replaces the slip rings as shown.

When side *A* of the coil passes upward along the N pole, the induced current flows in the direction shown and is picked up by brush 1. The current in the external circuit is also shown.

When side *B* of the coil passes upward along the N pole, the induced current flows in the direction shown and is picked up by brush 1. The current in the external circuit is in the same direction as it was when *A* passed along the N pole. Thus, this is a direct current.

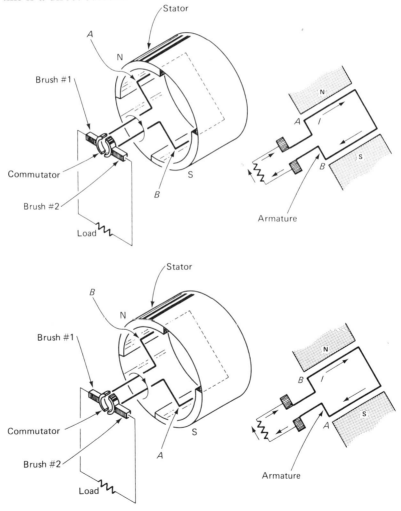

The current produced by this dc generator does not have the same value at all times. A graph of the induced current is shown.

Commercial dc generators that are used for industrial purposes contain many coils. The output current has almost the same value at all times due to the use of the large number of coils.

26.4 MOTORS

A *motor* is a mechanical–electrical device for converting electrical energy to mechanical energy. The principles involved were stated in Section 25.4. A current-carrying loop acts like a bar magnet; it has a magnetic moment. When a magnetic moment is placed in a magnetic field, a torque is exerted on the moment to align it in the same direction as the magnetic field. In the figure below, the magnetic moment associated with the loop of wire is shown in phantom. On an electric motor, the assembly containing the loop is called an armature. Notice that the direction of current flow is $B \rightarrow A$. The torque would tend to align the magnetic moment of the coil with the magnetic field, creating a counterclockwise rotation. If at just the time the magnetic moment is aligned with the magnetic field, the current stops, the inertia of the armature carries it past the alignment position. Thus, a current in the opposite direction $A \rightarrow B$ reverses the magnetic moment with respect to the loop. The new moment is in a position where the torque continues to force rotation in a counterclockwise direction. The device for reversing the current is called a *commutator* and is of the same type used with dc generators.

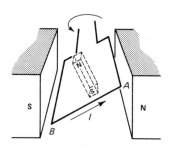

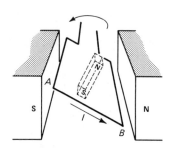

The commutator itself is mounted on the shaft of the armature and is simply a split ring. Contact is made between the dc source and the armature by brushes, which are usually spring-loaded carbon. If an ac source is used, it is possible to use slip rings and let the alternating current control the speed of the armature.

26.5 BACK EMF

The motor of Section 26.4 and the generator of Section 26.3 appear to be like devices. In fact, the only major difference that we see is that the current flows in the opposite direction in the generator to the current flow in the motor. The applied current is what produced the current flow in the motor and caused the armature to turn. Consistent with Faraday's law, the emf generated in the moving coil opposes the change. This induced emf is in the opposite direction to the applied emf and is called *back emf*. The back emf is in opposition to the applied emf. When an electric motor stalls, the armature is not rotating. There is no back emf, and the current through the armature can burn out the motor unless it is protected by an overload circuit breaker.

26.6 TRANSFORMERS

A *transformer* is an electrical device that changes the emf supplied by a source. The transformer may be used to raise voltage from a lower voltage to a higher voltage (step-up), or it may be used to lower voltage from a higher voltage to a lower voltage (step-down). A *step-down transformer* is used to lower the 120-V line that enters a television set to 6 V to heat the electron-gun filament in the picture tube, and a *step-up transformer* is used to produce the acceleration voltage of 25,000 V across the picture tube.

One of the reasons that it is more efficient for power companies to distribute ac power rather than dc power is that the power can be transmitted at high voltage and low current to minimize power loss in the lines. Transformers in the vicinity of the user can then reduce the voltage to the user's specifications.

A transformer utilizes the principle of induction. A primary coil is connected to an ac source. This coil produces a magnetic flux in a core. The core holds the magnetic flux in bounds and passes through a secondary coil. The flux changes at the same frequency as the primary source and induces an emf across the second coil. The induced emf has the same frequency as the original ac source. The voltage across each coil is proportional to the number of turns of each coil.

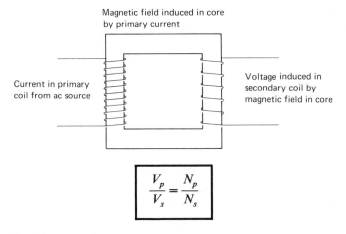

$$\frac{V_p}{V_s} = \frac{N_p}{N_s}$$

where V_p is the primary voltage,
$\quad V_s$ is the secondary voltage,
$\quad N_p$ is the number of primary turns, and
$\quad N_s$ is the number of secondary turns.

It follows that the step-down transformer has more turns in the primary coil than in the secondary coil and that the step-up transformer has more turns in the secondary than in the primary.

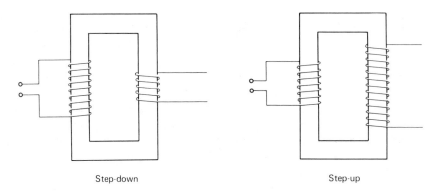

Step-down   Step-up

It must be emphasized that while the voltage in the secondary coil of a step-up transformer is higher than the voltage in the primary coil, there has been no creation of energy. In a 100 percent efficient transformer, the power supplied by the primary coil is exactly equal to the power of the secondary coil.

$$P_p = P_s$$
or
$$I_p V_p = I_s V_s$$

where P_p is the power (W) in primary coil,
$\quad$ P_s is the power (W) in secondary coil,
$\quad$ I_p is the current in primary coil,
$\quad$ I_s is the current in secondary coil,
$\quad$ V_p is the voltage of primary coil, and
$\quad$ V_s is the voltage of secondary coil.

It follows that the current relationship for a 100 percent efficient transformer is

$$\boxed{\frac{I_s}{I_p} = \frac{N_p}{N_s}}$$

where I_s is the current in secondary coil,
$\quad$ I_p is the current in primary coil,
$\quad$ N_p is the number of turns in primary, and
$\quad$ N_s is the number of turns in secondary.

Transformers do not create energy. Some energy is lost, however, during the change of voltage. Energy losses in transformers are of three types:

1. *Copper losses*—these result from the resistance of the copper wires in the coil and are unavoidable.

2. *Magnetic losses* (called *hysteresis losses*)—some energy is lost (turned into heat) by reversing the magnetism in the core.

3. *Eddy currents*—when a mass of metal (the core) is subjected to a changing magnetic field, currents are set up in the metal that do no useful work, waste energy, and produce heat. These losses can be lessened by *laminating* the core. Instead of using a solid block of metal for the core, thin sheets of metal with insulated surfaces are used, reducing these induced currents.

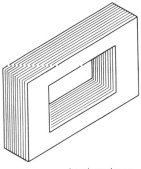

Laminated core

example A transformer on a television set has 62.0 turns on its primary coil and 15,$\overline{0}$00 turns on its secondary coil. If the voltage applied to the primary is 12$\overline{0}$ V, what is the secondary voltage?

DATA:

$$V_p = 12\bar{0} \text{ V}$$

$$N_p = 62.0 \text{ turns}$$

$$N_s = 15,\bar{0}00 \text{ turns}$$

$$V_s = ?$$

BASIC EQUATION:

$$\frac{V_p}{V_s} = \frac{N_p}{N_s}$$

WORKING EQUATION:

$$V_s = V_p \frac{N_s}{N_p}$$

SUBSTITUTION:

$$V_s = (12\bar{0} \text{ V})\left(\frac{15,\bar{0}00 \text{ turns}}{62.0 \text{ turns}}\right)$$

$$= 29,\bar{0}00 \text{ V}$$

EXERCISES

Sketch	Data	Basic Equation	Working Equation	Substitution
12 cm² $h = ?$ $b = 6.0$ cm	$A = 12 \text{ cm}^2$ $b = 6.0 \text{ cm}$ $h = ?$	$A = bh$	$h = \dfrac{A}{b}$	$h = \dfrac{12 \text{ cm}^2}{6.0 \text{ cm}}$ $= 2.0 \text{ cm}$

Find the effective value of voltage or current when the maximum value is:

1. $\mathcal{E}_{max} = 340$ V   **2.** $I_{max} = 7.1$ A

3. $I_{max} = 14$ A   **4.** $\mathcal{E}_{max} = 170$ V

5. $\mathcal{E}_{max} = 620$ V   **6.** $I_{max} = 21.1$ A

Find the maximum value of voltage or current when the effective value is:

7. $I = 1\bar{0}$ A   **8.** $\mathcal{E} = 115$ V

9. $\mathcal{E} = 240$ V   **10.** $I = 15$ A

11. $I = 5.0$ A   **12.** $\mathcal{E} = 80.0$ V

Solve for the indicated variable.

13. $V_p = 120$ V
$N_p = 2\bar{0}0$ turns
$N_s = 12$ turns
$V_s = ?$

14. $V_p = 120$ V
$N_p = 150$ turns
$N_s = 8\bar{0}0$ turns
$V_s = ?$

15. $V_p = 12\bar{0}$ V
$N_p = 45$ turns
$N_s = 9\bar{0}$ turns
$V_s = ?$

16. $V_p = 12\bar{0}$ V
$N_p = 10\bar{0}$ turns
$N_s = 10,\bar{0}00$ turns
$V_s = ?$

17. A transformer is used to reduce the line voltage of $12\bar{0}$ V to 6.30 V used by the electron gun of a television picture tube. The primary coil has 95.0 turns. How many turns does the secondary coil have?

18. A transformer is used to increase the line voltage of $12\bar{0}$ V to the $24,\bar{0}00$ V accelerating potential required by a color television picture tube. The primary coil has $10\bar{0}$ turns. How many turns does the secondary coil have?

19. A transformer has $10\bar{0}$ turns and $1\bar{0}$ turns, respectively, in its primary and secondary coils. When 5.0 A flows in the secondary circuit, what minimum current must exist in the primary coil?

20. A transformer has $1\bar{0}$ turns and $10\bar{0}$ turns, respectively, in its primary and secondary coils. When 5.0 A flows in the secondary circuit, what minimum current must exist in the primary coil?

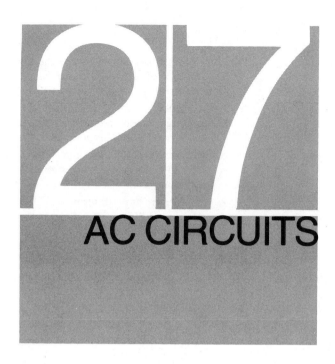

AC CIRCUITS

27.1 IMPEDANCE

Three components of electrical circuits were defined and their behavior in dc circuits explored in earlier chapters. The resistor simply dissipates energy as a heating element. The capacitor stores energy in an electric field, and time is required for a capacitor to charge or discharge through a resistance. The inductor tries to prevent changes in the current through the system and also acquires energy in its magnetic field. All three of these components tend to limit current flow in an ac circuit and are called *impedance*. Impedance is represented by the letter Z and is measured in ohms (Ω). The resistance is a resistive impedance in that it only dissipates energy and does not alter any relationship between potential differences and current. The capacitor and inductor provide reactive impedance because they do produce a phase difference between the voltage and current in an ac circuit. Specifically, the opposition to current flow by the capacitor is called *capacitive reactance* and the opposition to current flow by the inductor is called *inductive reactance*.

27.2 INDUCTIVE REACTANCE

Inductive reactance is measured in the same units as resistance and is given by

$$X_L = 2\pi f L$$

where X_L is the inductive reactance, in Ω,
  f is the frequency of the ac source, in Hz, and
  L is the inductance, in H.
Recall that when an inductor was placed in a dc circuit, it had a tendency to repress any change in current. In an ac circuit, the directions of both voltage and current are continually changing. The opposition to change of the inductor shows up as a current out of phase with and lagging behind the voltage by as much as 90°.

In a case where there is only inductance and a battery in a circuit, the current is written as

$$I = \frac{\mathcal{E}}{Z}$$

where I is the current, in A,
  $\mathcal{E}$ is the voltage, in V, and
  Z is the impedance, in Ω.
In this completely inductive circuit the current lags behind the voltage by a quarter of a cycle or 90°.

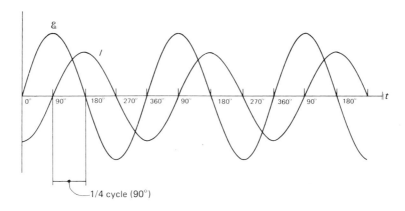

When a resistor is added to the inductor, the impedance becomes

$$Z = \sqrt{R^2 + X_L^2}$$

where Z is the impedance,
  R is the resistance, and
  X_L is the inductive reactance.
Note that the impedances of the two components do not add algebraically but rather add as vector components. The impedance can be represented graphically as the inductive reactance always being in the positive y-direction.

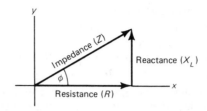

The resistance is always taken in the positive *x*-direction. The impedance *Z* is the vector sum of these components and the angle φ is the phase angle and is given by

$$\tan \phi = \frac{X_L}{R}$$

where φ is the phase angle,
  X_L is the inductive reactance, and
  R is the resistance.

The angle φ is the amount that the current lags behind the voltage. This angle was 90° for a pure inductive circuit and is 0° for a pure resistance.

example   An inductor of 0.175 H is connected to a 11̄0-V 60.0-Hz power supply.

(a) What is the inductive reactance of the inductor?
(b) When the inductor is placed in the circuit by itself, what current flows?
(c) If a 175-Ω resistor is connected in series with the inductor to the power supply, what is the impedance?
(d) What current flows in the circuit?
(e) What is the phase angle?

(a) DATA: $L = 0.175$ H

                                         $f = 60.0$ Hz

                                         $X_L = ?$

BASIC EQUATION: $X_L = 2\pi f L$

WORKING EQUATION: same

SUBSTITUTION: $X_L = 2\pi(60.0 \text{ Hz})(0.175 \text{ H})$

                                         $X_L = 66.0 \ \Omega$

(b) DATA: $\mathscr{E} = 11\bar{0}$ V

                                         $Z = 66.0 \ \Omega$ [from part (a)]

                                         $I = ?$

BASIC EQUATION: $I = \dfrac{\mathcal{E}}{Z}$

WORKING EQUATION: same

SUBSTITUTION: $I = \dfrac{11\bar{0} \text{ V}}{66.0 \text{ }\Omega}$

$= 1.67 \text{ A}$

(c) DATA: $R = 175 \text{ }\Omega$

$X_L = 66.0 \text{ }\Omega$ [from part (b)]

$Z = ?$

BASIC EQUATION: $Z = \sqrt{R^2 + X_L^2}$

WORKING EQUATION: same

SUBSTITUTION: $Z = \sqrt{(175 \text{ }\Omega)^2 + (66.0 \text{ }\Omega)^2}$

$= 187 \text{ }\Omega$

(d) DATA: $\mathcal{E} = 11\bar{0} \text{ V}$

$Z = 187 \text{ }\Omega$

$I = ?$

BASIC EQUATION: $I = \dfrac{\mathcal{E}}{Z}$

WORKING EQUATION: same

SUBSTITUTION: $I = \dfrac{11\bar{0} \text{ V}}{187 \text{ }\Omega}$

$= 0.588 \text{ A}$

(e) DATA: $R = 175 \text{ }\Omega$

$X_L = 66.0 \text{ }\Omega$ [from part (a)]

$\psi = ?$

BASIC EQUATION: $\tan \phi = \dfrac{X_L}{R}$

WORKING EQUATION: same

SUBSTITUTION: $\tan \phi = \dfrac{66.0 \text{ }\Omega}{175 \text{ }\Omega} = 0.3771$

$\phi = 21°$

The current lags the voltage by 21°.

27.3 CAPACITIVE REACTANCE

Capacitive reactance is measured in the same units as resistance and is given by

$$X_C = \frac{1}{2\pi f C}$$

Note that X_C as well as X_L is dependent upon the frequency of the circuit. In a dc circuit, no current will flow with a capacitor in series after the brief fraction of a second required to charge the capacitor.

In the RC circuit the capacitor has the ability to store energy, then supply a current when the applied potential is zero. When a resistor is added to the circuit, the impedance of the circuit is

$$Z = \sqrt{R^2 + X_C^2}$$

where Z is the impedance of circuit,
$\quad$ R is the resistance of circuit, and
$\quad$ X_C is the capacitive reactance (all in Ω).
The voltage will lag behind the current for this circuit. The impedances of the two components do not add algebraically but, rather, add vectorially. The impedance can be represented graphically as the capacitive reactance always being a negative y-component and the resistance a positive x-component. The phase angle ϕ is given by

$$\tan \phi = \frac{X_C}{R}$$

where ϕ is the phase angle,
$\quad$ X_C is the capacitive reactance, and
$\quad$ R is the resistance.

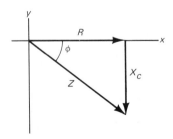

example$\quad$A capacitor of 2.65×10^{-6} F is connected to a $11\bar{0}$-V 60.0-Hz power supply.

(a) What is the capacitive reactance?
(b) When the capacitor is placed in the circuit, with a 175-Ω resistor, what is the impedance of the circuit?
(c) What effective current flows?
(d) What is the phase angle?

(a) DATA:

$$C = 2.65 \times 10^{-6} \text{ F}$$

$$f = 60.0 \text{ Hz}$$

$$X_C = ?$$

BASIC EQUATION:

$$X_C = \frac{1}{2\pi f C}$$

WORKING EQUATION: same

SUBSTITUTION:

$$X_C = \frac{1}{2\pi (60.0 \text{ Hz})(2.65 \times 10^{-6} \text{ F})}$$

$$= 10\bar{0}0 \ \Omega$$

(b) DATA:

$$R = 175 \ \Omega$$

$$X_C = 10\bar{0}0 \ \Omega \quad \text{[from part (a)]}$$

$$Z = ?$$

BASIC EQUATION:

$$Z = \sqrt{R^2 + X_C^2}$$

WORKING EQUATION: same

SUBSTITUTION:

$$Z = \sqrt{(175 \ \Omega)^2 + (10\bar{0}0 \ \Omega)^2}$$

$$= 1020 \ \Omega$$

(c) DATA:

$$\mathscr{E} = 11\bar{0} \text{ V}$$

$$Z = 1020 \ \Omega \quad \text{[from part (b)]}$$

$$I = ?$$

BASIC EQUATION:

$$I = \frac{\mathscr{E}}{Z}$$

WORKING EQUATION: same

SUBSTITUTION:

$$I = \frac{11\bar{0} \text{ V}}{1020 \ \Omega}$$

$$= 0.108 \text{ A}$$

(d) DATA:

$$R = 175 \ \Omega$$

$$X_C = 10\bar{0}0 \ \Omega$$

$$\phi = ?$$

BASIC EQUATION: $\tan\phi = \dfrac{X_C}{R}$

WORKING EQUATION: same

SUBSTITUTION: $\tan\phi = \dfrac{10\bar{0}0\ \Omega}{175\ \Omega} = 5.71$

$$\phi = 8\bar{0}°$$

The voltage lags behind the current by $8\bar{0}°$.

27.4 *LRC* CIRCUITS

When all three components are present in an ac circuit, the impedance of the circuit is given by

$$Z = \sqrt{R^2 + (X_L - X_C)^2}$$

where Z is the impedance of the circuit,
R is the resistance,
X_L is the inductive reactance, and
X_C is the capacitive reactance.
Notice that X_L and X_C add together to give the net y-component value, which adds vectorially to the resistance. The phase angle is now given by

$$\tan\phi = \dfrac{X_L - X_C}{R}$$

The angle ϕ is the phase angle and it corresponds to the angle by which the current lags behind the voltage. When the phase angle is positive, the current lags behind the voltage; but when ϕ is negative, the current leads the voltage.

example 1 An inductor of 0.175 H, a capacitor of 2.65×10^{-6} F, and a resistor of 175 Ω are connected in series to a $11\bar{0}$-V 60.0-Hz power supply.

(a) What is the impedance of the circuit?
(b) What effective current will flow?
(c) What is the phase angle?

(a) DATA: $L = 0.175$ H

$C = 2.65 \times 10^{-6}$ F

$R = 175\ \Omega$

$Z = ?$

BASIC EQUATIONS: $X_L = 2\pi fL$

$$X_C = \frac{1}{2\pi f C}$$

$$Z = \sqrt{R^2 + (X_L - X_C)^2}$$

WORKING EQUATION: Calculate X_L and X_C and substitute in the expression for Z.

SUBSTITUTIONS: $X_L = 2\pi(60.0 \text{ Hz})(0.175 \text{ H})$

$X_L = 66.0 \ \Omega$

$$X_C = \frac{1}{2\pi(60.0 \text{ Hz})(2.65 \times 10^{-6} \text{ C})}$$

$X_C = 10\bar{0}0 \ \Omega$

$$Z = \sqrt{(175 \ \Omega)^2 + (66.0 \ \Omega - 10\bar{0}0 \ \Omega)^2}$$

$= 95\bar{0} \ \Omega$

(b) DATA: $\mathscr{E} = 11\bar{0} \text{ V}$

$Z = 95\bar{0} \ \Omega$

$I = ?$

BASIC EQUATION: $I = \dfrac{\mathscr{E}}{Z}$

WORKING EQUATION: same

SUBSTITUTION: $I = \dfrac{11\bar{0} \text{ V}}{95\bar{0} \ \Omega}$

$= 0.116 \text{ A}$

(c) DATA: $X_L = 66.0 \ \Omega$ [from part (a)]

$X_C = 10\bar{0}0 \ \Omega$ [from part (a)]

$R = 175 \ \Omega$

$\phi = ?$

BASIC EQUATION: $\tan \phi = \dfrac{X_L - X_C}{R}$

WORKING EQUATION: same

SUBSTITUTION: $\tan \phi = \dfrac{66.0 \ \Omega - 10\bar{0}0 \ \Omega}{175 \ \Omega} = -5.34$

$\phi = -79°$

The voltage will lag behind current by 79°.

It will be noted that the inductive reactance and capacitive reactance may be equal in numerical value:

$$X_L = X_C$$

$$2\pi f L = \frac{1}{2\pi f C}$$

$$\boxed{f = \frac{1}{2\pi\sqrt{LC}}}$$

This value of f is called the resonance frequency. At this frequency the maximum energy is present in the circuit. The circuit responds most efficiently to voltages or currents of this frequency.

If no resistance were present in the circuit, the circuit would oscillate indefinitely with the energy alternating between the electric field in the capacitor and the magnetic field in the inductor. The frequency for resonance can be adjusted by changing either C or L. Most radios use a variable capacitor to tune the radio so that its frequency responds to the radio frequency of the desired station.

When resistance is added to the circuit, it may slightly change the resonance frequency, but its major effect is to dampen the oscillation of the circuit. That is, with each oscillation some energy is dissipated, with the result that the charge moving back and forth becomes less and less.

example 2 A variable capacitor can be adjusted from 10.0×10^{-12} F to 102×10^{-12} F. This capacitor is to be used to tune an AM radio (0.50×10^6 Hz to 1.60×10^6 Hz). What value of inductance should be used in series with the capacitor?

DATA:
$$C = 10.0 \times 10^{-12} \text{ F to } 102 \times 10^{-12} \text{ F}$$

$$f = 1.60 \times 10^6 \text{ Hz to } 0.50 \times 10^6 \text{ Hz}$$

$$L = ?$$

Because of the inverse proportion, the higher frequency corresponds to the lower capacitance and the lower frequency corresponds to the higher capacitance.

BASIC EQUATION:
$$f = \frac{1}{2\pi\sqrt{LC}}$$

WORKING EQUATION:
$$L = \frac{1}{(2\pi f)^2 C}$$

SUBSTITUTIONS:
$$L = \frac{1}{[2\pi(1.60 \times 10^6 \text{ Hz})]^2 (10.0 \times 10^{-12} \text{ F})}$$

$$=0.99 \times 10^{-3} \text{ H} = 99 \text{ mH}$$

$$L = \frac{1}{[2\pi(0.50 \times 10^6 \text{ Hz})]^2 (102 \times 10^{-12} \text{ F})}$$

$$=0.99 \times 10^{-3} \text{ H} = 99 \text{ mH}$$

27.5 RECTIFICATION

It is sometimes desirable to change ac into dc. This process is called *rectification*. Adapter-chargers for calculators, tape recorders, and portable radios take line ac voltage and reduce it and rectify it to charge and/or operate dc appliances. A device that permits current to pass in one direction but not in the other is called a *diode*. Alternating current is allowed to pass only in one direction and is thus changed into a direct current.

Additional circuit devices can be added to the rectifier which will smooth out the pulsed dc current somewhat.

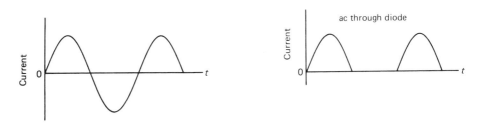

When simply a single diode is present in a circuit, the circuit is called *half-wave-rectified*. A *full-wave-rectified* circuit is shown on page 500. The transformer is centertapped at C and grounded. The points A and B are alternately going to be positively charged. When A is positive, the current will flow *CADC*. No current will pass through the branch *DB* of this circuit because the diode prevents that. One-half cycle later B is positively charged and the current flows *CBDC* with no current passing through the branch *DA* because of the diode. This type of circuit is a full-wave rectified circuit and gives the voltage pattern shown. By use of other components, it is possible to smooth the full-wave-rectified dc supply until it is almost a constant.

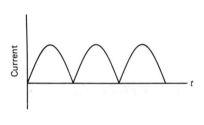

EXERCISES

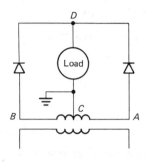

Sketch	Data	Basic Equation	Working Equation	Substitution
	$A = 12 \text{ cm}^2$	$A = bh$	$h = \dfrac{A}{b}$	$h = \dfrac{12 \text{ cm}^2}{6.0 \text{ cm}}$
	$b = 6.0 \text{ cm}$			
	$h = ?$			$= 2.0 \text{ cm}$

Find the inductive reactance of the following inductors at the given frequency.

1. $L = 4.00 \times 10^{-2}$ H, $f = 1.00 \times 10^4$ Hz

2. $L = 8.00 \times 10^{-1}$ H, $f = 10.0$ Hz

3. $L = 7.00 \times 10^{-4}$ H, $f = 90.0$ Hz

4. $L = 2.00 \times 10^{-2}$ H, $f = 1.00 \times 10^3$ Hz

5. $L = 5.00 \times 10^{-3}$ H, $f = 60.0$ Hz

Find the capacitive reactance of the circuits with the following characteristics.

6. $C = 6.00 \times 10^{-4}$ F, $f = 1.00 \times 10^5$ Hz

7. $C = 7.00 \times 10^{-3}$ F, $f = 1.00 \times 10^3$ Hz

8. $C = 3.00 \times 10^{-3}$ F, $f = 1.00 \times 10^4$ Hz

9. $C = 4.00 \times 10^{-5}$ F, $f = 1.00 \times 10^2$ Hz

10. $C = 8.00 \times 10^{-4}$ F, $f = 1.00 \times 10^2$ Hz

Find the impedance and phase angle of circuits with the following characteristics.

11. $R = 10\bar{0}0$ Ω, $L = 2.00 \times 10^{-3}$ H, $f = 40\bar{0}0$ Hz

12. $R = 2.00 \times 10^3$ Ω, $L = 7.00 \times 10^{-2}$ H, $f = 12\bar{0}$ Hz

13. $R = 15\bar{0}$ Ω, $L = 1.00 \times 10^{-2}$ H, $f = 15\bar{0}0$ Hz

14. $R = 3.00 \times 10^2$ Ω, $L = 3.00 \times 10^{-3}$ H, $f = 90\bar{0}$ Hz

15. $R = 20\bar{0}$ Ω, $L = 5.00 \times 10^{-2}$ H, $f = 50\bar{0}$ Hz

Find the impedance and phase angle of each circuit.

16. $R = 10\bar{0}0 \ \Omega$, $C = 1.00 \times 10^{-6}$ F, $f = 245$ Hz
17. $R = 25\bar{0} \ \Omega$, $C = 5.00 \times 10^{-6}$ F, $f = 52\bar{0}$ Hz
18. $R = 6.50 \ \Omega$, $C = 50.0 \times 10^{-6}$ F, $f = 10\bar{0}0$ Hz
19. $R = 15\bar{0} \ \Omega$, $C = 9.00 \times 10^{-6}$ F, $f = 10\bar{0}0$ Hz
20. $R = 12.0 \ \Omega$, $C = 4.50 \times 10^{-6}$ F, $f = 75.0$ Hz

Find the impedance and current in the following circuits.

21. $R = 45.0 \ \Omega$, $L = 0.500$ H, $C = 1.60 \times 10^{-6}$ F, $f = 60.0$ Hz, $\mathcal{E} = 5.00$ V
22. $R = 175 \ \Omega$, $L = 2.00 \times 10^{-2}$ H, $C = 22.0 \times 10^{-6}$ F, $f = 50\bar{0}$ Hz, $\mathcal{E} = 10.0$ V
23. $R = 95.0 \ \Omega$, $L = 1.00 \times 10^{-2}$ H, $C = 3.00 \times 10^{-6}$ F, $f = 260$ Hz, $\mathcal{E} = 15.0$ V
24. $R = 60.0 \ \Omega$, $L = 7.00 \times 10^{-3}$ H, $C = 2.60 \times 10^{-9}$ F, $f = 60.0$ Hz, $\mathcal{E} = 20.0$ V
25. $R = 162 \ \Omega$, $L = 6.30 \times 10^{-3}$ H, $C = 30.0 \times 10^{-9}$ F, $f = 12\bar{0}$ Hz, $\mathcal{E} = 25.0$ V

Find the resonant frequencies for the following circuits.

26. $C = 63.0 \times 10^{-9}$ F, $L = 0.100 \times 10^{-3}$ H
27. $C = 12.5 \times 10^{-12}$ F, $L = 0.550 \times 10^{-3}$ H
28. $C = 0.570 \times 10^{-6}$ F, $L = 1.50 \times 10^{-6}$ H
29. $C = 1.33 \times 10^{-6}$ F, $L = 3.70 \times 10^{-9}$ H
30. $C = 0.750 \times 10^{-12}$ F, $L = 15.0 \times 10^{-3}$ H

28.1 ELECTROMAGNETIC SPECTRUM

Electromagnetic radiation may be produced by an oscillatory electric circuit as described in Section 27.4. Electromagnetic radiation may come from a light bulb heated to incandescence, from an X-ray machine in a hospital, from a television station, or as a form of cosmic radiation from outer space. These are all part of the electromagnetic spectrum. They come from many different sources but are all transverse electromagnetic waves and differ from each other only in their frequency of oscillation. It is by this frequency that we classify these radiations into categories as shown in the figure below.

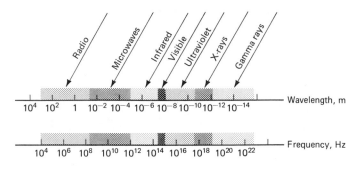

The Electromagnetic Spectrum

The velocity of electromagnetic radiation in a vacuum is a constant and is the same for all frequencies. This velocity is commonly referred to as the speed of light and is represented by the letter c.

$$c = 2.99 \times 10^8 \text{ m/s}$$

The frequency and the wavelength of a wave are related by the expression

$$\boxed{c = \lambda f}$$

where λ is the wavelength,

f is the frequency, and

c is the speed of light.

It is common to talk about a radiation in terms of its frequency or wavelength. A high-frequency radiation has a correspondingly short wavelength while a lower-frequency radiation has a longer wavelength. Because our eyes are able to perceive electromagnetic radiation of about 4×10^{-7} m to 7×10^{-7} m, this small region of the electromagnetic spectrum is called the *visible region* and we refer to it as *light*. The visible region will serve as our model for discussion of the properties of the entire spectrum.

As with matter waves, electromagnetic waves transmit energy. Electromagnetic waves are the means by which energy is transmitted from the sun to the earth. Light has the same characteristics that matter waves have:

1. Reflection at the surface of a medium.

2. Refraction when passing from one medium to another.

3. Interference (constructive and destructive) when two waves are properly superimposed.

4. Diffraction (bending) when the waves pass through an aperture or around an obstacle. When light is absorbed, emitted, or scattered by matter, it behaves as if it were a particle.

example 1   The distance from the earth to the moon is 3.80×10^8 m. How long does it take for a light signal to travel this distance?

DATA:

$s = 3.80 \times 10^8$ m

$v = 2.99 \times 10^8$ m/s

$t = ?$

BASIC EQUATION:

$s = vt$

WORKING EQUATION:

$t = \dfrac{s}{v}$

SUBSTITUTION:

$t = \dfrac{3.80 \times 10^8 \text{ m}}{2.99 \times 10^8 \text{ m/s}}$

$= 1.27$ s

example 2   The wavelength of yellow light is 5.80×10^{-7} m. What is the frequency of yellow light?

DATA:

$$\lambda = 5.80 \times 10^{-7}\,\text{m}$$

$$c = 2.99 \times 10^{8}\,\text{m/s}$$

$$f = ?$$

BASIC EQUATION:

$$c = \lambda f$$

WORKING EQUATION:

$$f = \frac{c}{\lambda}$$

SUBSTITUTION:

$$f = \frac{2.99 \times 10^{8}\,\text{m/s}}{5.80 \times 10^{-7}\,\text{m}} \qquad \boxed{\frac{\text{m/s}}{\text{m}} = \frac{1}{\text{s}} = \text{Hz}}$$

$$= 5.16 \times 10^{14}\,\text{Hz}$$

28.2   ELECTROMAGNETIC WAVES

The oscillating *LC* circuit of Section 27.4 serves as a source of electromagnetic waves when coupled with an antenna as shown below.

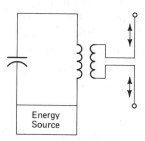

Energy Source

Charges oscillating back and forth in the antenna produce varying electric and magnetic fields which are propagated away from the antenna with a velocity *c*. These varying fields are the *electromagnetic radiation*. In fact, the field strength is the displacement of the electromagnetic waves. The field is a function of both position and time and can be superimposed to produce both constructive and destructive interference. These interference patterns can be observed in thin films such as soap bubbles and oil slicks. Interference effects are the wave nature of light.

28.3   LIGHT AS A PARTICLE

Light sometimes behaves as if it were a particle with directed momentum. Scattering of light by free electrons and the ejection of electrons when light strikes certain metal surfaces are examples of the particle nature of light. Each particle of light is called a photon and has an energy E given by

$$E = hf$$

where f is the frequency and
$\quad$ $h = 6.62 \times 10^{-34}$ J s (Planck's constant).
$\quad$ Photons may be thought of as light particles with specific energies that travel with the speed of light.

example$\quad$ What is the energy of a photon of yellow light ($\lambda = 5.80 \times 10^{-7}$ m)?

DATA:
$$\lambda = 5.80 \times 10^{-7} \text{ m}$$
$$c = 2.99 \times 10^8 \text{ m/s}$$
$$h = 6.62 \times 10^{-34} \text{ J s}$$
$$E = ?$$

BASIC EQUATIONS:
$$c = \lambda f$$
$$E = hf$$

WORKING EQUATION:
$$E = \frac{hc}{\lambda}$$

SUBSTITUTION:
$$E = \frac{(6.62 \times 10^{-34} \text{ J s})(2.99 \times 10^8 \text{ m/s})}{5.80 \times 10^{-7} \text{ m}}$$
$$= 3.41 \times 10^{-19} \text{ J}$$

28.4   ILLUMINATION

The determination of the necessary light sources for proper illumination in homes, business, and industry is often a matter of concern for engineering technicians. We will consider here some simple problems of this type.
$\quad$ The intensity (strength), I, of a light source is measured in terms of candles or candelas (cd). The early use of certain candles for standards of illumination led to the name of the unit. We now use a platinum source at a certain temperature as the standard for comparison. Another unit, the lumen (lm) is often used for the measurement of the intensity of a source. The conversion factor between candelas and lumens is

$$1 \text{ cd} = 4\pi \text{ lm}$$

Thus, a certain 40-W light bulb, which is rated at 35 cd, would have a rating of 440 lm.

$$35 \text{ cd} \times \frac{4\pi \text{ lm}}{1 \text{ cd}} = 440 \text{ lm}$$

When a surface is illuminated by a light source, the intensity of the illumination decreases as the distance between the source and the surface increases. If the source radiates light uniformly in all directions, the light is uniformly distributed over a spherical surface centered at the source. Since the surface area of a sphere is $4\pi r^2$, the intensity of illumination, E, at the surface is given by

$$E = \frac{I}{4\pi r^2}$$

where I is the intensity of the source in lumens and
   r is the distance between the source and the illuminated surface.

example   Find the intensity of illumination on a surface located 3.00 m from a source of $25\bar{0}$ lm.

DATA: $I = 25\bar{0}$ lm

                                        $r = 3.00$ m

                                        $E = ?$

BASIC EQUATION: $E = \dfrac{I}{4\pi r^2}$

WORKING EQUATION: same

SUBSTITUTION: $E = \dfrac{25\bar{0} \text{ lm}}{4\pi(3.00 \text{ m})^2}$

                                        $= 2.21 \dfrac{\text{lm}}{\text{m}^2}$

                                        $= 2.21$ lx

The unit used for intensity of illumination is lm/m^2 and is defined as a lux (lx) in the SI system.

28.5   PINHOLE CAMERA

The pinhole camera will be used to define terms and to illustrate a simple optical system. Refer to the figure below.

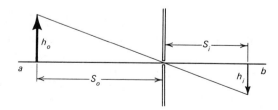

The pinhole camera uses a small round aperture as its optical element. The line *ab*, which passes through the center of the aperture and is perpendicular to the plane of the aperture, is called the *optical axis*. An object of length h_o is placed a distance S_o from the aperture. The distance is called the *object distance*. A ray of light traced from each point on the object through the aperture will strike a screen at S_i. For each point on the object there is a corresponding point on the screen forming an image of height h_i. The distance from the aperture to the screen is called the *image distance*. The ratio of the size of the image to the size of the object is called the *magnification*. It can be seen by similar triangles that

$$\frac{h_i}{h_o} = \frac{S_i}{S_o}$$

We can express the magnification as

$$m = -\frac{S_i}{S_o}$$

where *m* is the magnification,
 S_i is the distance from the optical element to the image, and
 S_o is the distance from the optical element to the object.
 The negative sign signifies that the image is inverted with respect to the object.
 One would think that the smaller the aperture, the better defined each point on the image would be. This statement is not correct for small apertures. Rays of light passing through an aperture tend to spread out, with each point across the aperture serving as a point source. This spreading of waves is called *diffraction* and the smaller the ratio of the diameter of the aperture to the wavelength of the light, the wider the spread.

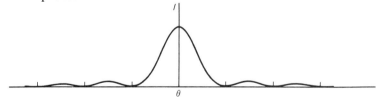

In the figure the central maximum is widened, and on either side, we find other maxima. These second, third, and so on maxima are interference effects from points across the aperture. The equation is

$$\frac{\Delta d}{d} = 1.22\frac{\lambda}{a}$$

where λ is the wavelength of light,

      a is the diameter of the aperture,

      d is the distance from the aperture to the object, and

      Δd is the minimum distance between two points of the object that can be resolved.

The criterion above gives the minimum distance that two points can be apart on the object and still be identified as two points on the image. This criterion is based on diffraction and limits the resolution of optical instruments such as telescopes and microscopes.

    example 1   A girl 1.50 m tall stands 5.00 m in front of a pinhole camera. The distance from the pinhole to the film is 10.0 cm.

(a) What is the magnification?

(b) How tall will her image be on the film?

(c) She is wearing a polka dot dress. The separation between polka dots is 2.00 cm. The pinhole is 2.0 mm in diameter. Assume the average wavelength of light to be 5.5×10^{-7} m. Will the polka dots be resolved in the picture?

(a) DATA:

$$S_o = 5.00 \text{ m}$$

$$S_i = 10.0 \text{ cm} = 0.100 \text{ m}$$

$$m = ?$$

BASIC EQUATION:

$$m = -\frac{S_i}{S_o}$$

WORKING EQUATION: same

SUBSTITUTION:

$$m = -\frac{0.100 \text{ m}}{5.00 \text{ m}}$$

$$= -0.0200$$

The negative sign signifies that the image on the film is inverted with respect to the girl.

(b) DATA:

$$h_o = 1.50 \text{ m}$$

$$m = -0.0200$$

$$h_i = ?$$

BASIC EQUATION:

$$\frac{h_i}{h_o} = m$$

WORKING EQUATION:

$$h_i = mh_o$$

SUBSTITUTION:

$$h_i = (-0.0200)(1.50 \text{ m})$$

$$= -0.0300 \text{ m}$$

The image will be 3.00 cm high and inverted.

(c) DATA:

$$d = 5.00 \text{ m}$$

$$\lambda = 5.50 \times 10^{-7} \text{ m}$$

$$a = 2.00 \times 10^{-3} \text{ m}$$

$$\Delta d = ?$$

BASIC EQUATION:

$$\frac{\Delta d}{d} = 1.22 \frac{\lambda}{a}$$

WORKING EQUATION:

$$\Delta d = 1.22 \frac{\lambda}{a} d$$

SUBSTITUTION:

$$\Delta d = (1.22) \left(\frac{5.50 \times 10^{-7} \text{ m}}{2.00 \times 10^{-3} \text{ m}} \right) (5.00 \text{ m})$$

$$= 1.68 \times 10^{-3} \text{ m} = 1.68 \text{ mm}$$

This answer tells us that two points 1.68 mm apart on the girl's dress would be resolved, so the polka dots would be clearly visible.

example 2 An aerial photographer is taking pictures of the ground from an altitude of 3.00 km. The aperture of the camera is 1.50 cm. Using the average wavelength of light as 5.50×10^{-7} m, find what distance on the earth will be resolved.

DATA:

$$d = 3.00 \times 10^3 \text{ m}$$

$$a = 1.50 \times 10^{-2} \text{ m}$$

$$\lambda = 5.50 \times 10^{-7} \text{ m}$$

$$\Delta d = ?$$

BASIC EQUATION:

$$\frac{\Delta d}{d} = 1.22 \frac{\lambda}{a}$$

WORKING EQUATION:

$$\Delta d = 1.22 \frac{\lambda}{a} d$$

SUBSTITUTION:

$$\Delta d = (1.22) \left(\frac{5.50 \times 10^{-7} \text{ m}}{1.50 \times 10^{-2} \text{ m}} \right) (3.00 \times 10^3 \text{ m})$$

$$= 0.134 \text{ m} - 13.4 \text{ cm}$$

Two points that are less than 13.4 cm apart could not be distinguished on the picture.

EXERCISES

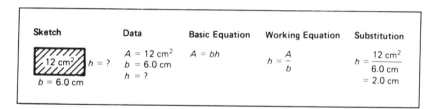

1. The mean Earth–sun distance is 1.49×10^{11} m. How long does it take for light from the sun to reach Earth?

2. How much time is required for a radiowave to travel around the earth? (Mean radius of the earth is 6.37×10^6 m.)

3. An X-ray tube is placed at a distance of 0.50 m from the film pack. How much time is required for the X-ray photon to travel this distance?

4. Find the distance traveled by a microwave in 5.0 s.

5. A television signal is sent to a communications satellite which is 32,000 km above a relay station. How long does it take for the signal to reach the satellite?

6. A radar wave that is bounced off an approaching storm returns to the radar transceiver in 3.0×10^{-5} s. How far is the storm from the radar station?

7. The wavelength of X-rays used for some radiography is 1.50×10^{-10} m.
   (a) What is the frequency of this radiation?
   (b) What energy is associated with a photon of this radiation?

8. A certain infrared source has an average wavelength of 8.0×10^{-6} m.
   (a) What is the frequency of this radiation?
   (b) What is the energy of a photon of this infrared light?

9. Find the intensity of the light source necessary to produce an illumination of 5.50 lx at a distance of 7.85 m from the source.

10. Find the intensity of illumination on a surface located 5.00 m from a source of 450 lm.

11. Find the intensity of the light source necessary to produce an illumination of 2.39 lx at a distance of 3.50 m from the source.

12. Find the intensity of the light source necessary to produce an illumination 5.28 lx at a distance of 6.50 m from the source.

13. The 1.00×10^{-10} m γ-rays given off by a certain radioactive sample are used to produce a picture of itself by means of a 1.0-mm aperture in a lead sheet. The sample is 10.0 cm from the pinhole and the film is placed 25.0 cm from the pinhole. What is the magnification of this arrangement?

14. A man 1.50 m tall is photographed with a pinhole camera from a distance of 3.50 m. The distance from the pinhole to the film is 20.0 cm. How tall is the man's image on the film?

15. A telescope with an objective lens that is 7.50 cm in diameter is used to spot on the rifle range. What is the minimum separation between bullet holes that can be detected on a target $50\overline{0}$ m away?

16. A pair of binoculars with 50.0-mm objective lenses is used to look at the moon. What is the smallest distance on the moon that can be resolved by the binoculars? (The mean Earth–moon distance is 3.80×10^5 km.)

REFLECTION

29.1 REFLECTION AND REFRACTION OF LIGHT

When light is incident upon an interface between two mediums, a portion of the light may enter the new medium. Light that does not enter the new medium is returned from the interface to the original medium. The portion of light entering the new medium changes its direction and is said to be *refracted*. The light that was returned from the interface is said to be *reflected*. Refracted light is discussed in Chapter 30. This chapter deals with reflected light.

The fraction of light that is reflected at an interface depends upon the materials on either side of the interface. In our discussion of reflection, we will consider a well-polished surface in air and all the light reflected at this surface. Such a surface is called a mirror. There are three types of mirrors to consider: plane, concave, and convex. The *plane mirror* is the most common—it gives an image that is the same size as the object. Plane mirrors are commonly found in our homes and businesses. A *concave mirror*, sometimes called a converging mirror, is commonly used as a shaving mirror or makeup mirror. A *convex mirror*, sometimes called a diverging mirror, is used as, for example, the wide-angle rearview mirror on a car or truck. Stores use convex mirrors to give a wide view of their aisles to serve as a security precaution.

29.2   LAW OF REFLECTION

The *law of reflection* requires that the angle of reflection be equal to the angle of incidence. The *angle of incidence*, θ_i, is the angle that the incoming light rays make with a normal to the mirror surface. The *angle of reflection*, θ_r, is the angle the reflected ray makes with the normal to the mirror surface. That is,

$$\theta_i = \theta_r$$

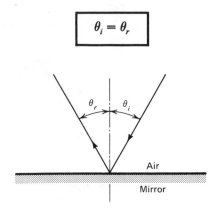

29.3   IMAGES FORMED BY A PLANE MIRROR

A candlestick is placed in front of a plane mirror. The mirror forms an image of the candlestick which appears as far behind the mirror as the candlestick is in front. The image of the candlestick seems to be the same size as the candlestick itself.

In the drawing below, the distance that the object is in front of the mirror is called the *object distance* and is denoted as S_o.

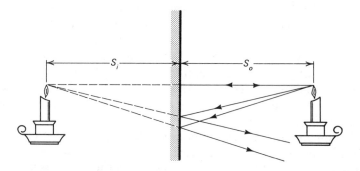

The distance that the image appears from the mirror is called the *image distance* and denoted S_i. In the figure, three rays of light are shown emerging from the tip of the flame. Each of these rays strikes the mirror and is reflected. The law of reflection demands that the angle of reflection be the same as the angle of incidence. The ray that comes to the mirror normal to the surface is reflected at the same angle. All the reflected rays of light that originated at the actual tip of the flame appear to come from a common point: namely, the tip of the flame in the image. Of course, these rays of reflected light are never at the image but rather diverge from the surface in such a pattern that they appear to come from a common point behind the mirror. This statement could be made for each point on the candlestick and each corresponding point on the image.

An image, such as formed by the plane mirror, where the rays of light do not actually pass through a point but diverge from the reflecting surface is called a *virtual image*. An image where the rays leaving the surface of the mirror converge through a common point is called a *real image*.

The image formed by the plane mirror has the same spatial direction as the object, so the image is said to be *erect*. If the image is turned upside down, the image is said to be *inverted*.

Since the image and the object are the same size, we say that the magnification, m, is unity. Also, note that the image distance is the same as the object distance but is behind the mirror rather than in front of it.

In summary

> For a plane mirror:
> the image is erect,
> the image is virtual,
> $m = 1$, and
> $S_i = - S_o$

example   A person is 1.80 m tall. What is the smallest plane mirror in which the person can have a full-length view?
This problem divides itself into the following parts:

(a) Find the position of the top of the mirror in which the person can see the top of his or her head.
(b) Find the position of the bottom of the mirror in which the person can see the bottom of his or her feet.
(c) Find the length of the mirror.

To determine the position of the top of the mirror, make an assumption that the person's eyes are 0.10 m below the top of his or her head or at a height of 1.70 m above the ground.

SKETCH:

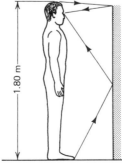

(a) DATA:

$$\text{top of the head} = 1.80 \text{ m}$$

$$\text{position of the eyes} = 1.70 \text{ m}$$

$$\text{top of the mirror} = ?$$

BASIC EQUATION:

$$\theta_r = \theta_i$$

This law of reflection requires that the "top of the mirror" (the point of reflection) be at the midpoint between the "top of the head" and the "position of the eyes."

WORKING EQUATION: top of the mirror =

$$\frac{(\text{top of the head}) + (\text{position of the eyes})}{2}$$

SUBSTITUTION:

$$\text{top of the mirror} = \frac{1.80 \text{ m} + 1.70 \text{ m}}{2}$$

$$= 1.75 \text{ m}$$

(b) DATA:

$$\text{position of the eyes} = 1.70 \text{ m}$$

$$\text{bottom of the feet} = 0.00 \text{ m}$$

$$\text{bottom of the mirror} = ?$$

BASIC EQUATION:

$$\theta_r = \theta_i$$

WORKING EQUATION: bottom of the mirror =

$$\frac{(\text{position of the eyes}) + (\text{bottom of the feet})}{2}$$

SUBSTITUTION:

$$\text{bottom of the mirror} = \frac{1.70 \text{ m} + 0.00 \text{ m}}{2}$$

$$= 0.85 \text{ m}$$

(c) DATA:

$$\text{top of the mirror} = 1.75 \text{ m} \qquad [\text{from part (a)}]$$

$$\text{bottom of the mirror} = 0.85 \text{ m} \qquad [\text{from part (b)}]$$

$$\text{length of the mirror} = ?$$

BASIC EQUATION: length of the mirror = top of the mirror − bottom of the mirror

SUBSTITUTION: length of the mirror = 1.75 m − 0.85 m

$$= 0.90 \text{ m}$$

Notice that the length of the mirror required for the person to just see a full-length view is one-half of his or her height.

29.4   IMAGES FORMED BY CONCAVE MIRRORS

A concave mirror is a mirror whose reflection surface has a center of curvature on the reflecting side of the mirror. A concave mirror is also called a converging mirror because when parallel light is incident upon the mirror, the reflected light converges as shown below.

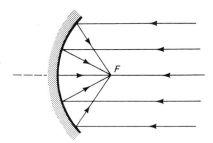

The reflected light passes through a small spot F at a distance, f, from the mirror. This distance is called the *focal length* and is a property of the mirror. If the concave surface of the mirror is parabolic, the spot that the rays converge through is a point, called the *focal point*. If the concave surface of the mirror is spherical but the dimensions of the mirror are small compared to the radius of the sphere, the spot is very small and may be regarded as a point. The focal length for the spherical region is

$$f = \frac{R}{2}$$

where f is the focal length and
$\qquad$ R is the radius of curvature.

The figure below shows a concave mirror and a source (arrow). The position of the focal point on a line, called the *optical axis*, through the center of each mirror is shown along with a point R that corresponds to the center of curvature for a spherical mirror. (Recall that $R = 2f$.)

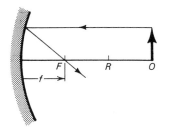
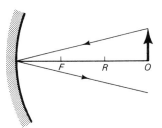
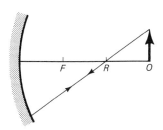

In each drawing, the path of a ray is shown. The first ray shown travels parallel to the optical axis to the mirror and is then reflected through the focal point, F. This behavior follows from the definition of focal length. The second ray shown travels from the object to the point where the optical axis intersects the surface of the mirror. The law of reflection requires that this ray reflect at an angle equal to the angle of incidence. The third ray shown passes from the object through the center of curvature to the mirror. This ray travels along the radius and is reflected back along the radius.

The three rays all intersect at a common point. This common point is the image, and the intersections of two of these rays determine its position. The three rays are superimposed in the figure below and it can be seen that the point on the image is defined.

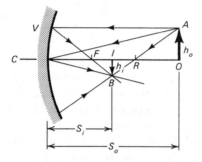

The triangles ACO and BCI are similar triangles and

$$\frac{h_i}{h_o} = \frac{S_i}{S_o}$$

where h_i is the height of the image,

$\quad h_o$ is the height of the object,

$\quad S_i$ is the distance from the mirror to the image, and

$\quad S_o$ is the distance from the mirror to the object.

The triangles CVF and IBF are similar and $CV = h_o$ when it is assumed the CV is small compared to the radius of curvature of the mirror. Thus, the corresponding sides are proportional. That is,

$$\frac{h_o}{f} = \frac{h_i}{S_i - f}$$

Since $\dfrac{h_i}{h_o} = \dfrac{S_i}{S_o}$, we have

$$\frac{S_i}{S_o} = \frac{S_i - f}{f}$$

or,

$$\frac{S_i}{S_o} = \frac{S_i}{f} - 1$$

Divide both sides by S_i

$$\frac{1}{S_o} = \frac{1}{f} - \frac{1}{S_i}$$

which gives

$$\frac{1}{S_o} + \frac{1}{S_i} = \frac{1}{f}$$

where f is the focal length of the mirror. For a concave mirror a graphical location of the position and size of the image may be determined by following these steps:

1. Draw the mirror and its optical axis. Make points on the optical axis at the focal point, F, and at the center of the curvature, R. Locate the object, O, on the figure.

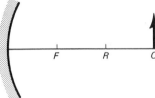

2. Draw a line from the tip of the object to the mirror and through the focal point, as in the figure below.

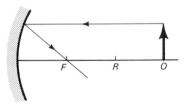

3. Draw a line that passes through the tip of the object and the center of curvature, R, and extend this line until it intersects the line drawn in step 2. This intersection is the tip of the image and determines the position of the image.

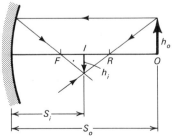

example 1   A concave mirror has a radius of curvature of 0.50 m. An object is placed 0.80 m from the mirror.

(a) What is the focal length of the mirror?
(b) What is the position of the image?
(c) What is the magnification?
(d) Is the image real?
(e) Is the image erect?

(a) DATA: $\qquad\qquad\qquad\qquad\qquad\qquad\quad r = +0.50$ m

$$f = ?$$

BASIC EQUATION: $\qquad\qquad\qquad\qquad\quad f = \dfrac{R}{2}$

WORKING EQUATION: $\qquad\qquad\qquad$ same

SUBSTITUTION: $\qquad\qquad\qquad\qquad\quad f = \dfrac{+0.50 \text{ m}}{2}$

$$= +0.25 \text{ m}$$

(b) DATA: $\qquad\qquad\qquad\qquad\qquad\qquad\quad S_o = 0.80$ m

$$f = 0.25 \text{ m} \qquad \text{[from part (a)]}$$

$$S_i = ?$$

BASIC EQUATION: $\qquad\qquad\qquad\qquad\quad \dfrac{1}{S_o} + \dfrac{1}{S_i} = \dfrac{1}{f}$

SUBSTITUTION: $\qquad\qquad\qquad\qquad\quad \dfrac{1}{0.80 \text{ m}} + \dfrac{1}{S_i} = \dfrac{1}{0.25 \text{ m}}$

$$S_i = 0.36 \text{ m}$$

(c) DATA: $\qquad\qquad\qquad\qquad\qquad\qquad\quad S_o = 0.80$ m

$$S_i = 0.36 \text{ m}$$

$$m = ?$$

BASIC EQUATION: $\qquad\qquad\qquad\qquad\quad m = -\dfrac{S_i}{S_o}$

WORKING EQUATION: $\qquad\qquad\qquad$ same

SUBSTITUTION: $\qquad\qquad\qquad\qquad\quad m = -\dfrac{0.36 \text{ m}}{0.80 \text{ m}}$

$$= -0.45$$

(d) Yes. The image is real because the image distance, S_i, is positive. A positive value of S_i ensures that the image is formed by converging rays and passes through a common point.

(e) No. The image is inverted. The negative value of the magnification means that the image is inverted.

Example 1 is shown schematically below. In this figure only two rays are used: one passes from the mirror through the focal point, F; the other passes through the center of curvature, R.

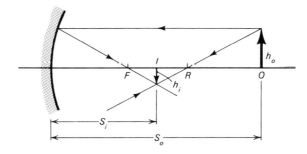

example 2 An object is placed 0.15 m in front of a concave mirror, whose focal length is 0.20 m.

(a) What is the position of the image?
(b) What is the magnification?
(c) Is the image real?
(d) Is the image erect?

(a) DATA:

$$f = 0.20 \text{ m}$$

$$S_o = 0.15 \text{ m}$$

$$S_i = ?$$

BASIC EQUATION:

$$\frac{1}{S_o} + \frac{1}{S_i} = \frac{1}{f}$$

SUBSTITUTION:

$$\frac{1}{0.15 \text{ m}} + \frac{1}{S_i} = \frac{1}{0.20 \text{ m}}$$

$$S_i = -0.60 \text{ m}$$

(b) DATA:

$$S_o = 0.15 \text{ m}$$

$$S_i = -0.60 \text{ m} \qquad \text{[from part (a)]}$$

$$m = ?$$

BASIC EQUATION:

$$m = -\frac{S_i}{S_o}$$

WORKING EQUATION: same

SUBSTITUTION:

$$m = -\frac{-0.60 \text{ m}}{0.15 \text{ m}}$$

$$= 4.0$$

(c) No. The image is virtual, (S_i is negative).
(d) Yes. The image is erect, (m is positive).

Example 2 is shown schematically below.

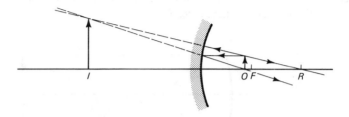

29.5   IMAGES FORMED BY CONVEX MIRRORS

A convex mirror is a mirror whose center of curvature is on the side away from the reflecting surface of the mirror. A convex mirror is also called a diverging mirror because when parallel light is incident upon the mirror, the reflected light diverges from the surface. The figure below shows that the divergent light rays appear to be coming from a point behind the mirror.

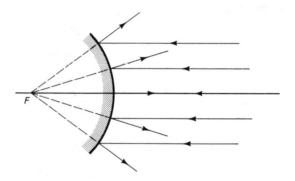

This point, F, is called the focal point of the mirror and the distance from the mirror to this point is called the focal length, f. This focal point is a virtual focal point and the focal length is negative because it is not on the reflecting side of the mirror. As with a concave mirror, the focal length of a small convex mirror is half the radius of curvature, that is, $f = R/2$.

A graphical construction of the image can be made by a convex mirror by adapting the rules set down for the concave mirror.

1. Draw the mirror and its optical axis. Make points on the optical axis at the focal point, F, and the center of curvature, R. Locate the object, O, on the figure.

2. Draw a line from the tip of the object to the mirror. Reflect the ray from the mirror so that its extension passes through the focal point (top figure page 521).

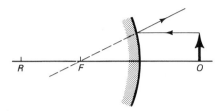

3. Draw a line that passes through the tip of the object and the center of curvature. The intersection of this line and the line in step 2 determines the position of the image.

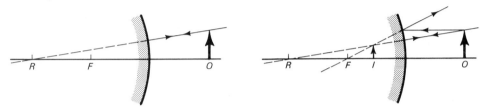

In the figure below, we have drawn the line in step 2 which passes through the focal point and also a ray that strikes the convex mirror at the intersection of the mirror and its optical axis and is reflected at the same angle.

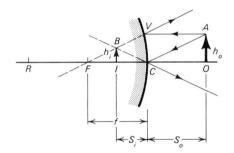

The reflected ray is extended to intersect the ray passing through the focal point. This intersection determines the position of the image. When the mirror is small compared to its radius of curvature, the line VC is equal to h_o and triangles VCF and BIF are similar. Triangles BIC and AOC are also similar; thus,

$$\frac{f - S_i}{f} = \frac{h_i}{h_o} = \frac{S_i}{S_o}$$

or

$$\frac{1}{S_o} + \frac{1}{-S_i} = \frac{1}{-f}$$

This expression is the same as the one found for the concave mirror if one recalls that the image distance and the focal length were pointed out to be negative at the beginning of this section. That is,

$$\frac{1}{S_o} + \frac{1}{S_i} = \frac{1}{f}$$

A positive focal length and a positive radius of curvature are always associated with a concave mirror. A negative focal length and a negative radius of curvature are always associated with a convex mirror. If the rays of light converge through a point, the image is real and can be displayed on a screen. If the rays of light diverge from the reflective surface and their extension converges behind the mirror, the image is virtual and the image distance is negative.

example   A convex mirror has a focal length of -0.25 m. An object is placed at a distance of 0.35 m from the mirror.

(a) What is the position of the image?
(b) What is the magnification?
(c) Is the image real?
(d) Is the image erect?

(a) DATA:
$$f = -0.25 \text{ m}$$
$$S_o = 0.35 \text{ m}$$
$$S_i = ?$$

BASIC EQUATION:
$$\frac{1}{S_o} + \frac{1}{S_i} = \frac{1}{f}$$

SUBSTITUTION:
$$\frac{1}{0.35 \text{ m}} + \frac{1}{S_i} = \frac{1}{-0.25 \text{ m}}$$
$$S_i = -0.15 \text{ m}$$

(b) DATA:
$$S_o = 0.35 \text{ m}$$
$$S_i = -0.15 \text{ m} \qquad \text{[from part (a)]}$$
$$m = ?$$

BASIC EQUATION:
$$m = -\frac{S_i}{S_o}$$

WORKING EQUATION: same

SUBSTITUTION:
$$m = -\frac{-0.15 \text{ m}}{0.35 \text{ m}}$$
$$= 0.43$$

(c) No. The image is virtual. (S_i is negative.)
(d) Yes. The image is erect. (m is positive.)

A schematic drawing of this example is shown below.

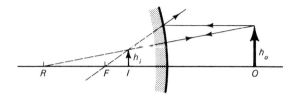

29.6 THE MIRROR FORMULA

The equations

$$\frac{1}{S_o} + \frac{1}{S_i} = \frac{1}{f}$$

$$m = -\frac{S_i}{S_o}$$

where S_o is the object distance,
S_i is the image distance,
f is the focal length, and
m is the magnification,
have been demonstrated for concave and convex mirrors.

These equations are also true for plane mirrors. The radius of curvature for the plane mirror is infinite and its focal length is also infinite. The mirror equation reduces to

$$\frac{1}{S_o} + \frac{1}{S_i} = 0$$

or

$$S_i = -S_o$$

and the magnification is

$$m = -\frac{S_i}{S_o} = 1$$

which tells that the image is virtual, erect, and the same size as the object.

EXERCISES

Sketch	Data	Basic Equation	Working Equation	Substitution
$b = 6.0$ cm	$A = 12$ cm^2 $b = 6.0$ cm $h = ?$	$A = bh$	$h = \dfrac{A}{b}$	$h = \dfrac{12 \text{ cm}^2}{6.0 \text{ cm}}$ $= 2.0$ cm

Determine the values of f, S_i, and m in Exercises 1–12.

1. $R = 0.50$ m
$S_o = 0.60$ m

2. $R = 0.50$ m
$S_o = 0.50$ m

3. $R = 0.50$ m
$S_o = 0.30$ m

4. $R = 0.50$ m
$S_o = 0.20$ m

5. $R = -0.50$ m
$S_o = 0.35$ m

6. $R = -0.50$ m
$S_o = 0.20$ m

7. $R = 0.40$ m
$S_o = 0.40$ m

8. $R = 0.40$ m
$S_o = 0.60$ m

9. $R = 0.40$ m
$S_o = 0.20$ m

10. $R = 0.40$ m
$S_o = 0.10$ m

11. $R = -0.40$ m
$S_o = 0.15$ m

12. $R = -0.40$ m
$S_o = 0.30$ m

13. A child standing 3.00 m in front of a plane mirror wishes to take a "self-portrait" in the mirror. At what range should the camera be set for sharp focus?

14. A camera and an object are 3.00 m apart. Both camera and object are 2.00 m in front of a plane mirror. What is the correct value of the range setting on the camera for a sharp focus?

15. A man has a shaving mirror whose focal length is 0.15 m. When he shaves, he places his face 0.12 m from the mirror.

(a) Where is the image of the man's face?
(b) What is the magnification?
(c) Is the image erect?
(d) Is the image real?

16. A woman has a makeup mirror whose focal length is 0.18 m. She places her face 0.15 m in front of the mirror when applying lipstick.

(a) Where is the image of the woman's lips?
(b) What is the magnification?
(c) Is the image erect?
(d) Is the image real?

17. At what distance in front of the shaving mirror in Exercise 15 must the man place his face for the image to appear four times larger?

18. At what distance in front of the makeup mirror in Exercise 16 must the woman place her face for the image to appear three times larger?

19. A convex mirror forms an image that is one-half the size of the object. If the focal length of the mirror is -0.10 m, where is the object placed?

20. An object is placed 0.40 m from a convex mirror. It is desired for the image to be one-third the height of the object. What focal-length mirror should be used?

REFRACTION

30.1 REFRACTION OF LIGHT

When light is incident upon an interface between two mediums, a portion of light may enter the new medium and be transmitted through it. This portion of light entering the new medium changes its direction and is said to be refracted. We use the refraction of light to construct prisms and converging and diverging lenses just as we used the reflection of light to make various mirrors. The refraction of light at an interface is shown below. Refraction obeys *Snell's law*:

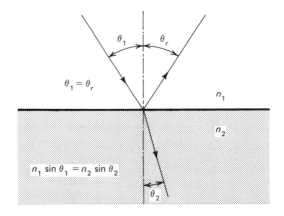

$$n_1 \sin \theta_1 = n_2 \sin \theta_2$$

where n_1 is the index of refraction of medium 1,

n_2 is the index of refraction of medium 2,

θ_1 is the angle the light ray in medium 1 makes with the normal to the interface, and

θ_2 is the angle the light ray in medium 2 makes with the normal to the interface.

The *index of refraction* of a medium is

$$n_1 = \frac{c}{v_1}$$

where n_1 is the index of refraction of medium 1 with respect to a vacuum,

c is the speed of light in a vacuum, and

v_1 is the speed of light in medium 1.

Recall that the speed of light in a vacuum, c, is a constant for all of the electromagnetic spectrum and regardless of the relative motion of the source and observer. The speed of light in a typical medium differs with wavelength and will vary by 2 to 3 percent across the visible region of the electromagnetic spectrum. It is this wavelength variation of the index of refraction that produces the color spectrum seen in rainbows, irregular pieces of glass, prisms, and thin films.

example 1 A ray of light is incident upon a piece of plate glass ($n = 1.65$) at an angle of $3\bar{0}°$. What is the angle of refraction?

SKETCH:

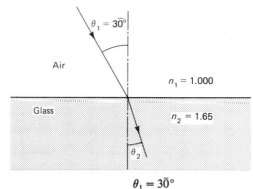

DATA:

$$\theta_1 = 3\bar{0}°$$

$$n_1 = 1.000$$

$$n_2 = 1.65$$

$$\theta_2 = ?$$

BASIC EQUATION: $$n_1 \sin \theta_1 = n_2 \sin \theta_2$$

WORKING EQUATION: $$\sin \theta_2 = \frac{n_1}{n_2} \sin \theta_1$$

SUBSTITUTION: $$\sin \theta_2 = \frac{1.000}{1.65} \sin 3\bar{0}° = 0.303$$

$$\theta_2 = 18°$$

example 2 A sea shell is at a depth of 2.00 m below the surface of clear water ($n = 1.33$). At what depth would the shell appear to an observer directly above the water?

We will first set up the problem for the general case of the view at one side, then restrict the problem to the special case of the observer directly overhead.

SKETCH:

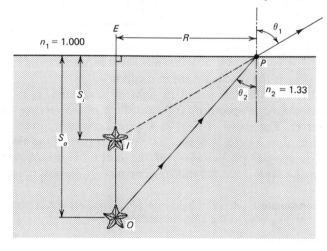

DATA:

$$n_1 = 1.000$$

$$n_2 = 1.33$$

$$S_o = 2.00 \text{ m}$$

$$S_i = ?$$

BASIC EQUATION: $$n_1 \sin \theta_1 = n_2 \sin \theta_2$$

Let R be the distance from the point where the line OE passes through the water surface to the point P, where the refracted ray passes through the surface. Then

$$R = S_i \tan \theta_1 = S_o \tan \theta_2$$

As the viewer moves to a position directly above the shell, both angle θ_1 and angle θ_2 approach zero. Then, as $\theta_1 \rightarrow 0$ and $\theta_2 \rightarrow 0$ and

$$\tan \theta_1 \rightarrow \theta_1$$

$$\tan \theta_2 \rightarrow \theta_2$$

So, $S_i\theta_1 = S_o\theta_2$ (1)

Similarly as $\theta_1 \to 0$ and $\theta_2 \to 0$,

$$\sin \theta_1 \to \theta_1 \text{ and}$$
$$\sin \theta_2 \to \theta_2$$

Then, Snell's law becomes

$$n_1\theta_1 = n_2\theta_2$$ (2)

From equation (1), we have

$$\frac{\theta_1}{\theta_2} = \frac{S_o}{S_i}$$

From equation (2), we have

$$\frac{\theta_1}{\theta_2} = \frac{n_2}{n_1}$$

Thus,

$$\frac{S_o}{S_i} = \frac{n_2}{n_1}$$

WORKING EQUATION: $S_i = \dfrac{n_1}{n_2} S_o$

SUBSTITUTION: $S_i = \dfrac{1.000}{1.33}(2.00 \text{ m})$

$$= 1.50 \text{ m}$$

The sea shell appears to be only 1.50 m below the surface, while its true depth is 2.00 m.

example 3 A prism is in the shape of an equilateral triangle. A ray of white light is incident at an angle of 50.00°.

(a) At what angle does the violet light ($\lambda = 4.000 \times 10^{-7}$ m and $n = 1.535$) emerge from the prism?
(b) At what angle does red light ($\lambda = 7.000 \times 10^{-7}$ m and $n = 1.518$) emerge from the prism?
(c) What is the angular separation between the violet and red light?

SKETCH:

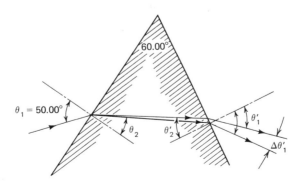

(a) DATA:

$$n_1 = 1.000$$
$$n_2 = 1.535$$
$$\theta_1 = 50.00°$$
$$\theta_2' = ?$$

First find θ_2, then θ_2', and finally θ_1'.

BASIC EQUATION:

$$n_1 \sin \theta_1 = n_2 \sin \theta_2$$

WORKING EQUATION:

$$\sin \theta_2 = \frac{n_1}{n_2} \sin \theta_1$$

SUBSTITUTION:

$$\sin \theta_2 = \frac{1.000}{1.535} (\sin 50.00°) = 0.4991$$
$$\theta_2 = 29.94°$$

Reference to the sketch above shows that the angles

$$\theta_2 + \theta_2' = 60.00°$$

so

$$\theta_2' = 30.06°$$

To find θ_1':

WORKING EQUATION:

$$\sin \theta_1' = \frac{n_2}{n_1} \sin \theta_2'$$

SUBSTITUTION:

$$\sin \theta_1' = \frac{1.535}{1.000} (\sin 30.06°) = 0.7689$$
$$\theta_1' = 50.26°$$

(b) DATA:

$$n_1 = 1.000$$
$$n_2 = 1.518$$
$$\theta_1 = 50.00°$$
$$\theta_1' = ?$$

First, find θ_2, then θ_2', and finally θ_1'.

BASIC EQUATION:

$$n_1 \sin \theta_1 = n_2 \sin \theta_2$$

WORKING EQUATION:

$$\sin \theta_2 = \frac{n_1}{n_2} \sin \theta_1$$

SUBSTITUTION:

$$\sin \theta_2 = \frac{1.000}{1.518} (\sin 50.00°) = 0.5046$$
$$\theta_2 = 30.30°$$

Reference to the sketch shows that the angles

$$\theta_2 + \theta_2' = 60.00°$$

so

$$\theta_2' = 29.70°$$

To find θ_1':

WORKING EQUATION:

$$\sin \theta_1' = \frac{n_2}{n_1} \sin \theta_2'$$

SUBSTITUTION:

$$\sin \theta_1' = \frac{1.518}{1.000} (\sin 29.70°) = 0.7521$$

$$\theta_1' = 48.77°$$

(c) DATA:

for violet light $\theta_1' = 50.26°$

for red light $\theta_1' = 48.77°$

$$\Delta\theta_1' = 1.49°$$

The angle $\Delta\theta_1' = 1.49°$ is the angular separation between the colors of light at the ends of the visible spectrum for this prism. The index of refraction $n_1 = c/v_1$ is the index of refraction of medium 1 with respect to a vacuum. The index of refraction of air with respect to a vacuum for visible light differs from unity by less than 0.0005. When four or less significant digits are used, the index of refraction of a material with respect to air and the index of refraction of the same material with respect to a vacuum are the same.

The index of refraction of a medium with respect to a vacuum is given by

$$n_1 = \frac{c}{v_1}$$

It follows that

$$\boxed{n_{12} = \frac{n_1}{n_2} = \frac{c/v_1}{c/v_2} = \frac{v_2}{v_1}}$$

where n_{12} is the index of refraction of medium 1 with respect to medium 2,
n_1 is the index of refraction of medium 1 with respect to a vacuum,
n_2 is the index of refraction of medium 2 with respect to a vacuum,
v_1 is the speed of light in medium 1, and
v_2 is the speed of light in medium 2.

example 4 At what angle would the red light of Example 3 emerge from the prism if the prism were immersed in water ($n = 1.330$)?

SKETCH:

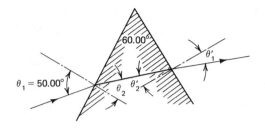

DATA:

$$n_1 = 1.330$$
$$n_2 = 1.518$$
$$\theta_1 = 50.00°$$
$$\theta_1' = ?$$

First find θ_2, then θ_2', and finally θ_1'.

BASIC EQUATION: $n_1 \sin \theta_1 = n_2 \sin \theta_2$

WORKING EQUATION: $\sin \theta_2 = \dfrac{n_1}{n_2} \sin \theta_1$

SUBSTITUTION: $\sin \theta_2 = \dfrac{1.330}{1.518} (\sin 50.00°) = 0.6712$

$$\theta_2 = 42.16°$$

Reference to the sketch shows that the angles

$$\theta_2 + \theta_2' = 60.00°$$

so

$$\theta_2' = 17.84°$$

To find θ_1':

WORKING EQUATION: $\sin \theta_1' = \dfrac{n_2}{n_1} \sin \theta_2'$

SUBSTITUTION: $\sin \theta_1' = \dfrac{1.518}{1.330} (\sin 17.84°) = 0.3497$

$$\theta_1' = 20.47°$$

30.2   TOTAL INTERNAL REFLECTION

If the source of light is in the medium having the larger index of refraction (referring to the figure on page 533, $n_2 > n_1$), the angle θ_1 in the new or exterior medium will be larger than the angle θ_2 in the original medium. Snell's law,

$$n_1 \sin \theta_1 = n_2 \sin \theta_2$$

will be obeyed up to the limit that $\sin \theta_1$ cannot be greater than unity or θ_1 cannot be greater than 90°. The angle θ_2 for which θ_1 becomes 90° is called the *critical angle θ_c*.

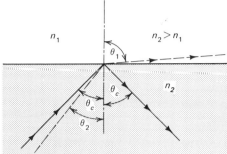

At any angle larger than the critical angle, the light cannot enter the external medium and is internally reflected. Since all the light is reflected, it is called *total internal reflection*:

$$n_1 \sin 90° = n_2 \sin \theta_c$$

$$\boxed{\sin \theta_c = \frac{n_1}{n_2}}$$

where θ_c is the critical angle,
    n_1 is the index of refraction of exterior medium, and
    n_2 is the index of refraction of medium.
When the exterior medium is air, $n = 1.000$ and

$$\boxed{\sin \theta_c = \frac{1}{n}}$$

where n is the index of refraction of the medium with respect to air.

example Total internal reflection in a 90°- 45°- 45° prism is to be used to reflect a light ray as shown in the figure below. What is the minimum acceptable value of the index of refraction of the glass in this prism?

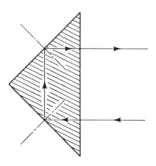

DATA: $\theta_c = 45.0°$

                                             $n = ?$

BASIC EQUATION: $\sin \theta_c = \dfrac{1}{n}$

WORKING EQUATION: $n = \dfrac{1}{\sin \theta_c}$

SUBSTITUTION: $n = \dfrac{1}{\sin 45.0°}$

                                             $= 1.41$

The prism must be constructed of glass of index of refraction greater than 1.41. Prisms are commonly used in binoculars to effect great path lengths by reflecting light over the same path several times.

30.3   TYPES OF LENSES

Materials that transmit light can be used to make light rays converge or diverge to form images. In the figure below, it can be seen that a transparent object whose

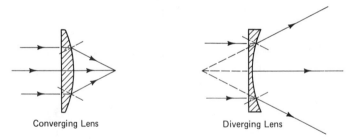

Converging Lens   Diverging Lens

central portion is thicker than its edges will cause parallel rays to converge; and a transparent object whose central portion is thinner than its edges will cause parallel rays to diverge.

For our discussion, the surfaces will be portions of spheres and the restriction from Chapter 29 will still be imposed—that the dimensions of the transparent object are small compared to the radii of curvature of its surface. A second condition that the distances between the spherical surfaces be small is also imposed here. The transparent object is called a "thin lens." If the lens causes parallel light to converge, we call it a "converging lens" and the region where all the rays cross is called the focal point. The distance from the lens to the focal point is called the focal length. For a converging lens that distance has a positive value. If the lens causes parallel light to diverge, we call it a "diverging lens" and the region in which the diverging rays would cross if extended is the focal point and the distance from the lens to the focal point is its focal length. For a diverging lens that distance has a negative value. It will be noticed that for both lenses and mirrors, a positive value of the focal length indicates a converging optical element and a negative value indicates a diverging optical element.

30.4 IMAGES FORMED BY CONVERGING LENSES

The definition of focal length from Section 30.3, together with the fact that a ray of light passing through the optical center of the lens is not deviated, can be used to construct a diagram to find the image of an object produced by a converging lens. With reference to the figure below,

h_0 is the height of object,
h_i is the height of image,
S_o is the distance from the lens to the object,
S_i is the distance from the lens to the image, and
f is the focal length of the lens.

A ray $AVFB$ is a ray that is parallel to the optical axis of the lens and is refracted through the focal point. The ray ACB strikes the center of the lens and is transmitted without deviation.

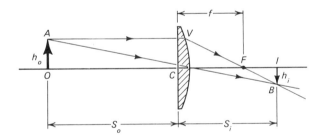

Triangles AOC and BIC are similar, so

$$\frac{h_i}{h_o} = \frac{S_i}{S_o}$$

Triangles VCF and BIF are similar, so

$$\frac{h_i}{h_o} = \frac{S_i - f}{f}$$

Eliminating the ratio h_i/h_o from the equations above in the manner shown in Section 29.4 gives

$$\boxed{\frac{1}{S_o} + \frac{1}{S_i} = \frac{1}{f}}$$

The *magnification* is defined as the ratio of the height of the image to the height of the object:

$$m = -\frac{S_i}{S_o}$$

where m is the magnification,

S_i is the distance from lens to the image, and

S_o is the distance from lens to the object.

When the magnification is negative, the image is inverted; when the magnification is positive, the image is erect. As with mirrors, when the light from the line converges for a real image, S_i has a positive value. When the light from the lens is diverging as if from a point behind the lens, the image is virtual and the value of S_i is negative.

example 1   An object 0.060 m tall is 0.80 m from a converging lens of focal length 0.25 m.

(a) Where is the image formed?

(b) What is the height of the image?

(a) DATA:

$$S_o = 0.80 \text{ m}$$

$$f = 0.25 \text{ m}$$

$$S_i = ?$$

BASIC EQUATION:

$$\frac{1}{S_o} + \frac{1}{S_i} = \frac{1}{f}$$

SUBSTITUTION:

$$\frac{1}{0.80 \text{ m}} + \frac{1}{S_i} = \frac{1}{0.25 \text{ m}}$$

$$S_i = 0.36 \text{ m}$$

(b) DATA:

$$S_o = 0.80 \text{ m}$$

$$S_i = 0.36 \text{ m}$$

$$h_o = 0.060 \text{ m}$$

$$h_i = ?$$

BASIC EQUATIONS:

$$m = -\frac{S_i}{S_o}$$

$$\frac{h_i}{h_o} = m$$

WORKING EQUATION:

$$h_i = -\frac{S_i}{S_o} h_o$$

SUBSTITUTION:

$$h_i = -\frac{0.36 \text{ m}}{0.80 \text{ m}}(0.060 \text{ m})$$

$$= -0.027 \text{ m}$$

The image is located 0.36 m from the lens and is inverted and 0.027 m high. This optical arrangement is shown schematically below.

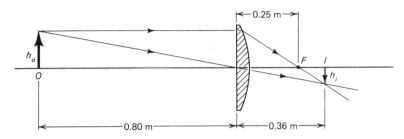

example 2 An object viewed through a converging lens has an upright image 4.0 times the size of the object. Assume that the viewer focused his or her eye at a distance 0.24 m behind the lens. Find the focal length of the lens.

SKETCH:

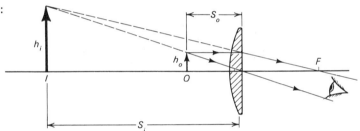

DATA: $S_i = -0.24 \text{ m}$ (*Note*: virtual image)

$$m = 4.0$$

$$f = ?$$

BASIC EQUATIONS: $m = -\dfrac{S_i}{S_o}$

$$\frac{1}{S_o} + \frac{1}{S_i} = \frac{1}{f}$$

Solving the second equation for f, we have

$$f = \frac{S_o S_i}{S_o + S_i}$$

Divide the numerator and the denominator by S_o.

$$f = \frac{S_i}{1 + \dfrac{S_i}{S_o}}$$

Since $\dfrac{S_i}{S_o} = -m$, we have

WORKING EQUATION:

$$f = \frac{S_i}{1-m}$$

SUBSTITUTION:

$$f = \frac{-0.24 \text{ m}}{1-4.0}$$

$$= 0.080 \text{ m}$$

In this example, the converging lens is being used as a simple microscope or magnifying glass.

30.5   IMAGES FORMED BY DIVERGING LENSES

An image may be constructed for a diverging lens in the same manner as the image was constructed for the converging lens. The figure below shows a diverging lens. Two rays from the top of an object O are traced.

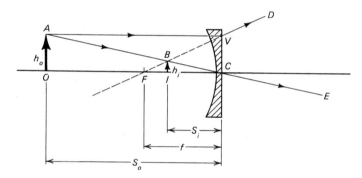

The first ray, AVD, is the ray approaching the lens parallel to the optical axis; after passing through the lens, it appears to come through the focal point, F. The second ray, ACE, passes through the optical center of the lens and is not deviated. The point where the two lines cross, point B, is the image of the point A. This image is virtual and the distance, S_i, is on the opposite side of the lens from the transmitted rays and is taken to be negative.

In the figure, triangles AOC and BIC are similar giving

$$\frac{h_i}{h_o} = \frac{S_i}{S_o}$$

and triangles VCF and BIF are similar, giving

$$\frac{h_i}{h_o} = \frac{f - S_i}{f}$$

Combining these two equations gives

$$\frac{1}{S_o} + \frac{1}{-S_i} = \frac{1}{-f}$$

Recall that the focal length for a diverging lens is negative and the value of S_i for a virtual image is negative:

$$\frac{1}{S_o} + \frac{1}{S_i} = \frac{1}{f}$$

$$m = -\frac{S_i}{S_o}$$

The two equations above are applicable to concave, plane, and convex mirrors as well as converging and diverging lens. Care must be exercised to use and interpret negative values within the conventions stated.

example An object that is 0.60 m from a lens is viewed through a diverging lens of -0.10 m focal length. What is the magnification of this system?

DATA:

$$S_o = 0.60 \text{ m}$$

$$f = -0.10 \text{ m}$$

$$m = ?$$

BASIC EQUATIONS:

$$\frac{1}{S_o} + \frac{1}{S_i} = \frac{1}{f}$$

$$m = -\frac{S_i}{S_o}$$

Solving the first equation for S_i, we have

$$S_i = \frac{S_o f}{S_o - f}$$

Divide both sides by S_o.

$$\frac{S_i}{S_o} = \frac{f}{S_o - f}$$

Then,

$$m = -\frac{S_i}{S_o} = \frac{f}{-(S_o - f)} = \frac{f}{f - S_o}$$

WORKING EQUATION:

$$m = \frac{f}{f - S_o}$$

SUBSTITUTION:

$$m = \frac{-0.10 \text{ m}}{-0.10 \text{ m} - 0.60 \text{ m}}$$

$$= 0.14$$

The image is upright and only one-seventh of the size of the object. This type of converging lens is used as the "visitor–viewer," so a person can be identified without opening the door, and is shown schematically below.

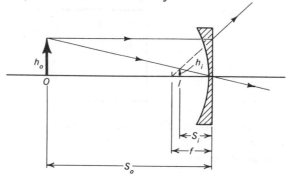

30.6   THE LENS-MAKER FORMULA

The focal length of spherical lenses depends on the material of which the lens is constructed and the geometry of the lens. The *lens-maker formula* is given as

$$\frac{1}{f} = (n - 1)\left(\frac{1}{R'} - \frac{1}{R''} \right)$$

where f is the focal length,
      n is the index of refraction of lens,
      R' is the radius of curvature of first surface, and
      R'' is the radius of curvature of second surface.
A lens is reversible and has the same focal length regardless of the direction in which light travels through it. For consistency, it is best to adopt a convention. If the center of the curvature is on the same side of the lens as light that has passed through the lens, then that radius of curvature is positive. If the center of curvature is on the same side of the lens as light that is incident upon the lens, then that radius of curvature is negative.

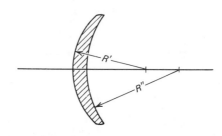

example 1 Find the focal length of a thin converging lens; the surface is convex with a radius of curvature of 0.20 m, and the other surface is concave with a radius of curvature of 0.40 m. The index of refraction of the glass is 1.50.

SKETCH:

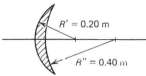

DATA:

$$R' = 0.20 \text{ m}$$

$$R'' = 0.40 \text{ m}$$

$$n = 1.50$$

$$f = ?$$

BASIC EQUATION:

$$\frac{1}{f} = (n - 1)\left(\frac{1}{R'} - \frac{1}{R''} \right)$$

WORKING EQUATION: same

SUBSTITUTION:

$$\frac{1}{f} = (1.50 - 1)\left(\frac{1}{0.20 \text{ m}} - \frac{1}{0.40 \text{ m}} \right)$$

$$f = 0.80 \text{ m}$$

example 2 Find the focal length of a thin diverging lens. One surface is convex with a radius of curvature of 0.40 m, and the other surface is concave with a radius of curvature of 0.20 m. The index of refraction of the lens is 1.50.

SKETCH:

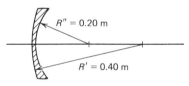

DATA:

$$R' = 0.40 \text{ m}$$

$$R'' = 0.20 \text{ m}$$

$$n = 1.50$$

BASIC EQUATION:

$$\frac{1}{f} = (n - 1)\left(\frac{1}{R'} - \frac{1}{R''} \right)$$

WORKING EQUATION: same

SUBSTITUTION:

$$\frac{1}{f} = (1.50 - 1)\left(\frac{1}{0.40 \text{ m}} - \frac{1}{0.20 \text{ m}} \right)$$

$$f = -0.80 \text{ m}$$

30.7 THE SIMPLE MICROSCOPE

A single converging lens may be used as a simple microscope. It is sometimes called a magnifier and sometimes called a reading glass. A schematic drawing for the simple microscope is shown.

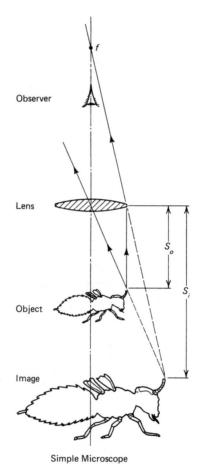

Simple Microscope

The object is placed inside the focal point in a position so that a virtual image is formed at the distance of most distinct vision for the observer. The distance of most distinct vision varies from person to person. An average value of 0.25 m will be used here. A person can easily determine his or her distance of most distinct vision by trying to read some fine print, such as the mint mark on a dime. The distance from the person's eye to the fine print is this distance. The basic equation for this lens is

$$\frac{1}{S_o} + \frac{1}{S_i} = \frac{1}{f}$$

This relation may be rewritten as

$$-\frac{S_i}{S_o} = -\frac{S_i}{f} + 1$$

The term $-(S_i/S_o)$ is the magnification m, and the image distance S_i is taken as -0.25 m (the negative sign because the image is virtual), the distance of most distinct vision. Making these substitutions gives

$$\boxed{m = \frac{0.25}{f} + 1}$$

where m is the magnification of the simple microscope and
$\quad f$ is the focal length of the lens in metres.
It should be noted that the image is virtual and erect.

30.8 THE COMPOUND MICROSCOPE

A compound microscope is shown schematically. It consists of two lenses and may have a calibrated reticule which permits measuring the size of objects.

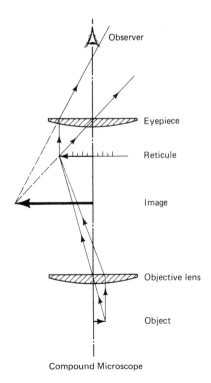

Compound Microscope

The objective lens has a very short focal length. The object to be observed is placed just outside the focal length of the objective lens, producing a real image in the plane of the calibrated reticule. The eyepiece serves as a simple microscope to magnify the real image superimposed on the calibrated reticule.

The position of the reticule is fixed in the microscope and the observer adjusts the eyepiece position to find the best image of the calibrated reticule. The distance from the reticule to the objective lens is fixed by the construction of the microscope. This microscope is moved to give the best image of the object in the plane of the reticule, which, of course, gives the best image to the observer. Since the dimensions of the microscope are fixed, especially the distance from the plane of the reticule to the objective lens, the magnification for that lens in the microscope can be stated and the lens labeled. The objective lenses for a microscope are marked $10 \times$, $30 \times$, and so on. The eyepieces are labeled with the magnification as described in Section 30.7 on the simple microscope. The product of the magnifications of the eyepiece and the objective lens gives the magnification for the microscope.

Continued use of a microscope as described here would soon cause the observer to become tired because his or her eyes would always be strained from viewing at the distance of most distinct vision. An experienced observer would adjust the eyepiece such that the eye focuses as in viewing a distant object. The light rays are parallel from the eyepiece to the observer, or the image is said to be an infinite distance from the observer. In this case, magnification (size of image/ size of object) loses definition and the term *magnifying power* is introduced. Magnifying power of an optical instrument is the ratio of the angle subtended by the image to the angle subtended by the object. In the case of the compound microscope, the magnifying power is essentially the same as the magnification of the microscope, as described above.

30.9   THE SIMPLE TELESCOPE

A simple telescope, sometimes called the astronomical telescope, uses two converging lenses: an objective and an eyepiece. A telescope is used to observe distant

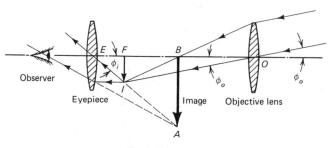

Simple Telescope

objects. The object distance is great, so the image formed by the objective is at the focal point of the lens. This image is then magnified by the eyepiece, which serves as a simple magnifier here just as it did in the microscope. Magnification is of little value. The object is so far from the objective lens that its image is very much smaller than the object. Rather, we shall use magnifying power (the ratio of the angle subtended by the image to the angle subtended by the object). In the preceding figure, the magnifying power is

$$M.P. = \frac{\phi_i}{\phi_o}$$

where M.P. is the magnifying power of the telescope,
ϕ_i is the angle subtended by the image, and
ϕ_o is the angle subtended by the object.

The angle *IOF* is the angle subtended by the object and is ϕ_o. The angle *AEB* is the angle subtended by the image and is ϕ_i. Angle *FEI* is equal to angle *AEB*. The image *FI* formed by the objective lens is subtended by both ϕ_o and ϕ_i. The diameters of the lenses are small when compared to their focal lengths. That is, θ_i and θ_o are small. Thus,

$$\phi_i \cong \tan \phi_i = \frac{FI}{EF} = \frac{FI}{f_e}$$

and

$$\phi_o \cong \tan \phi_o = \frac{FI}{OF} = \frac{FI}{f_o}$$

where   *FI*   is the image formed by the objective and lies in the focal plane of
both the objective lens and the eyepiece,
$EF = f_e$ is the focal length of the eyepiece, and
$OF = f_o$ is the focal length of the objective lens.
The magnifying power of this telescope is

$$M.P. = \frac{f_o}{f_e}$$

It should be noted that this image is inverted and used mostly for astronomical observations.

An innovation which changes this telescope to one that gives an erect image is shown on page 546.

A third converging lens, called an inverting lens, is added. The lens is positioned so that the image formed by the objective lens is just exactly two focal

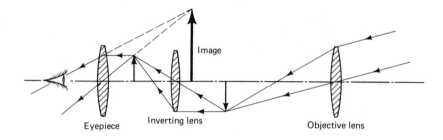

lengths from the inverting lens, thus reproducing the image erect at a point two focal lengths on the other side of the lens. This lens has only inverted the image and has in no way magnified or diminished the image.

The use of this innovation increases the length of the telescope by four focal lengths, producing a long instrument called a spy glass.

EXERCISES

Sketch	Data	Basic Equation	Working Equation	Substitution
$h = ?$ $b = 6.0$ cm, 12 cm²	$A = 12$ cm² $b = 6.0$ cm $h = ?$	$A = bh$	$h = \dfrac{A}{b}$	$h = \dfrac{12 \text{ cm}^2}{6.0 \text{ cm}}$ $= 2.0$ cm

1. The index of refraction for water is 1.33. What is the speed of light in water?

2. A certain glass has an index of refraction of 1.50. What is the speed of light in this glass?

3. The speed of light in a brand name of oil is 2.5×10^8 m/s. What is the index of refraction of this oil?

4. The speed of light in a soft glass is 2.1×10^8 m/s. What is the index of refraction of this glass?

5. Light is incident upon a piece of glass ($n = 1.50$) at an angle of 40.0°. At what angle will the light be refracted?

6. Light is incident on a piece of glass ($n = 1.60$) at an angle of 45.0°. At what angle will the light be refracted?

7. A ray of light traveling in glass ($n = 1.50$) is incident upon the surface at an angle of 40.0°. What angle will the ray make with the normal to the glass surface when it emerges into air?

8. A ray of light traveling in a glass ($n = 1.60$) is incident upon the surface at an angle of 45.0°. What angle will the ray make with the normal to the glass surface when it emerges into air?

9. A fish is at a depth of 4.00 m below the surface of fresh water ($n = 1.33$). At what depth will the fish appear to an observer directly above the fish?

10. When looking straight down into a pool of water ($n = 1.33$), the bottom appears at 10.0 m below the surface. How deep is the pool?

11. A ray of white light is incident upon a plane glass surface at an angle of 56.00°. If the index of refraction for violet light is 1.562 and for red light is 1.542, what is the angular separation between the ends of the spectrum after refraction at this surface?

12. An equilateral prism is used to separate two wavelengths of light; 5.550×10^{-7} m and 5.577×10^{-7} m. Their respective indices of refraction are 1.460 and 1.458. What is their angular separation if the incident ray makes an angle of 48.00° with one face of the prism?

13. At what angle would the 5.550×10^{-7} m light in Exercise 12 emerge from the prism if the prism were immersed in water? ($n = 1.330$)

14. A flood lamp is 2.00 m beneath the surface of water. The lamp produces a bright circle of light on the water surface. What is the diameter of the circle of light? ($n = 1.33$)

15. The critical angle for a piece of glass in air is 40.0°. What is the critical angle for the same piece of glass when immersed in water? ($n = 1.33$)

16. A 90°- 45°- 45° prism is to be used to reflect light by total internal reflection, as shown. The prism will be mounted in a fluid whose index of refraction is 1.16. What is the minimum value of the index of refraction that is acceptable for the prism?

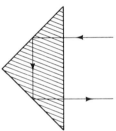

In Exercises 17–34, solve for the indicated quantities.

17. $S_o = +0.40$ m
$f = +0.20$ m
$S_i = ?$

18. $S_o = +0.15$ m
$f = +0.20$ m
$S_i = ?$

19. $S_o = +0.40$ m
$f = -0.20$ m
$S_i = ?$

20. $S_o = +0.15$ m
$f = -0.20$ m
$S_i = ?$

21. $S_o = 0.10$ m
$f = +0.30$ m
$S_i = ?$

22. $S_o = +0.60$ m
$f = +0.30$ m
$S_i = ?$

23. $S_o = +0.50$ m
$S_i = +0.50$ m
$f = ?$

24. $S_o = +0.50$ m
$S_i = -0.50$ m
$f = ?$

25. $S_o = +0.20$ m
$S_i = -0.40$ m
$f = ?$
$m = ?$

26. $S_o = +0.40$ m
$S_i = -0.20$ m
$f = ?$
$m = ?$

27. $S_o = +0.10$ m
$S_i = +0.10$ m
$f = ?$
$m = ?$

28. $S_o = +0.30$ m
$S_i = -0.15$ m
$f = ?$
$m = ?$

29. $n = 1.50$
$R' = +0.20$ m
$R'' = -0.40$ m
$f = ?$

30. $n = 1.50$
$R' = +0.20$ m
$R'' = +0.40$ m
$f = ?$

31. $n = 1.50$
$R' = -0.30$ m
$R'' = -0.30$ m
$f = ?$

32. $n = 1.50$
$R' = -0.30$ m
$R'' = +0.30$ m
$f = ?$

33. $n = 1.50$
$R' = +0.30$ m
$R'' = +0.60$ m
$f = ?$

34. $n = 1.50$
$R' = +0.60$ m
$R'' = +0.30$ m
$f = ?$

35. A simple microscope lens is to be used as a magnifier of $5\times$. The image is to be upright and located 0.25 m from the lens. What is the focal length of the lens?

36. A converging lens is used in a spy glass to invert the image of the objective lens without changing its size. A maximum distance of 0.160 m may be used ($S_o + S_i = 0.160$ m) by the inversion lens. What is the maximum focal length that this lens can have?

37. An object is placed 0.15 m in front of a lens that has two concave sides, each with a radius of curvature of 0.80 m. The index of refraction of the glass is 1.60.

(a) Where is the image formed?
(b) Is the image real?
(c) Is the image erect?

38. Lenses are often described by the term *diopter*. A diopter is the reciprocal of the focal length expressed in metres. Given a piece of glass ($n = 1.50$), design a + 2.00-diopter lens.

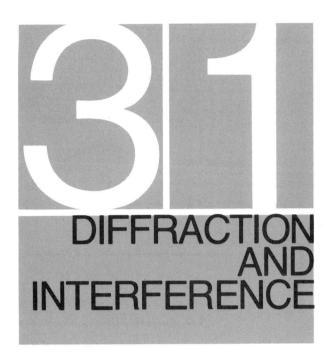

31
DIFFRACTION AND INTERFERENCE

31.1   DIFFRACTION OF WAVES

When a collimated beam of light passes by a sharp edge, around a fine wire, or through a narrow slit, the beam tends to spread or flare out. This behavior is called *diffraction*. Diffraction is not restricted to light but is true of any wave phenomenon. A person can observe the divergence of water waves when they pass through a narrow restriction. The diffraction of sound waves at doorways and other openings makes it possible for us to "hear around corners."

In the description of diffraction effects, it is best to use monochromatic light or light of a single wavelength. The figure on page 550 shows the intensity as a function of angle of divergence from the forward direction for a beam of monochromatic light. The slit widths, a, are stated in terms of the wavelength and are plotted for widths of $\lambda/3$, λ, 3λ, and 10λ. When the slit is small compared to the wavelength, the spread of light is much greater. For the case where the slit width is one-third of the wavelength, the light diffracted at an angle of 45° is 90 percent as intense as the light in the forward direction, and even at a diffraction angle of 90° the diffracted light is more than two-thirds of the intensity of the light in the forward direction. When the slit is the same size as the wavelength, the intensity spread is not as severe. At a diffraction angle of 45° the intensity is only one-eighth of the forward intensity and gradually falls to zero at 90° from the forward beam. A slit width of three wavelengths causes the diffracted intensity to fall to a zero value at less than 20° and shows other weak maxima inside of 90°. The intensity curve for a slit of 10 wavelengths reaches its first zero-intensity value at less

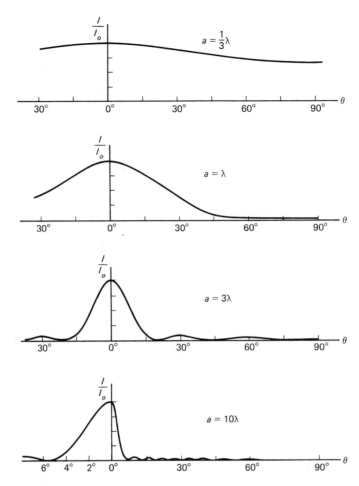

than 6° and exhibits nine zero values inside of 90° with very weak maxima. The maxima and minima in the diffraction pattern result from the interference between light rays passing through different points across the width of the slit.

The *intensity distribution* from a single slit is

$$I = I_0\left(\frac{\sin \alpha}{\alpha}\right)^2$$
$$\alpha = \frac{\pi a \sin \theta}{\lambda}$$

where I is the diffracted intensity at an angle θ from the forward direction,
I_0 is the intensity in the forward direction,
θ is the diffraction angle,
a is the width of the slit, and
λ is the wavelength of light.

Diffraction was the source of the broadening of points on the image in our earlier discussion of the pinhole camera.

example   Collimated light of 5.5×10^{-7} m wavelength (green) passes through a 0.050-mm slit. At what deviation from the forward direction will the first minimum of intensity be observed?

DATA:

$$\lambda = 5.5 \times 10^{-7} \text{ m}$$

$$a = 0.050 \text{ mm} = 5.0 \times 10^{-5} \text{ m}$$

$$\theta = ?$$

BASIC EQUATIONS:

$$I = I_0 \left(\frac{\sin \alpha}{\alpha} \right)^2$$

$$\alpha = \frac{\pi a \sin \theta}{\lambda}$$

The intensity will first be zero when

WORKING EQUATIONS:

$$\alpha = \pi$$

$$\alpha = \pi = \frac{\pi a \sin \theta}{\lambda}$$

$$\sin \theta = \frac{\lambda}{a}$$

SUBSTITUTION:

$$\sin \theta = \frac{5.5 \times 10^{-7} \text{ m}}{5.0 \times 10^{-5} \text{ m}} = 0.011$$

$$\theta = 0.63°$$

31.2   INTERFERENCE FROM TWO SOURCES

In the figure on page 552, two sources of electromagnetic radiation are shown separated by a distance d. These two sources could be from any part of the electromagnetic spectrum, but we will assume two radio sources, S_1 and S_2, separated by a distance d. These sources emit radiation equally in all directions. We will observe the waves from these sources at point P, which lies the distance r from the midpoint of these two sources. The distance r is much greater than the separation, d, between the two sources and makes an angle θ with a perpendicular line through the line joining S_1 and S_2. The sources S_1 and S_2 are two antennae whose radiation is generated by a common source and, consequently, the radiations leaving these two antennae are of exactly the same frequency and wavelength and phase of oscillation. The radiation arriving at P from S_2 will have traveled a distance $r_2 - r_1$ farther than the radiation from S_1. This path difference must be an integer number of wavelengths in order for the interference to be constructive and produce a maximum.

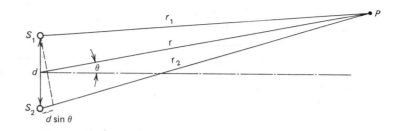

$$r_2 - r_1 = d \sin \theta = n\lambda, \text{ or}$$
$$n\lambda = d \sin \theta \qquad \text{maxima}$$

where $r_2 - r_1$ is the path difference from sources S_1 and S_2 to the observer at P,
$\quad$ d is the separation between the two sources,
$\quad$ λ is the wavelength of radiation, and
$\quad$ n is an integer and its value is called order.
It can also be seen that the interference produces a null (minimum) when

$$\left(n + \tfrac{1}{2}\right)\lambda = d \sin \theta \qquad \text{minimum}$$

where the notation is the same as above.

The foregoing relations give discrete values for the positions of maximum and minimum intensities of the radiation observed from the two sources. A continuous value of the intensity is

$$I = I_m \cos^2 \beta$$
$$\beta = \frac{\pi d}{\lambda} \sin \theta$$

where I is the intensity at an angle θ from the forward direction,
$\quad$ I_m is the intensity in the forward direction,
$\quad$ θ is the angular distance between observer and forward direction,
$\quad$ d is the distances between sources, and
$\quad$ λ is the wavelength of the radiation.

In the figure below a plot is shown of the intensity of radiation as a function of the angle β.

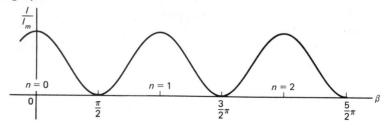

It should be noted that this plot shows equal peak intensities regardless of the order of interference n. The central peak corresponds to both rays of radiation from S_1 and S_2, which traveled the same distance. These rays would have zero phase difference. For the peaks immediately on either side of the $n = 0$ peak, $n = 1$, and the path difference is exactly one wavelength. Similarly, $n = 2$ corresponds to a two-wavelength path difference, $n = 3$ to a three-wavelength path difference, and so on.

> **example** Two microwave sources of 2.50-cm wavelength are separated by 14.0 cm.
>
> (a) Find the angle from the perpendicular bisector of the line joining these sources to the first intensity maximum.
> (b) How many maxima can be observed on either side of the central maximum?

(a) DATA:

$\lambda = 2.50 \times 10^{-2}$ m

$d = 0.140$ m

$n = 1$

$\theta = ?$

BASIC EQUATION:

$n\lambda = d \sin \theta$

WORKING EQUATION:

$\sin \theta = \dfrac{n\lambda}{d}$

SUBSTITUTION:

$\sin \theta = \dfrac{1(2.5 \times 10^{-2} \text{ m})}{0.140 \text{ m}} = 0.1786$

$\theta = 10.3°$

(b) DATA:

$\lambda = 2.50 \times 10^{-2}$ m

$d = 0.140$ m

$\sin \theta = 1$ (maximum value)

$n = ?$ (an integer)

BASIC EQUATION:

$n\lambda = d \sin \theta$

WORKING EQUATION:

$n = \text{*INT}\left[\dfrac{d \sin \theta}{\lambda} \right]$

SUBSTITUTION:

$n = \text{INT}\left[\dfrac{(0.140 \text{ m})(1)}{2.50 \times 10^{-2} \text{ m}} \right] = \text{INT}[5.6]$

$= 5$ maxima

*This notation refers to finding the largest integer less than or equal to the quantity in the brackets.

31.3 INTERFERENCE FROM SEVERAL SOURCES

If the two antennae were increased to three or more antennae, each separated by a distance d, each powered by the same common generator and in phase, the radiation observed at an angle θ would be modified. Intensity distributions are shown below for two, three, or five sources.

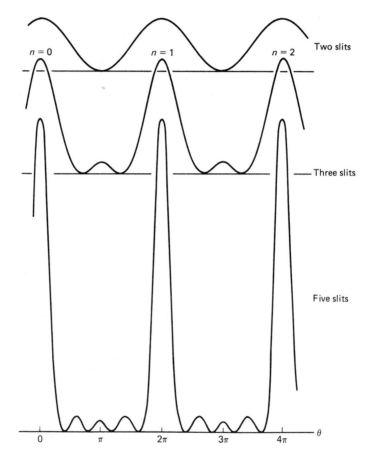

Notice that for the two sources the values of intensity follow a smooth cosine-squared curve with the intensity peaking at integer values of n with only a single minimum between the orders. When three antennae are used to transmit, strong interference peaks occur at the same values where the peaks occurred for a two-antenna system with the same distance d between the antennae. These intensity peaks, however, become sharper and a weaker satellite peak appears midway between the strong peaks. These strong peaks satisfy the equation for a maxima of the two-source system:

$$n\lambda = d \sin \theta$$

where n is the order of the interference peak,
   λ is the wavelength of the radiation,
   d is the distance between adjacent sources of radiation, and
   θ is the angle between the forward direction and interference peak.
   When the number of sources is increased to five, each separated by a distance, d, strong diffraction peaks occur at the position satisfied by the expression

$$n\lambda = d \sin \theta$$

There are now three small satellite peaks that appear between each pair of the strong order peaks. It should also be noted that if we assume all of our radiation sources to be of equal intensity, I_0, the peak values of intensity are not just the number of sources times the intensity of a single source but, rather, the number squared times the intensity of a single source.

$$I_N = N^2 I_0$$

where I_N is the peak intensity of a principal diffraction peak with
   N sources,
   I_0 is the intensity from a single source, and
   N is the number of sources.
The peak intensities in our drawing on page 554 are $4I_0$, $9I_0$, and $25I_0$ rather than $2I_0$, $3I_0$, and $5I_0$ because the amplitudes of the waves superimpose and the intensity is proportional to the square of the amplitudes. In general, as more and more sources are added, maintaining the spacing, d, between each pair, the principal diffraction peak becomes sharper and stronger and the satellite peaks increase in number and tend to disappear into each other, producing a weak background to the strong principal peaks.

31.4 INTERFERENCE FROM TWO SLITS

Earlier we saw that light passing through a slit diffracted or flared out and that the slit acted as a light source where angular intensity distribution depended upon the slit width, a. We also found that two sources radiating uniformly in all directions produced an interference pattern as a function of angle. Two slits with light from a common source such as another slit or a laser beam will superimpose to give

$$I_\theta = I_m \cos^2 \beta \left(\frac{\sin \alpha}{\alpha} \right)^2$$

$$\alpha = \frac{\pi a \sin \theta}{\lambda}$$

$$\beta = \frac{\pi d \sin \theta}{\lambda}$$

where θ is the diffraction angle,
   λ is the wavelength of light,
   a is the width of a single slit,
   d is the separation between slits,
   I_m is the intensity in the forward direction, and
   I_θ is the intensity at the angle θ.

The interference term will have maximum values when $\beta = 0, \pi, 2\pi, \ldots, n\pi$, where n is an integer.

$$n\pi = \beta = \frac{\pi d \sin \theta}{\lambda}$$

or

$$n\lambda = d \sin \theta \qquad \text{maxima}$$

This expression is, of course, the same as developed for a system of slits earlier. The diffraction term $(\sin \alpha / \alpha)^2$ will have minima values for $\alpha = \pi, 2\pi, \ldots, n'\pi$, where n' is an integer.

$$n'\pi = \alpha = \frac{\pi a \sin \theta}{\lambda}$$

or

$$n'\lambda = a \sin \theta \qquad \text{minima}$$

For a given wavelength and slit system, the angular positions of the constructive interference lines are determined by the spacing between the slits, d. The relative intensities of these lines depend upon the width of the slits, a.

The figure on page 557 shows an intensity distribution curve for a double-slit arrangement where the slit separation is three times the slit width ($d = 3a$). Curves of $\cos^2 \beta$ and $(\sin \alpha / \alpha)^2$ are drawn such that $n = 3$ coincides with $n' = 1$. The two curves are multiplied, giving the third curve, which corresponds to the intensity distribution to be observed from the double slit.

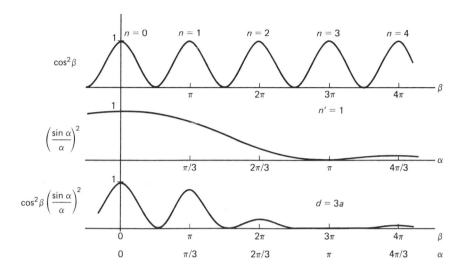

example   In a double-slit diffraction pattern the fringes of red light ($\lambda = 6.30 \times 10^{-7}$ m) are observed at 16.5°, 34.6°, and 58.4°.

(a) What is the separation between slits?

(b) At what angles would blue light ($\lambda = 4.80 \times 10^{-7}$ m) show maxima?

(a) DATA:

$$\lambda = 6.30 \times 10^{-7} \text{ m}$$

$$\theta = 16.5°, 34.6°, 58.4°$$

$$n = 1, 2, 3$$

$$d = ?$$

BASIC EQUATION:

$$n\lambda = d \sin \theta$$

WORKING EQUATION:

$$d = \frac{n\lambda}{\sin \theta}$$

SUBSTITUTIONS:

$$d = \frac{1(6.30 \times 10^{-7} \text{ m})}{\sin 16.5°} = 2.22 \times 10^{-6} \text{m}$$

$$d = \frac{2(6.30 \times 10^{-7} \text{ m})}{\sin 34.6°} = 2.22 \times 10^{-6} \text{m}$$

$$d = \frac{3(6.30 \times 10^{-7} \text{ m})}{\sin 58.4°} = 2.22 \times 10^{-6} \text{m}$$

(b) DATA:
$$\lambda = 4.80 \times 10^{-7} \text{ m}$$
$$n = 1, 2, 3$$
$$d = 2.22 \times 10^{-6} \text{ m [from part (a)]}$$
$$\theta = ?$$

BASIC EQUATION:
$$n\lambda = d \sin \theta$$

WORKING EQUATION:
$$\sin \theta = \frac{n\lambda}{d}$$

SUBSTITUTIONS:
$$\sin \theta_1 = \frac{1(4.80 \times 10^{-7} \text{ m})}{2.22 \times 10^{-6} \text{ m}} = 0.2162$$

$$\sin \theta_2 = \frac{2(4.80 \times 10^{-7} \text{ m})}{2.22 \times 10^{-6} \text{ m}} = 0.4324$$

$$\sin \theta_3 = \frac{3(4.80 \times 10^{-7} \text{ m})}{2.22 \times 10^{-6} \text{ m}} = 0.6486$$

$$\theta_1 = 12.5°$$
$$\theta_2 = 25.6°$$
$$\theta_3 = 40.4°$$

31.5   THE DIFFRACTION GRATING

The diffraction grating is a scientific instrument which embodies the principles that we have discussed. A diffraction grating has many slits. A typical grating may have 20,000 or more slits. Recall that when more sources were added, the result was an interference pattern with increased intensities in the principal peaks, sharper peaks, and a relatively lower background between the peaks. A diffraction grating with its many slits produces interference maxima which are sharp lines and not broad peaks. Each angular position corresponds to a specific wavelength of radiation that satisfies the interference and diffraction conditions.

Materials are composed of atoms which are bound together in many complex ways. When a material is furnished energy by an electric spark or flame, it releases this energy in the form of radiation. Each material radiates a light wavelength distribution which is characteristic of that material. This distribution of radiation is called a *spectrum* and is a property of that material and that material alone. The spectrum may be considered as that material's fingerprint and the material can be identified by its spectrum. The diffraction grating is the tool that displays a material's spectrum. Each wavelength of radiation from the material satisfies the interference conditions of a specific grating for only discrete angular positions. Many materials have had their spectra classified. Many volumes of books, decks of cards, and computer tapes are available listing these spectra and using aids to rapidly identify materials from their spectrum.

The drawing below illustrates a spectrum from atomic hydrogen, showing only principal lines in the visible light region. You will note that the intensities of the lines are not uniform but vary from line to line. This intensity variation is part of the spectrum and is a valuable aid in identifying materials.

The diffraction grating is also a research tool for using the molecular spectra of compounds to determine bonding energies.

EXERCISES

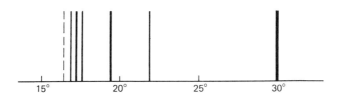

1. What slit width is necessary to produce a minimum at a diffraction angle of 10.0° with green light ($\lambda = 5.50 \times 10^{-7}$ m)?

2. A double slit produces a separation of 1.2° between the first and second minima in the diffraction pattern of yellow light ($\lambda = 5.90 \times 10^{-7}$ m). What is the slit width?

3. A 12.0-cm microwave source is used for demonstration. What slit separation would produce a third-order maximum at 90°?

4. An AM radio station uses a frequency of $55,\overline{0}00$ Hz. Two antennae are separated by $10\overline{0}0$ m. How many minimum intensities will be observed on a smooth, continuous path enclosing these two sources?

5. A diffraction grating has $30\overline{0}0$ lines/cm. What would be the angular range covered by the second-order diffraction pattern of visible spectra (4.00×10^{-7} m to 7.00×10^{-7} m)?

32.1 DISCRETE PHYSICAL PROPERTIES

Some experiments conducted during the last part of the nineteenth century and early twentieth century discovered phenomena that could not be explained by classical mechanics and electromagnetism. To explain these observed effects, it was necessary to make several assumptions regarding subatomic systems. It was necessary to postulate that electromagnetic waves produced by vibrating atoms could have only certain discrete values of energy. These values were integer multiples of an energy increment that was proportional to the waves' frequencies. Atoms were shown to have only discrete energy values and not a continuum of energy. Materials were found to absorb certain energies of radiation more readily than other energies. A system that can have only certain limited values of a physical property is said to be *quantized*. An atom that may be said to have only certain values of angular momentum or other physical property is also said to be quantized. In the following sections several experiments will be presented which require the concept of quanta to describe. These experiments not only played major roles in the development of quantum physics but are also commonly used in the tools of today's technology.

32.2 RADIATION FROM A BLACK BODY

A solid object gives off electromagnetic radiation. The intensity and energy distribution of that radiation depends upon the temperature of the object and the material of the object. Some materials absorb and emit electromagnetic radiation well, whereas others absorb and emit radiation poorly. If an object is a good absorber of a radiation, it is an equally good emitter of that radiation. If it were possible to have an object that absorbed well but emitted poorly, the object could be placed in an enclosed box where it would continue to absorb radiation from the box faster than it would emit radiation. The object would get hotter and hotter, as it would absorb more energy than it would emit. If the object were a poor absorber but a good emitter of radiation, it would emit radiation faster than it would absorb radiation when placed in the enclosed box. Now the object would get colder and colder, as it would emit more radiation than it would absorb. We know that if an object is placed in a box, the object and the box will come to the same temperature. When the two are at the same temperature, they are in thermal equilibrium and they radiate and absorb radiation at the same rate. A material which absorbs all the radiation that falls on it is not only an ideal absorber but is also an ideal emitter. Such a substance is called a *black body*, because all radiation that strikes it is absorbed and none is reflected. Regardless of the material, it is possible to construct an ideal absorber by making a pinhole entrance to a cavity, as shown in the illustration. The pinhole acts as a black body, in that essentially all radiation which strikes it is absorbed. If the object is heated until it glows, the pinhole will be brighter than the area surrounding it because it is also an ideal emitter. The distribution of radiation from an ideal emitter depends only upon the temperature of the walls of the cavity and not on the materials of its construction.

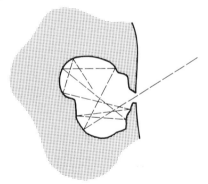

The figure below shows the spectral emissivity $\mathcal{R}_T(\lambda)$ of a black body as a function of wavelength for a specific temperature of 5700 K. The product of the spectral

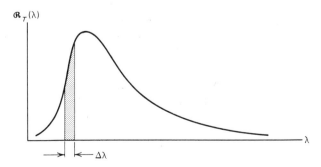

emissivity $\mathscr{R}_T(\lambda)$ and a small wavelength increment $\Delta\lambda$ is $\mathscr{R}_T(\lambda)\ \Delta\lambda$ and gives the power radiated per unit area of emitter surface in the wavelength range from λ to $\lambda + \Delta\lambda$. The figure below shows the spectral emissivity for several different temperatures.

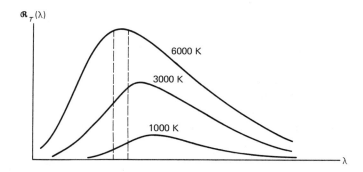

Note that as the temperature increases, the power per unit area radiated increases for all wavelengths, but the peak of the curve shifts toward shorter wavelengths. The wavelength limits of the visible spectrum are indicated on the figure. Comparison of the spectral emissivity at different temperatures with the wavelength sensitivity of the human eye shows why stars have a red, white, or blue appearance, dependent upon their temperatures.

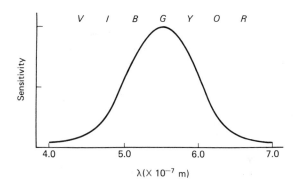

An instrument for measuring furnace temperature, called an optical pyrometer, uses the object's color. The pyrometer is simply a telescope that has a tungsten filament at the focal spot of the eyepiece. The tungsten filament is calibrated for different currents heating it. The current is adjusted until the filament and the image of the furnace are the same color.

Max Planck obtained a mathematical expression for the spectral emissivity curve which agrees with experiment. He assumed that the electrons in the walls of the cavity vibrate because of thermal energy. He assumed that the energy from these oscillating electrons were discrete, that is, could have only certain values, and that the electromagnetic waves given off could have only discrete energies. He further assumed that the energies of the oscillating electrons were proportional to their frequencies:

$$E = hf$$

where E is the energy of the oscillating electrons,
    h is the constant of proportionality, and
    f is the frequency of the oscillating electron.

When Planck introduced this concept of discrete values of energy proportional to the vibration frequencies of the electrons, it was regarded by many as a theoretical aid. However, it was the beginning of quantum physics. The only unknown in Planck's equation was h and it was evaluated to make the mathematical expression fix the experimental data.

$$h = 6.62 \times 10^{-34} \text{ J s} \quad \text{(see Section 28.3)}$$

The peak of the spectral emissivity curve occurs at different wavelengths as a function of temperature:

$$\lambda_{\max} T = b$$

where $\lambda_{\max}$ is the wavelength where the spectral emissivity is a maximum,
    T is the absolute temperature of the emitting body, and
    $b = 2.898 \times 10^{-3}$ m K (a constant).

This expression is called *Wein's displacement law* and the constant is known from experiment. Planck's theory provides a mathematical expression for b in terms of known physical constants and h. Solving for h confirms the value from the distribution curve.

The area beneath the spectral distribution curve corresponds to the total power per unit area emitted by a black body.

$$R_T = \sigma T^4$$

where R_T is the radiancy, in W/m^2,
    $\sigma = 5.67 \times 10^{-8} \text{ W}/(\text{m}^2 \text{ K}^4)$, and
    T is the absolute temperature of the black body.

This law is called *Stefan's law* and the constant is Stefan's constant. The constant is known from experiment and can be determined analytically from Planck's theory in terms of known physical constants and h. Solving for h confirms the value from the distribution curve.

example 1 Light from the sun is most intense at a wavelength of 5.10×10^{-7} m (green light). What is the temperature of the sun?

DATA:
$$\lambda_{max} = 5.10 \times 10^{-7} \text{ m}$$
$$b = 2.898 \times 10^{-3} \text{m K}$$
$$T = ?$$

BASIC EQUATION:
$$\lambda_{max} T = b$$

WORKING EQUATION:
$$T = \frac{b}{\lambda_{max}}$$

SUBSTITUTION:
$$T = \frac{2.898 \times 10^{-3} \text{ m K}}{5.10 \times 10^{-7} \text{ m}}$$
$$= 5680 \text{ K}$$

example 2   If the temperature of the surface of the sun is $57\overline{0}0$ K, what is the total energy radiated by the sun each second? The radius of the sun is 7.0×10^9 m.

DATA:
$$T = 57\overline{0}0 \text{ K}$$
$$\sigma = 5.67 \times 10^{-8} \text{ W/(m}^2 \text{ K}^4)$$
$$R_T = ?$$

BASIC EQUATION:
$$R_T = \sigma T^4$$

WORKING EQUATION: same

SUBSTITUTION:
$$R_T = (5.67 \times 10^{-8} \text{ W/(m}^2 \text{ K}^4))(57\overline{0}0 \text{ K})^4$$
$$= 5.99 \times 10^7 \text{ W/m}^2$$

The power radiated by the sun will be the product of R and the area of the sun's surface: $P = R_T A$.

DATA:
$$R_T = 5.99 \times 10^7 \text{ W/m}^2 \qquad \text{(from above)}$$
$$r_s = 7.0 \times 10^9 \text{ m}$$
$$P = ?$$

BASIC EQUATIONS:
$$P = R_T A$$
$$A = 4\pi r_s^2$$

WORKING EQUATION:
$$P = R_T (4\pi r_s^2)$$

SUBSTITUTION:
$$P = (5.99 \times 10^7 \text{ W/m}^2)[4\pi(7.0 \times 10^9 \text{ m})^2]$$
$$= 3.7 \times 10^{28} \text{ W}$$

32.3   THE PHOTOELECTRIC EFFECT

The photoelectric effect is the physical basis for some control devices and detection systems in common usage in our homes today. Many homeowners use photoelectric devices to control convenience and security lights around their property.

Photoelectric cells are used to open doors as a courtesy to customers at stores. Many security alarm systems depend upon photoelectric cells to detect intruders.

A circuit for the photoelectric effect is shown below. Light is incident upon a metallic surface. The term "light" is used here in its broad sense as electromagnetic radiation. The energy from the absorbed light gives enough energy to electrons to permit them to escape from the metal.

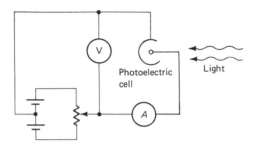

When certain light is incident upon the metal, electrons are emitted. When the light is turned off, no electrons are emitted. It is important to stress that there is no time delay. As soon as the light strikes the metal, electrons are emitted and when the light is removed, no more electrons escape. Not all light will produce the emission of electrons from a given metal plate. Each material has a cutoff frequency and light of lower frequency has no effect on that plate.

Refer to the circuit shown. When light is incident upon the photoelectric surface, electrons are emitted and attracted to the positive electrode. This permits a current to flow. A plot of that current versus the accelerating voltage is presented below.

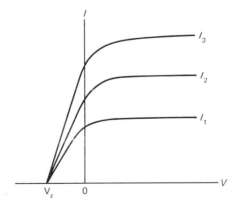

When the voltage is positive, a saturation current flows. As zero accelerating voltage is approached, the current flow drops; but even when the voltage is negative and repels the electrons, some electrons have enough kinetic energy to reach the electrode. For a given material and a given frequency of light, this value of repelling potential, called *stopping voltage*, is always the same and corresponds to the maximum kinetic energy that the photoemitted electrons have when leaving the surface of the metal. Please note that increased intensities of the same light increase the saturated current but do not change the stopping voltage. When light of a different frequency is incident upon the metal, a different stopping potential is obtained. Different light frequencies give different maximum kinetic energies to the emitted electrons. The maximum kinetic energy of the emitted electrons, in terms of the stopping potential, is

$$K_{\max} = eV_s$$

where $K_{\max}$ is kinetic energy, in joules,
  e is charge of an electron, in coulombs, and
  V_s is the stopping potential, in volts.

   A plot of stopping potential versus frequency for a material is shown below. When V_s is zero, the line intersects the frequency axis at f_0. This frequency is the cutoff frequency for that material, and light of lesser energy (lower frequency) produces no effect. The energy of light of this frequency is just equal to the energy that is required to free the electron from the photoelectric material. This energy is called the *photoelectric work function.*

$$\phi = hf_0$$

where ϕ is the photoelectric work function, in joules,
  h is Planck's constant, and
  f_0 is the cutoff frequency.

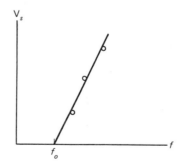

   Recall the instant nature of the emission of electrons when the plate is struck by light, plus the fact that an increased intensity of a given light source does not increase the maximum kinetic energy of electrons, only the number of emitted electrons. The interpretation is made that light may be considered as composed of particles called photons and each photon has an energy proportional to its frequency (Section 28.3). The photons exist only as electromagnetic radiation, travel with the speed of light, have both energy and momentum, and do not have a rest mass. The interaction between photon and the photoelectron is a quantum one in which the photon gives all of its energy to the electron and ceases to exist. The electron, having acquired this energy, is now able to escape from the material. Many electrons that receive energy from photons lose some or all of their energy in internal collisions with other electrons and nuclei. In fact, a very small percentage escape with their maximum kinetic energy.

$$eV_s = hf - \phi$$

where eV_s is the maximum kinetic energy of a photoelectron,
 f is the frequency of the incident light,
 h is Planck's constant, and
 ϕ is the photoelectric work function.

It is common practice to present work functions in the energy unit of electron volts (eV). An *electron volt* is the energy an electron receives when accelerated through a potential difference of 1 V (1.60×10^{-19} J $= 1.00$ eV). This equation states that the maximum kinetic energy of a photoelectron is the difference between the energy of the incident photon and the work function of the material.

The photoelectric effect was given a quantum interpretation by Albert Einstein early in the twentieth century and was a very important step toward our modern quantum concepts of physics.

example   The stopping potentials for a certain material are 0.88 V and 1.42 V when light of wavelengths 3.2×10^{-7} m and 2.8×10^{-7} m is incident upon it. What is the work function of the material?

DATA:
$$eV_s = 0.88 \text{ eV}$$
$$eV_s' = 1.42 \text{ eV}$$
$$\lambda = 3.2 \times 10^{-7} \text{ m}$$
$$\lambda' = 2.8 \times 10^{-7} \text{ m}$$
$$c = 2.99 \times 10^8 \text{ m/s}$$
$$h = (6.62 \times 10^{-34} \text{ Js}) \left(\frac{1.00 \text{ eV}}{1.60 \times 10^{-19} \text{ J}} \right)$$
$$= 4.14 \times 10^{-15} \text{ eVs}$$
$$\phi = ?$$

BASIC EQUATIONS:
$$eV_s = hf - \phi$$
$$f = \frac{c}{\lambda}$$

WORKING EQUATION:
$$\phi = \frac{hc}{\lambda} - eV_s$$

SUBSTITUTIONS:
$$\phi = \frac{(4.14 \times 10^{-15}\text{eVs})(2.99 \times 10^8 \text{ m/s})}{3.2 \times 10^{-7} \text{ m}} - 0.88 \text{ eV} = 3.0 \text{ eV}$$

$$\phi' = \frac{(4.14 \times 10^{-15}\text{eVs})(2.99 \times 10^8 \text{ m/s})}{2.8 \times 10^{-7} \text{ m}} - 1.42 \text{ eV} = 3.0 \text{ eV}$$

32.4   X-RAY PRODUCTION

The production of X-radiation was discovered by Wilhelm Conrad Roentgen at the end of the nineteenth century when he was experimenting with electrical discharges through gases at very low pressures. The early X-ray tubes consisted of two electrodes: the positive electrode was called the target and the negative electrode was pointed so that electrons would be pulled from it easily by high electric fields. In X-ray tubes of later design, the pointed electrode was replaced by a heated electrode and electrons received enough energy from heat to escape from the electrode. This process is called *thermionic emission*.

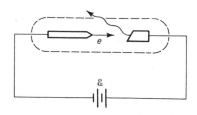

The production of X-rays is produced by accelerating electrons into the positive electrode or target. When the electrons are slowed down or scattered, they give off energy in the form of electromagnetic radiation. The figure on page 569 shows a plot of the X-ray spectrum as a function of wavelength for several different voltages ($V_3 > V_2 > V_1$). This spectrum of radiation is called the *continuous* X-ray spectrum and its shape depends upon the applied voltage between the two electrons. It will be noted that as the voltage is increased, the intensity is greater for all wavelengths. Also, the shortest wavelength is sharply defined and becomes less for a higher applied voltage. The X-ray photons corresponding to λ_0 have received all of the kinetic energy from a single electron accelerated through the applied voltage. The energy of the X-ray is just equal to the maximum energy acquired by an electron:

$$\boxed{hf_0 = \frac{hc}{\lambda_0} = eV}$$

where h is Planck's constant,
$\quad f_0$ is the frequency corresponding to λ_0,
$\quad \lambda_0$ is the short-wavelength cutoff,

c is the speed of light,

e is charge on an electron, and

V is the voltage applied to the X-ray tube.

This equation is often expressed as the following working equation:

$$\lambda_0 V = \frac{hc}{e} = 12.4 \text{ angstrom kilovolt}$$

where λ_0 is the short-wavelength cutoff, in angstroms, and

V is the applied voltage, in kilovolts.

One angstrom unit (Å) is 10^{-10} m.

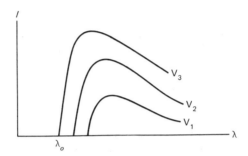

The continuous X-ray spectrum does have an efficiency of production directly proportional to the atomic number of the target material, but its distribution and cutoff wavelength are dependent upon the applied voltage to the X-ray tube. Longer wavelengths in the distribution are the results of multiple acceleration of electrons; that is, an electron in losing its kinetic energy may produce more than one X-ray photon.

The continuous X-ray spectrum may have superimposed on it spikes at certain wavelengths dependent upon the material of which the target is constructed and the applied tube voltage. Because these spikes are characteristic of the target material, they are called the characteristic spectrum and will be discussed in Chapter 33.

example   What is the value of the short-wavelength cutoff when an X-ray tube is operated at $10\bar{0},000$ V?

DATA: $V = 10\bar{0} \text{ kV}$

                                           $\lambda_0 = ?$

BASIC EQUATION: $V\lambda_0 = 12.4 \text{ Å kV}$

WORKING EQUATION: $\lambda_0 = \dfrac{12.4 \text{ Å kV}}{V}$

SUBSTITUTION: $\lambda_0 = \dfrac{12.4 \text{ Å kV}}{10\bar{0} \text{ kV}}$

                                           $= 0.124 \text{ Å or } 1.24 \times 10^{-11} \text{ m}$

EXERCISES

Sketch	Data	Basic Equation	Working Equation	Substitution
(sketch: $12\ cm^2$, $h = ?$, $b = 6.0\ cm$)	$A = 12\ cm^2$ $b = 6.0\ cm$ $h = ?$	$A = bh$	$h = \dfrac{A}{b}$	$h = \dfrac{12\ cm^2}{6.0\ cm}$ $= 2.0\ cm$

1. A star has a temperature of 8600 K. At what wavelength is there a maximum of power emitted by the star? What color does the star appear to be?

2. A star has a temperature of 6800 K. At what wavelength is there a maximum of power emitted? What color does the star appear to be?

3. What is the surface power density of radiation emitted by a star at 8600 K?

4. What is the surface power density of radiation emitted by a star at 6800 K?

5. Energy from the sun arrives at the surface of Earth at the rate of 1.42×10^3 W/m^2. How much energy is received from the sun each hour? (The radius of the earth is 6.38×10^6 m.)

6. What total power is radiated by the sun if 1.42×10^3 W/m^2 reaches the surface of Earth? (The mean sun–Earth distance is 1.49×10^{11} m.)

7. The stopping potential for a certain salt is 0.34 V when radiated by 5800-Å light. What is the work function of this salt?

8. A metal whose work function is 2.8 eV produces electrons which require 1.4 V of stopping potential to stop current. What is the wavelength of the incident radiation?

9. What applied voltage is required to produce 1.00-Å X-rays?

10. Some industrial radiographic X-ray units operate at 225,000 V. What is the shortest radiation generated?

11. A machine called a betatron is used to accelerate electrons through a potential of 1.00 million volts to generate X-radiation for therapeutic purposes. What is the shortest wavelength produced by this machine?

12. Long-wavelength X-radiation is sometimes used for the treatment of certain skin ailments. These units are commonly operated at 20$\bar{0}$0 V. What is the shortest wavelength of X-rays produced by this machine?

33.1 ATOMIC MODELS

Several early atomic models were proposed in an attempt to describe and predict line spectra. The first successful model was that of Niels Bohr, which explained the atomic spectra of hydrogen. It is important to realize that the Bohr atom is clearly limited in scope but was a forerunner to modern quantum theory. Contemporary atomic quantum theory with its mathematical formalism is beyond the scope of this book.

33.2 THE BOHR ATOMIC MODEL

Bohr proposed a planetary model where an electron moves in a stable orbit around a central positively charged nucleus. The nucleus of the hydrogen atom is a proton. Coulomb attraction between the orbiting electron and the proton provides the force necessary for a stable orbit. The total energy of the atom is quantized:

$$E_n = -2.176 \times 10^{-18}\left(\frac{1}{n^2}\right)\text{J}$$

or

$$E_n = -13.6\left(\frac{1}{n^2}\right)\text{eV}$$

where E_n is the energy of the nth state and
     n is an integer.
The coefficients of -2.176×10^{-18} J and -13.6 eV are equivalent. The energy unit electron volt, eV, is commonly used in describing atomic energy states. The coefficient of $1/n^2$ can be derived and its value is determined by several physical constants, including Planck's constant, h. This relationship gives the energy of a hydrogen atom and has only discrete values, depending upon the value of the integer n. The integer n is called a *quantum number* and describes the state of the atom. When $n = 1$, the atom is in its lowest energy state, which is called the *ground state*. The energy term is negative, which means that the nucleus and electron are bound together and as n is increased the binding becomes less and less and the atom is said to be *excited*. An excited atom may decay to a lower energy state by emitting electromagnetic radiation. The energy of the emitted radiation photon is the difference between the energy of the initial excited state and the lower state to which it has decayed.

$$hf = -(E_m - E_n)$$
$$hf = (-13.6 \text{ eV})\left(\frac{1}{m^2} - \frac{1}{n^2}\right)$$

where f is the frequency of the emitted photon,
     h is Planck's constant,
     n is an integer whose value identifies the final energy state, and
     $m = n + 1, n + 2, n + 3, \ldots$ is an integer whose value identifies the initial energy state.
Not only is the energy of the atom quantized, but other physical quantities, such as the radius of the orbit, become quantized. For the hydrogen atom,

$$r_n = a_0 n^2$$

where r_n is the radius of the electron orbit corresponding to the nth energy state,
     n is the quantum number, and
     $a_0 = 5.29 \times 10^{-11}$ m, a constant called the Bohr radius.
The following figure shows a plot of energy and orbit radius of the hydrogen atom. The smooth curve is the energy–radius relationship from classical physics. The quantum condition permitted only discrete energies and discrete radii. The four lowest energy levels are shown on the drawing.
     As the energies become higher, they remain discrete but are much closer together. If the atom were raised to such a high energy level that its total energy is greater than zero, the electron would have enough energy to break free of the nucleus, giving a free electron and a proton. The proton is the nucleus of the hydrogen atom. It is the positively charged nucleus of a hydrogen atom and may be called a hydrogen ion.

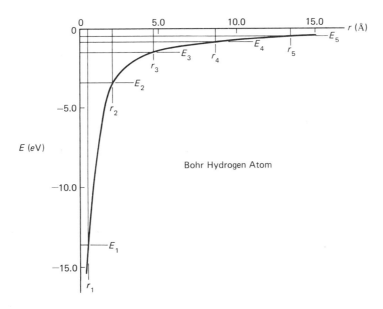

33.3   ENERGY LEVELS

The figure below is an energy-level diagram for hydrogen, giving only the five lowest energy levels. The emission spectrum for these five levels is shown. The series of the spectral lines for hydrogen which have the common terminal energy state $n = 1$ is called the Lyman series; $n = 2$, the Balmer series; $n = 3$, the Paschen

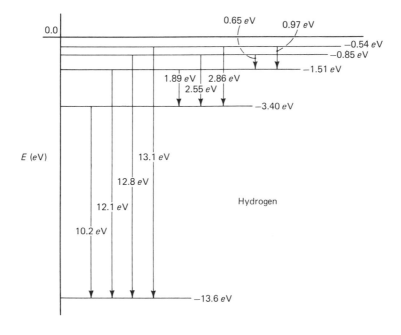

series; $n = 4$, the Brackett series; and $n = 5$, the Pfund series. The Lyman series is all in the ultraviolet region of the electromagnetic spectrum. The Balmer series lies in the visible spectrum; all the others are in the infrared region.

example   In the hydrogen spectrum one series of spectral lines is called the Balmer series ($n = 2$) and lies in the visible region.

(a) Calculate the frequencies and wavelengths of the first four lines of the Balmer series.

(b) What is the shortest wavelength in the Balmer series? This wavelength is called the *series limit*.

(a) DATA:
$$n = 2$$
$$m = 3, 4, 5, 6$$
$$h = 4.14 \times 10^{-15} \text{ eVs} \qquad \text{(See example, p.567)}$$
$$c = 2.99 \times 10^8 \text{ m/s}$$

BASIC EQUATIONS:
$$hf = (-13.6 \text{ eV})\left(\frac{1}{m^2} - \frac{1}{n^2}\right)$$
$$c = f\lambda$$

WORKING EQUATIONS:
$$f = \frac{13.6 \text{ eV}}{h}\left(\frac{1}{n^2} - \frac{1}{m^2}\right)$$
$$\lambda = \frac{c}{f}$$

SUBSTITUTIONS:
$$f_{3-2} = \frac{13.6 \text{ eV}}{4.14 \times 10^{-15}\text{eVs}}\left(\frac{1}{2^2} - \frac{1}{3^2}\right) = 4.56 \times 10^{14} \text{ Hz}$$

$$f_{4-2} = \frac{13.6 \text{ eV}}{4.14 \times 10^{-15}\text{eVs}}\left(\frac{1}{2^2} - \frac{1}{4^2}\right) = 6.16 \times 10^{14} \text{ Hz}$$

$$f_{5-2} = \frac{13.6 \text{ eV}}{4.14 \times 10^{-15}\text{eVs}}\left(\frac{1}{2^2} - \frac{1}{5^2}\right) = 6.90 \times 10^{14} \text{ Hz}$$

$$f_{6-2} = \frac{13.6 \text{ eV}}{4.14 \times 10^{-15}\text{eVs}}\left(\frac{1}{2^2} - \frac{1}{6^2}\right) = 7.30 \times 10^{14} \text{ Hz}$$

$$\lambda_{3-2} = \frac{2.99 \times 10^8 \text{ m/s}}{4.56 \times 10^{14} \text{ Hz}} = 6.56 \times 10^{-7} \text{ m or } 6560 \text{ Å}$$

$$\lambda_{4-2} = \frac{2.99 \times 10^8 \text{ m/s}}{6.16 \times 10^{14} \text{ Hz}} = 4.85 \times 10^{-7} \text{ m or } 4850 \text{ Å}$$

$$\lambda_{5-2} = \frac{2.99 \times 10^8 \text{ m/s}}{6.90 \times 10^{14} \text{ Hz}} = 4.33 \times 10^{-7} \text{ m or } 4330 \text{ Å}$$

$$\lambda_{6-2} = \frac{2.99 \times 10^8 \text{ m/s}}{7.30 \times 10^{14} \text{ Hz}} = 4.10 \times 10^{-7} \text{ m or } 41\overline{0}0 \text{ Å}$$

(b) Data: $n = 2$

$m = \infty$

$h = 4.14 \times 10^{-15} \text{ eVs}$

$c = 2.99 \times 10^8 \text{ m/s}$

Basic Equations: $hf = (-13.6 \text{ eV})\left(\dfrac{1}{m^2} - \dfrac{1}{n^2}\right)$

$c = f\lambda$

Working Equations: $f = \dfrac{13.6 \text{ eV}}{h}\left(\dfrac{1}{n^2} - \dfrac{1}{m^2}\right)$

$\lambda = \dfrac{c}{f}$

Substitutions: $f = \dfrac{13.6 \text{ eV}}{S4.14 \times 10^{-15}\text{eVs}}\left(\dfrac{1}{2^2} - 0\right)$

$= 8.21 \times 10^{14} \text{ Hz}$

Then, $\lambda = \dfrac{2.99 \times 10^8 \text{ m/s}}{8.21 \times 10^{14} \text{ Hz}}$

$= 3.64 \times 10^{-7} \text{ m or } 3640 \text{ Å}$

33.4   WAVES AND ELECTRONS

An electron in orbit about the nucleus is in continuous centripetal acceleration and by classical physics would be continually losing energy and spiraling in toward the nucleus. The quantum hypothesis, of course, says that electrons can only lose or gain energy in discrete amounts. A physicist, Louis de Broglie, proposed that all matter had a wave aspect associated with it and that wavelength was given by

$$\lambda = \frac{h}{mv}$$

where λ is the wavelength associated with a particle,
$\quad h$ is Planck's constant,
$\quad m$ is the mass of the particle, and
$\quad v$ is the velocity of the particle.

When an electron is in a stable orbit about the hydrogen nucleus, the atom's energy has a discrete value and the orbit radius has a discrete value. There can be only one discrete value of the velocity for that orbit. In fact, it follows that the circumference of each orbit corresponds to integer multiples of the wavelength of the electron for that discrete velocity. When $n = 1$, the orbit has a circumference of one wavelength; $n = 2$, the orbit has a circumference of two wavelengths; and so on. The interpretation is given that the electron orbits must meet the conditions necessary to form standing waves to be stable just as a jump rope must meet length and tension conditions to form standing waves.

33.5 LIMITATIONS OF THE BOHR ATOM

The Bohr atomic model had many successes for hydrogen and hydrogenlike atoms (one electron); however, it also is quite limited. First, it is applicable only to a two-body atom and cannot successfully cope with atoms with more than one electron. The orbit described by only one quantum number must be circular, so n became known as the principal quantum number. A second quantum number, ℓ, the orbital quantum number, is introduced to give eccentricity to the orbit. When placed in an applied magnetic field, the projections of angular momentum are quantized. Another quantum number, m, is introduced to describe this discrete projection. Each electron may have one or two angular momentum values in a magnetic field; so a spin quantum number, s, is introduced, which has only two values. No two electrons in a vicinity are permitted to have the same quantum numbers, so each electron in an atom has a wave function that is dependent upon the quantum numbers n, ℓ, m, and s. The atom has an energy state that is a combination of these wave functions, and radiation is emitted or absorbed by transferring from one energy state to another energy state.

33.6 X-RAY SPECTRA NOTATION

Spectral emission in the X-ray region occurs when one of the tightly bound electrons is removed from an inner orbit of an atom. The notation of K has been given to the electrons with principal quantum number $n = 1$, L for $n = 2$, M for $n = 3$, and so on. An energy-level diagram is shown on page 577 for the purpose of illustrating X-ray notation of spectral lines. If the transition giving off a photon ends at an energy level, where $n = 1$, the transition is called a K line; if the transition ends at a level where $n = 2$, the transition is called an L line, and so on. If the transition originated at a level, $n + 1$, the subscript α is added; if the transition originated at a level $n + 2$, the subscript β is added; and so on.

In the figure, transitions for K_α, K_β, K_γ, K_δ, L_α, L_β, L_γ, M_α, M_β are shown.

The energy levels shown are degenerate; for example, the L state has eight different distinct energy levels, but a neutral atom in the absence of a magnetic field has only three energy values. Probability of decay from an L state to a K state

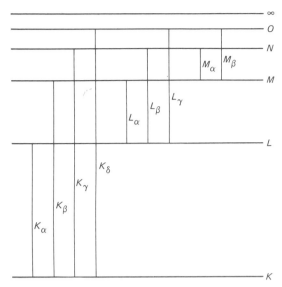

depends upon the wave functions of the states, and the more probable the decay, the greater that spectral line's intensity. The transition may even be prohibited. There are more than one K_α or K_β, and so on, for any atom. The more intense lines were discovered first, so further subscripts 1, 2, 3, and so on, are added and indicate only intensity. For example, L_{β_1}, L_{β_2}, L_{β_3} all represent transitions from a state with principal quantum number, n, of 4 to a state with principal quantum number, n, of 2; but the additional subscript only tells us that L_{β_1} is more intense than L_{β_2}.

Optical spectra come from the higher, more complex energy states, which may be dependent upon chemical combination. The X-ray spectrum comes from the lower-energy states and is altered little by chemical combinations. The X-ray spectrum is characteristic of each element and a valuable tool for analysis of elements present in a material.

EXERCISES

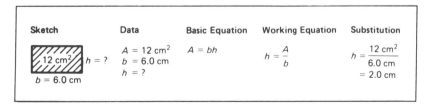

Sketch	Data	Basic Equation	Working Equation	Substitution
$A = 12 \text{ cm}^2$, $b = 6.0 \text{ cm}$, $h = ?$	$A = 12 \text{ cm}^2$ $b = 6.0 \text{ cm}$ $h = ?$	$A = bh$	$h = \dfrac{A}{b}$	$h = \dfrac{12 \text{ cm}^2}{6.0 \text{ cm}}$ $= 2.0 \text{ cm}$

1. The Lyman series for hydrogen has $n = 1$ as its terminal state. (a) Calculate the frequencies and wavelengths of the first four lines of the Lyman series. (b) What is the series limit of the Lyman series?

2. The Paschen series for hydrogen has $n = 3$ as its terminal state. Calculate the longest- and shortest-wavelength limits for this series.

3. The Brackett series has hydrogen for $n = 4$ as its terminal state. Calculate the longest- and shortest-wavelength limits for this series.

4. The Cu K_α-radiation has a wavelength of 1.54×10^{-10} m. What is the energy difference between the copper K and L shells?

5. The series limit for K-radiation is 1.38×10^{-10} m. What energy is required to ionize a copper atom?

6. Iron has a K-series limit of 1.74×10^{-10} m. What energy is required to ionize an iron atom?

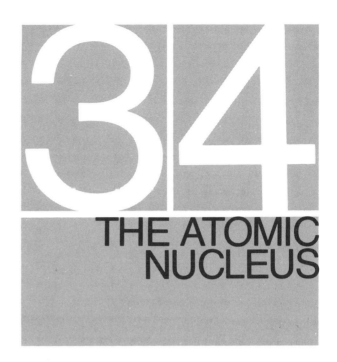

THE ATOMIC NUCLEUS

34.1   THE NUCLEUS

The atom has a massive nucleus surrounded by a number of electrons. In Chapter 33 we saw that the electrons obey rules for their orbits and that the energy states of the atom can have only discrete values.

The nucleus of the atom is composed of two principal components. One component is called protons. The second is called neutrons. Collectively, they are referred to as nucleons.

We recall that the mass of the proton and the mass of the neutron differ slightly from each other, but both are about 1.67×10^{-27} kg. The mass of an electron is only 9.1×10^{-31} kg, so the mass of the atom is concentrated in the nucleus. The size of an atom is generally considered to be the space occupied by the electrons. The radius of a stable, uncharged atom is of the order of 1×10^{-10} m, but the radius of the nucleus is of the order of 1×10^{-15} m. The radius of the nucleus is about 1/100,000 of the radius of the uncharged atom and contains almost all the mass of the atom. The atomic mass is proportional to the number of nucleons. The number of nucleons in an atom is called the *mass number* of that atom. In addition to giving mass to an atom, protons also have a positive charge. The charge on a proton has the same distinct value as the charge of an electron, but is positive. The nucleus is positively charged and a neutral atom will have the same number of electrons in orbit as there are protons in the nucleus. The electronic structure determines how atoms combine with other atoms or give chemical properties to an atom. The atomic number of an atom is the number of protons in the nucleus. All atoms with the same atomic number are the same element and have the same chemical properties. They may have different mass numbers but still have the same number of protons. For example, one atom may have 28 protons and 32 neutrons, and a second may have 28 protons and 30 neutrons. The first atom is nickel because of the 28 protons. It has a mass number of 60. The second atom is nickel because of the 28 protons, but it has a mass number of 58. These atoms are isotopes of nickel. The first would be written $^{60}_{28}$Ni, and the second $^{58}_{28}$Ni, or simply ^{60}Ni and ^{58}Ni, and would be called "nickel 60" and "nickel 58."

We found that the electronic structures of atoms were quantized with a number of discrete energy levels. The same is true for the nucleus itself. The nucleus has a complex energy-state array. The atom may be in a stable state or may be in an excited state.

34.2   RADIOACTIVITY

Atoms that are in excited states will have a tendency to decay to lower energy configurations. Nuclei decay to lower energy configurations by giving off electromagnetic radiation, a process called γ decay; or by permitting a negative charge from the nucleus, a process called β decay; or by emitting a helium nucleus (2 protons + 2 neutrons) from the nucleus, a process called α decay.

The helium nucleus is called an α particle. Alpha emission is more common among the elements with large mass numbers. When an atom emits an α particle, it becomes a different element whose atomic number is two less than its initial atomic number and whose mass number is four less than the initial atom. The α particle will leave the nucleus with kinetic energy, but because of its charge and mass, α-radiation is not a very penetrating radiation.

The β-radiation is composed of electrons that have been ejected from the nucleus. The β particles have kinetic energy, but because of their charge cannot penetrate through much material. The β decay is found mostly in materials that have been created artificially. The decay process takes a negative quantum of charge from the nucleus and does not change the mass number. Beta decay produces a product that is one atomic number greater than and of the same mass number as the initial atom.

Gamma decay is the emission of a high-energy photon from the nucleus. This radiation is called γ-radiation and except for its origin has the identical properties of the artificially produced X-radiation. This electromagnetic radiation is very penetrating and can be difficult to shield against. The emission of γ-radiation does not change either the mass number or the atomic number of the element, but only leaves the nucleus in a lower energy state. Gamma decay usually follows some other reaction, such as β decay, when the atom is left in an excited state.

34.3   NUCLEAR FISSION

It was discovered that one isotope of uranium, ^{235}U, when bombarded with low-energy neutrons (slow neutrons), would split into two or more elements toward the center of the atomic table and yield one or more neutrons as well as γ-radiation. This process, when a neutron is captured by an atomic nucleus and that nucleus almost immediately splits, is called *nuclear fission*. The fragments from the fission have high kinetic energy. As a result, the vicinity about the fission region becomes hot.

34.4   CHAIN REACTIONS

The nucleus undergoing fission produces one or more neutrons. These neutrons are of high kinetic energy (fast neutrons) and, consequently, are not likely to be captured by a fissionable nucleus. If, however, some material is present that can take this kinetic energy from the fast neutron, the neutron becomes a slow neutron and is much more likely to produce a fission from another nucleus. The material that produces slow neutrons from fast neutrons is called a moderator and is best made of light elements to take maximum energy from a neutron on collision. Moderators are usually water, paraffin, carbon, or materials composed of light atoms.

When an appropriate concentration of fissionable atoms is placed in a moderating material, the neutrons produced by a fission can be slowed enough to be captured by other fissionable atoms and produce more fissions. This pyramiding process is called a *chain reaction*. When fissionable materials and moderating materials are assembled in the proper geometry so as to sustain a chain reaction, the assembly is said to be critical. To prevent an explosion from a critical assembly, control of the rate of the chain reaction is needed.

34.5   NUCLEAR REACTORS

A nuclear reactor adds to the critical assembly a control element. This control element is a good absorber of slow neutrons. When an excess of the control element is in the assembly, the chain reaction stops because of a deficiency of neutrons. As the control element is withdrawn from the critical assembly, the chain reaction continues. In a nuclear reactor the rate at which the nuclear reaction proceeds is controlled by increasing or decreasing the amount of the control element in the critical assembly. A schematic drawing of a nuclear reactor is shown on page 582.

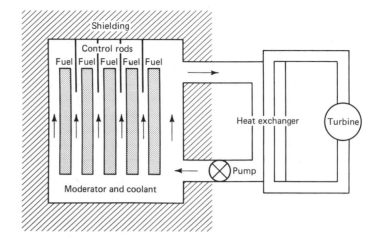

The fuel elements are composed of fissionable materials, either uranium ^{235}U, or plutonium ^{239}Pu, which may be alloyed with other metals or may be in a ceramic form. This fuel is in a metal jacket which bonds tightly to the fuel for heat transfer. The metal jacket contains the fissionable material. These fuel elements are mounted in a moderator which takes energy from the neutrons. The reactor must be cooled. This is usually done with water, which also serves as the moderator, or it may be done with liquid metals (Na and K). Thus heat, of course, is a product of power reactors and may be pumped through a heat exchanger as shown here, where a secondary system drives a generator or turbine. In the fission process, a number of elements are produced which are radioactive, and the reactor requires massive shielding against high-energy γ-radiation and against the fast and slow neutrons. The reactor is surrounded by a building, called a containment shell, which meets stringent strength and leakage requirements.

There are two main types of reactors. One is designed primarily to produce neutrons for scientific study and is called a research reactor. The other is designed to produce heat efficiently and is called a power reactor. Power reactors are of many designs; some are used to produce fissionable material when in operation and are called breeder reactors. Reactors, in addition to providing neutrons for research and energy for power, are used to produce isotopes of elements useful for diagnosis and treatment.

APPENDIX A REVIEW OF ALGEBRA

A.1 EXPONENTS AND RADICALS

$$a^m \cdot a^n = a^{m+n}$$

$$\frac{a^m}{a^n} = a^{m-n} \qquad (a \neq 0)$$

$$(a^m)^n = a^{mn}$$

$$(ab)^n = a^n b^n$$

$$\left(\frac{a}{b}\right)^n = \frac{a^n}{b^n} \qquad (b \neq 0)$$

$$a^{-n} = \frac{1}{a^n} \qquad \text{and} \qquad \frac{1}{a^{-n}} = a^n$$

$$a^0 = 1$$

$$a^{1/n} = \sqrt[n]{a}$$

$$a^{m/n} = \sqrt[n]{a^m} = \left(\sqrt[n]{a}\right)^m$$

$$\sqrt[n]{a} \cdot \sqrt[n]{b} = \sqrt[n]{ab}$$

$$\frac{\sqrt[n]{a}}{\sqrt[n]{b}} = \sqrt[n]{\frac{a}{b}} \qquad (b \neq 0)$$

$$\sqrt[m]{\sqrt[n]{a}} = \sqrt[mn]{a}$$

A.2   QUADRATIC FORMULA

The solutions of the quadratic equation

$$ax^2 + bx + c = 0 \qquad (a \neq 0)$$

are

$$x = \frac{-b \pm \sqrt{b^2 - 4ac}}{2a}$$

A.3   THE ADDITION–SUBTRACTION METHOD

To solve a pair of linear equations by the addition–subtraction method:

1. If necessary, multiply each side of one or both equations so that the numerical coefficients of one of the variables are of equal absolute value.
2. If these coefficients of equal absolute value have like signs, subtract one equation from the other. If they have unlike signs, add the equations. That is, do whatever is necessary to eliminate that variable.
3. Solve the resulting equation for the remaining variable.
4. Substitute the solution for the variable found in step 3 in either of the original equations and solve this resulting equation for the second variable.

example   Solve:

$$2F_1 + 3F_2 = 180$$
$$\underline{F_1 - 2F_2 = -50} \quad \text{(multiply each side of this}$$

equation by 2 so that the

numerical coefficients of F_1

will be of equal absolute value)

$$2F_1 + 3F_2 = 180$$
$$\underline{2F_1 - 4F_2 = -100} \quad \text{(subtract the equations)}$$
$$7F_2 = 280$$
$$F_2 = 40$$

Substitute $F_2 = 40$ in either original equation.

$$F_1 - 2F_2 = -50$$
$$F_1 - 2(40) = -50$$
$$F_1 = 30$$

The solution is $F_1 = 30$ and $F_2 = 40$.

A.4   THE SUBSTITUTION METHOD

To solve a pair of linear equations by the substitution method:

1. From either of the two given equations, solve for one variable in terms of the other.
2. Substitute this resultant value of the variable, from step 1, into the remaining equation.
3. Solve the equation obtained from step 2 for its variable.
4. From the equation in step 2, substitute the solution for the variable found in step 3 and solve this resulting equation for the second variable.

example   Solve:

$$3F_1 - 4F_2 = 100$$
$$F_1 - 2F_2 = \ \ 0$$

Solve the second equation for F_1:

$$F_1 - 2F_2 = \ \ 0$$
$$F_1 = 2F_2$$

Substitute $F_1 = 2F_2$ into the first equation.

$$3F_1 - 4F_2 = 100$$
$$3(2F_2) - 4F_2 = 100$$
$$6F_2 - 4F_2 = 100$$
$$2F_2 = 100$$
$$F_2 = \ \ 50$$

Substitute $F_2 = 50$ into the second equation:

$$F_1 = 2F_2$$
$$F_1 = 2(50)$$
$$F_1 = 100$$

The solution is $F_1 = 100$ and $F_2 = 50$.

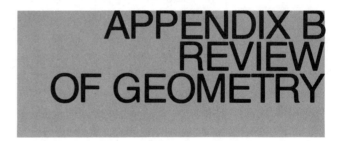

APPENDIX B
REVIEW
OF GEOMETRY

B.1   INTRODUCTION

A basic knowledge of fundamentals of geometry is necessary to solve technical problems as well as to understand mathematical discussions and developments. We will briefly review the most basic, most often used, formulas and theorems.

B.2   ANGLES AND LINES

1. A *right angle* is an angle whose measure is 90°.
2. An *acute angle* is an angle whose measure is less than 90°.
3. An *obtuse angle* is an angle whose measure is more than 90° but less than 180°.
4. Two *vertical angles* are the opposite angles formed by two intersecting lines. Angles m and n are vertical angles as are angles p and q as shown below.

5. Vertical angles are equal.
6. Two angles are *supplementary* when their sum is 180°.
7. Two angles are *complementary* when their sum is 90°.
8. Two lines are *perpendicular* when they form a right angle.
9. The *shortest* distance between a point and a line is the *perpendicular* distance between them.

10. Two lines are *parallel* if they lie in the same plane and do not intersect.

11. When two parallel lines are intersected by a third line called a transversal,

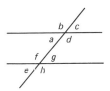

   (a) the alternate interior angles are *equal*.

$$(\angle a = \angle g \quad \text{and} \quad \angle d = \angle f)$$

   (b) the corresponding angles are *equal*.

$$(\angle a = \angle e, \angle b = \angle f, \angle c = \angle g, \angle d = \angle h)$$

   (c) the interior angles on the same side of the transversal are *supplementary*.

$$(\angle a + \angle f = 180° \quad \text{and} \quad \angle d + \angle g = 180°)$$

B.3 TRIANGLES

1. A *polygon* is a closed figure whose sides are all straight line segments.

2. A *triangle* is a polygon with three sides.

3. A *scalene* triangle is a triangle having *no two* sides equal.

4. An *isosceles* triangle is a triangle having *two* sides equal.

5. An *equilateral* triangle is a triangle having *all three* sides equal.

6. An *acute* triangle is a triangle having *all three* angles acute.

7. An *obtuse* triangle is a triangle having *one* obtuse angle.

8. A *right* triangle is a triangle having *one* right angle.

9. An *oblique* triangle is a triangle that does not contain a right angle.

10. In a right triangle, the side opposite the right angle is called the *hypotenuse, c,* and the other two sides are called the *legs, a* and *b*, as shown below.

11. *Pythagorean theorem*: The square of the hypotenuse of a right triangle is equal to the sum of the squares of the legs: $c^2 = a^2 + b^2$.

12. A *median* of a triangle is a line segment joining any vertex to the *midpoint* of the opposite side.

13. An *altitude* of a triangle is a *perpendicular* line segment from any vertex to the opposite side.

14. An *angle bisector* of a triangle is a line segment that bisects any angle and intersects the opposite side.

15. The *sum* of the interior angles of *any* triangle is 180°.

16. The *area* of a triangle equals one-half the base times the height ($A = \frac{1}{2}bh$).

17. In a 30° − 60° − 90° triangle,
    (a) the side opposite the 30° angle equals one-half the hypotenuse, and
    (b) the side opposite the 60° angle equals $\sqrt{3}/2$ times the length of the hypotenuse.

18. Triangles or any other polygons are *similar* if they have their corresponding angles equal or their corresponding sides proportional.

19. Triangles or any other polygons are *congruent* if they have their corresponding angles and sides equal.

B.4 QUADRILATERALS

1. A *quadrilateral* is a polygon with four sides.

2. A *parallelogram* is a quadrilateral having two pairs of parallel sides.

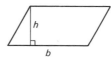

3. The *area* of a parallelogram equals the length of the base times the height ($A = bh$).

4. The opposite sides and the opposite angles of a parallelogram are equal.

5. A *rectangle* is a parallelogram with right angles.

6. A *square* is a rectangle having equal sides.

7. A *rhombus* is a parallelogram having equal sides.

8. A *trapezoid* is a quadrilateral having only one pair of parallel sides.

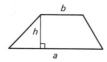

9. The area of a trapezoid is given by the formula: $A = \frac{1}{2}h(a + b)$.

B.5 CIRCLES

1. A *circle* is the set of all points on a curve which are equidistant from a given point called the center.

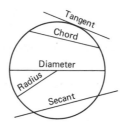

2. A *radius* is a line segment joining the center and any point on the circle.

3. A *chord* is a line segment joining any two points on the circle.

4. A *diameter* is a chord passing through the center.

5. A *tangent* is a line intersecting a circle at only one point.

6. A *secant* is a line intersecting a circle in two points.

7. The *area* of a circle is given: $A = \pi r^2$, where r is the radius.

8. The *circumference* of a circle is given: $C = 2\pi r$, where r is the radius or by $C = \pi d$, where d is the diameter.

B.6   AREAS AND VOLUMES OF SOLIDS

1. The *lateral surface area* of a solid is the sum of the areas of the sides excluding the area of the bases.

2. The *total surface area* of a solid is the sum of the lateral surface area plus the area of the bases.

3. The *volume* of a solid is the number of cubic units of measure contained in the solid.

Instead of verbally describing each of the following solids, we will name, draw, and give the formula for the volume and lateral surface area of each. We shall use B, r, and h as the area of the base, length of radius, and height, respectively.

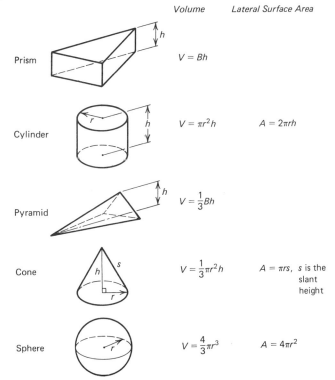

	Volume	*Lateral Surface Area*
Prism	$V = Bh$	
Cylinder	$V = \pi r^2 h$	$A = 2\pi rh$
Pyramid	$V = \frac{1}{3} Bh$	
Cone	$V = \frac{1}{3}\pi r^2 h$	$A = \pi rs$, s is the slant height
Sphere	$V = \frac{4}{3}\pi r^3$	$A = 4\pi r^2$

APPENDIX C
REVIEW OF
TRIGONOMETRY

C.1 DEFINITIONS

$$\sin A = \frac{\text{side opposite } A}{\text{hypotenuse}} = \frac{a}{h} \text{ ; } \sin B = \frac{b}{h}$$

$$\cos A = \frac{\text{side adjacent } A}{\text{hypotenuse}} = \frac{b}{h} \text{ ; } \cos B = \frac{a}{h}$$

$$\tan A = \frac{\text{side opposite } A}{\text{side adjacent } A} = \frac{a}{b} \text{ ; } \tan B = \frac{b}{a}$$

C.2 LAW OF SINES

$$\frac{a}{\sin A} = \frac{b}{\sin B} = \frac{c}{\sin C}$$

C.3 LAW OF COSINES

$$a^2 = b^2 + c^2 - 2bc \cos A$$
$$b^2 = a^2 + c^2 - 2ac \cos B$$
$$c^2 = a^2 + b^2 - 2ab \cos C$$

C.4 TRIGONOMETRIC IDENTITIES

1. $\tan x = \dfrac{\sin x}{\cos x}$

2. $\cot x = \dfrac{\cos x}{\sin x}$

3. $\sec x = \dfrac{1}{\cos x}$

4. $\csc x = \dfrac{1}{\sin x}$

5. $\cot x = \dfrac{1}{\tan x}$

6. $\sin^2 x + \cos^2 x = 1$

7. $1 + \tan^2 x = \sec^2 x$

8. $1 + \cot^2 x = \csc^2 x$

9. $\sin(x + y) = \sin x \cos y + \cos x \sin y$

10. $\sin(x - y) = \sin x \cos y - \cos x \sin y$

11. $\cos(x + y) = \cos x \cos y - \sin x \sin y$

12. $\cos(x - y) = \cos x \cos y + \sin x \sin y$

13. $\tan(x + y) = \dfrac{\tan x + \tan y}{1 - \tan x \tan y}$

14. $\tan(x - y) = \dfrac{\tan x - \tan y}{1 + \tan x \tan y}$

15. $\sin(-x) = -\sin x$

16. $\cos(-x) = \cos x$

17. $\tan(-x) = -\tan x$

18. $\cot(-x) = -\cot x$

19. $\sec(-x) = \sec x$

20. $\csc(-x) = -\csc x$

21. $\sin 2x = 2 \sin x \cos x$

22. $\cos 2x = \cos^2 x - \sin^2 x$
$$= 1 - 2 \sin^2 x$$
$$= 2 \cos^2 x - 1$$

23. $\tan 2x = \dfrac{2 \tan x}{1 - \tan^2 x}$

24. $\sin \dfrac{x}{2} = \pm \sqrt{\dfrac{1 - \cos x}{2}}$

25. $\cos\dfrac{x}{2} = \pm\sqrt{\dfrac{1 + \cos x}{2}}$

26. $\tan\dfrac{x}{2} = \pm\sqrt{\dfrac{1 - \cos x}{1 + \cos x}} = \dfrac{1 - \cos x}{\sin x} = \dfrac{\sin x}{1 + \cos x}$

27. $\sin x + \sin y = 2\sin\dfrac{x + y}{2}\cos\dfrac{x - y}{2}$

28. $\sin x - \sin y = 2\cos\dfrac{x + y}{2}\sin\dfrac{x - y}{2}$

29. $\cos x + \cos y = 2\cos\dfrac{x + y}{2}\cos\dfrac{x - y}{2}$

30. $\cos x - \cos y = -2\sin\dfrac{x + y}{2}\sin\dfrac{x - y}{2}$

C.5   THE GREEK ALPHABET

Capital	Lowercase	Name
A	α	alpha
B	β	beta
Γ	γ	gamma
Δ	δ	delta
E	ε	epsilon
Z	ζ	zeta
H	η	eta
Θ	θ	theta
I	ι	iota
K	κ	kappa
Λ	λ	lambda
M	μ	mu
N	ν	nu
Ξ	ξ	xi
O	o	omicron
Π	π	pi
P	ρ	rho
Σ	σ	sigma
T	τ	tau
Υ	υ	upsilon
Φ	ϕ	phi
X	χ	chi
Ψ	ψ	psi
Ω	ω	omega

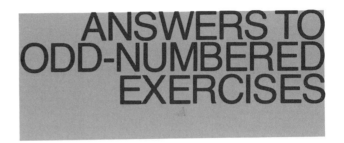

ANSWERS TO
ODD-NUMBERED
EXERCISES

CHAPTER 1

Pages 15–17

1. 10^8 **3.** 10^{12} **5.** 10^{-9} **7.** 10^5 **9.** 10^{16} **11.** 10^9
13. 5.06×10^4 **15.** 2.4×10^{-3} **17.** 8.5×10^6 **19.** 1×10^{-10} **21.** 4100
23. 0.000632 **25.** 1,050,000,000 **27.** 0.0000000004 **29.** 10^9 **31.** 10^{-9} **33.** 10^{-18}
35. (a) 10^{30} (b) 10^{23} (c) 10^{24} (d) 10^{25} (e) 10^{24} (f) 10^{27} (g) 10^{27} (h) 10^{26} (i) 10^{26} (j) 10^{24}
37. (a) 10^2 (b) 10^3 (c) 10 (d) 1 (e) 10
39. (a) 1836.11 (b) 1838.63 (c) 1.00138 (d) 6.56×10^{54}

Page 20

1. 2.5 km **3.** 1.3 g **5.** 650 m **7.** 60,000 m/s **9.** 1.8×10^7 m/h **11.** 16.7 m/s **13.** 12 cm
15. 80 Ω

Pages 24–25

1. 3; 1 km **3.** 2; 1000 L **5.** 2; 1 mg **7.** 2; 0.0001 V **9.** 3; 0.0001 g
11. 78,200 m **13.** 19 V **15.** 8600 km **17.** 1,100,000 V **19.** 901 m^2
21. 21,380,000 cm^3 or 2.138×10^7 cm^3 **23.** $3\bar{0}$ m **25.** 340 V/A
27. 31,000 Nm/s or 3.1×10^4 Nm/s **29.** $3\bar{0}0$ V^2/Ω

Pages 28–29

1. Direct; $k = \frac{3}{4}$ **3.** Inverse; $k = \frac{3}{2}$ **5.** Neither **7.** $y = kz$ **9.** $a = kbc$
11. $r = ks/\sqrt{t}$ **13.** $f = kgh/j^2$ **15.** $k = \frac{1}{3}$; 12 **17.** $k = 54$; 3 **19.** $k = 6$; 36
21. $k = 32$; $\frac{64}{3}$ **23.** $k = 6$; 1800 cm^3 **25.** $k = \frac{8}{9}$; 430 W

CHAPTER 2

Pages 42–43

1. 59 km at 53° north of east (53°)   **3.** 26$\overline{0}$ km at 23° south of west (203°)
5. 58.5 km at 52.6° north of east (52.6°)   **7.** 741 km at 49.1° south of east (310.9°)
9. 58.5 km at 52.6° north of east (52.6°)   **11.** 104 km at 34.9° north of west (145.1°)
13. 44.3 km at 8.8° south of west (188.8°)   **15.** 42 km at 51° north of east (51°)
17. 35$\overline{0}$ km at 49.6° north of east (49.6°)   **19.** 264 km at 49° south of west (229°)
21. 15 km   **23.** 180°   **25.** 0

CHAPTER 3

Pages 49–50

1. (a) 16.9 m^2 (b) 16.5 m   **3.** (a) 42,100 cm^3 (b) 8320 cm^2 **5.** 19,400 cm^2
7. 20$\overline{0}$ cm^2 **9.** (a) 15.0 m   (b) 13.6 m   **11.** 5.2 m   **13.** 24.8 cm

CHAPTER 4

Pages 79–82

1. 410 km   **3.** 72 km/h   **5.** (a) constant   (b) 6.5 m/s
7. (a) 16 km/h   (b) 24 km/h   **9.** 432 km/h at 1$\overline{0}$° west of north (10$\overline{0}$°)
11. 455 km/h at 9° west of south (261°)   **13.** (a) 53° upstream   (b) 4.2 min
15. 1.2 m/s^2 **17.** (a) 32.2 m/s   (b) 80.8 m   **19.** (a) constant   (b) 22 m/s^2
21. (a) 21 m/s   (b) 22 m   **23.** (a) 221 m/s   (b) 22.6 s   (c) 45.2 s
25. (a) 8.24 s   (b) 5.98 s   (c) 58.6 m/s   **27.** (a) 16.5 m   (b) 3.67 s
29. (a) 19.3 km   (b) 42.9 s   (c) 42$\overline{0}$ m/s   (d) 45$\overline{0}$ m/s
31. (a) 45°   (b) 7720 m   (c) 90°

CHAPTER 5

Pages 101–102

1. 5250 N   **3.** 6.00 kg   **5.** 10.5 m/s^2 **7.** (a) 4.62 s   (b) 58 m   (c) 14.2 m/s
9. (a) -1.44 m/s^2 (b) 3530 N   (c) 5620 N   (d) -0.905 m/s^2
11. (a) 1.42×10^4 N   (b) 2360 N   **13.** (a) 1.51 m/s^2 (b) 5390 N   (c) 40,400 N
15. (a) 1.8×10^4 kg   (b) 1.4×10^4 N   **17.** (a) 0.51 m/s^2 (b) 9.7 m/s^2

CHAPTER 6

Pages 118–122

1. $F_x = 134$ N, $F_y = 112$ N   **3.** $F_N = 980$ N, F = 690 N
5. (a) 30$\overline{0}$ N   (b) 7$\overline{0}$ N   **7.** 407 N at 10.6° north of east
9. 1030 N at 11.0° north of east   **11.** 407 N at 10.6° south of west
13. 1030 N at 11.0° south of west   **15.** $|F_1| = 70.7$ N, $|F_2| = 70.7$ N   **17.** $|F_1| = 820$ N, $|F_2| = 480$ N
19. $|F_1| = 577$ N, $|F_2| = 289$ N   **21.** $|T| = \overline{8}900$ N, $|C| = 12,600$ N   **23.** $|T_1| = |T_2| = 1750$ N
25. 110 N   **27.** $|C_1| = |C_2| = 8490$ N   **29.** $|T_1| = 13,000$ N, $|T_2| = 9500$ N
31. 970 N   **33.** 450 N   **35.** 8.5°

CHAPTER 7

Pages 136–139

1. 3.20×10^4 J **3.** 4410 J **5.** 0.30 N **7.** 1.45×10^5 J
9. (a) 670 N (b) 1.3×10^4 J **11.** (a) 58 N (b) 1500 J
13. (a) 57 N (b) 1100 J **15.** 39.2 kW **17.** 12.3 s **19.** 219 N **21.** 59.4 s
23. (a) 17 kW (b) 28 kW **25.** 4.2×10^6 J **27.** 4.6×10^{-17} J **29.** 79,100 J
31. (a) 2.45 MW (b) 1.47×10^8 J **33.** 8.85 m/s **35.** 72.7 m/s

CHAPTER 8

Pages 153–155

1. (a) 6.58×10^5 kg m/s (b) 562 m/s **3.** 0.103 m/s
5. (a) 15.7 m/s (b) $22\bar{0}0$ kg m/s (c) 1.73×10^4 J **7.** 4.50 m/s to the left
9. 9.75 m/s to the right **11.** 4.80 m/s to the right **13.** $A = 2.5$ m/s (left); $B = 2.5$ m/s (right)
15. $A = 2.66$ m/s (left); $B = 5.33$ m/s (right) **17.** 7.14 m/s (right) **19.** 29 km/h (east)
21. (a) 66.7 km/h (b) 96.0 km/h
23. (a) 8.3 m/s or $3\bar{0}$ km/h (b) 15 m/s or 54 km/h (c) 7 m/s or ~ 25 km/h (d) 21 m/s or 76 km/h

CHAPTER 9

Pages 185–191

1. (a) 13π rad (b) 2340° **3.** (a) 5 rev (b) 1800°
5. $18,\bar{0}00/\pi$ rpm or 5730 rpm **7.** (a) 6.7 rev/s (b) $4\bar{0}0$ rpm (c) 42 rad/s
9. (a) 68.6 rad/s (b) 1.23×10^4 rad (c) 157 cm (d) 3080 m (e) 1720 cm/s
11. (a) 50.5 rad/s (b) 1520 rad (c) $50\bar{0}$ m **13.** 3.93 m/s **15.** (a) 3.5 rad/s² (b) 2600 rad
17. (a) 0.909 rad/s² (b) 5.45 rad/s (c) 1.80 m/s (d) 6.48 km/h (e) 16.4 rad
19. (a) 16.7 rad/s² (b) 3.34 m/s² (c) 50.2 m/s (d) 12,600 m/s²
21. (a) 1.08×10^5 km/h (b) 77.2 km/h² or 5.96×10^{-3} m/s² (c) 0 (d) 0
23. 5430 N **25.** (a) 16° (b) 4200 N **27.** 845 rpm **29.** 7.00 m/s
31. 27,000 km/h **33.** 58.0 N **35.** 6.0×10^4 kg m²
37. (a) 8.72 rad/s² (b) $50\bar{0}$ rpm (c) 31.5 kg m²
39. 7.0 J **41.** 5.0 rad/s **43.** 350 kW

CHAPTER 10

Pages 208–213

1. $50\bar{0}$ N **3.** $F_1 = 455$ N, $F_2 = 395$ N **5.** 1.8×10^5 N **7.** $90,\bar{0}00$ N, 101,000 N
9. 8.3×10^4 N, 1.1×10^5 N **11.** 275 N down; 3.75 m from A **13.** (3.06 cm, 3.06 cm)
15. (4.1 cm, -0.10 cm) **17.** (0, 2.95 cm) **19.** (a) 9920 N (b) $A_x = 76\bar{0}0$ N; $A_y = 620$ N
21. (a) 18,500 N (b) $A_x = 17,400$ N, $A_y = 2520$ N **23.** (a) 15,800 N (b) $A_x = 14,800$ N, $A_y = 3450$ N
25. 6690 N; 10.3° north of west; 2.55 m from A

CHAPTER 11

Pages 234–238

1. (a) 2.7 (b) 3200 N **3.** (a) 6.0 (b) 130 N **5.** (a) 6.0 (b) 1950 N
7. (a) 3.35 (b) 2.99 m (c) 1.5 m **9.** 3 **11.** 4 **13.** 5 **15.** 7

17. **19.** **21.** **23.** (a) 6̄00 N (b) 5.0 m (c) 1.8 (d) 9̄0%

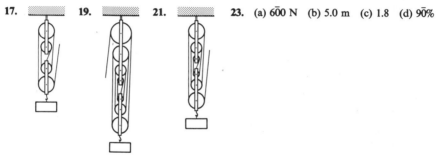

25. (a) (b) 4 (c) 10̄0 N; 12̄0 m (d) 25̄0 N; 52.0 m

27. (a) 9.86 m (b) 5.63 (c) 143̄0 N (d) 23̄0 N (e) 2.36
29. (a) 1.1 cm (b) 1.5×10^5 N (c) 12,000 N

CHAPTER 12

Pages 255–261

1. 92 teeth **3.** 192 rpm **5.** 32̄0 rpm **7.** 85 rpm **9.** 75 teeth
11. 69 teeth **13.** Counterclockwise **15.** Clockwise **17.** Clockwise
19. Clockwise **21.** Clockwise **23.** 1160 rpm **25.** 576 rpm **27.** 1480 rpm
29. 40 **31.** 20 **33.** 1140 rpm **35.** 30̄ cm **37.** 160 rpm
39. Clockwise **41.** Clockwise **43.** 320 cm **45.** 0.53 **47.** 2.50 **49.** 0.65
51. (a) 14.4 (b) 51̄0 N (c) 3680 J (d) 8160 J (e) 123 W (f) 272 W

CHAPTER 13

Pages 293–296

1. 34 kPa; 61 kPa; 110 kPa **3.** 20̄0 N/m **5.** 48.9 N
7. 90.0 cm **9.** (a) 0.025 cm (b) 1.6×10^6 N
11. (a) 110 MPa (b) 1.2×10^{-3} (c) 9.2×10^{10} Pa or 92 GPa
13. 1.2×10^8 Pa or 120 MPa **15.** 1.41×10^8 Pa or 141 MPa
17. (a) 7.1 mm (b) 13 mm (c) 2̄000 kg
19. (a) 1.3 L (b) 5.9×10^{-10}/Pa **21.** 2.1 MPa **23.** (a) 4.3 GPa (b) 0.61 mm (c) 2.9°
25. 0.070 rad or 4.0° **27.** 2800 kg/m³ **29.** 96 cm³ **31.** 1.0 m³ or 1.0×10^6 cm³
33. 2.5 m³ **35.** (a) 10̄00 L (b) 1500 L (c) 74 L
37. 2.7 **39.** 0.68 **41.** 0.917 **43.** Sinks **45.** Floats **47.** 1.2×10^{14}

CHAPTER 14

Pages 320–323

1. 176 kPa   **3.** (a) 114 MPa   (b) 5.2×10^8 N
5. 351 kPa   **7.** 12   **9.** 33 N
11. (a) 139 N   (b) 17.7 kPa   (c) 17.7 kPa   (d) 36.0
13. 6.8 N   **15.** (a) 1800 m^3 (b) 1.40×10^7 N   **17.** 89 percent   **19.** 11.4
21. 2.70   **23.** 0.21   **25.** 0.690   **27.** 340 L/min
29. (a) 9.6 cm   (b) 0.34 m/s   **31.** (a) 16.0 m/s   (b) 7540 L/min
33. (a) 9.90 m/s   (b) 0.891 L/min

CHAPTER 15

Pages 338–340

1. (a) 933 K   (b) 600 K   (c) 1808 K   (d) 234 K   (e) 273 K
3. 19,400 cal   **5.** 1.5×10^8 J or 150 MJ   **7.** 2.9×10^7 cal
9. 49°C   **11.** 1100 g   **13.** 24°C   **15.** 33°C   **17.** 460 J/kg C°
19. 355,000 J or 355 kJ   **21.** 203 kcal   **23.** 262 kW   **25.** 9.42 C°

CHAPTER 16

Pages 348–350

1. 0.064 m or 6.4 cm   **3.** 0.18 m or 18 cm   **5.** 200.10 m
7. 0.27 m or 27 cm   **9.** 2.004 cm   **11.** 0.37 cm^2 **13.** 348.6 cm^2
15. 60.10 cm^3 **17.** 653 L   **19.** 9.39 m^3 **21.** 1.23×10^5 L   **23.** 0.58 cm^3 **25.** $240

CHAPTER 17

Pages 361–362

1. (a) 202.64 kPa   (b) 506.60 kPa   (c) 20.26 kPa   **3.** 296 kPa   **5.** 83.3 kPa
7. 2.4 kg/m^3 **9.** 580 kPa   **11.** 17.3 m^3 **13.** (a) 54.6 percent   (b) 83.2 percent
15. 380 cm^3 **17.** 1400 L   **19.** 239 kPa   **21.** 295°C   **23.** 15.1 MPa
25. -13.8 kPa (87.5 kPa)

CHAPTER 18

Pages 373–374

1. 10,000 cal   **3.** 5.03×10^6 J or 5.03 MJ   **5.** 2.73×10^6 J or 2.73 MJ
7. 3.7×10^4 cal or 37 kcal   **9.** 18°C   **11.** 1900 kcal   **13.** 3.39×10^6 J or 3.39 MJ
15. 3.13×10^6 J or 3.13 MJ   **17.** 79.8°C   **19.** 147,000 cal

CHAPTER 19

Pages 384–385

1. (a) 250 s   (b) 0.0040 Hz   (c) 1.29 m   (d) 0.0052 m/s;   right to left   (e) 2.5 m   (f) 0.76 m
3. 1.25 m, 2.75 m, 4.25 m, 5.75 m, 7.25 m, 8.75 m
5. 1.00 m, 2.00 m, 3.00 m, 4.00 m, 5.00 m, 6.00 m, 7.00 m
7. (a) 0.64 W/m^2 (b) 0.040 W/m^2 **9.**   31.2 N

CHAPTER 20

Pages 396–398

1. 334 m/s   **3.**   347 m/s   **5.**   (2n + 1) 346 Hz for $n = 0, 1, 2, \ldots, 28$
7. None   **9.**   0.670 m   **11.**   (a) 0.250 m   (b) 343 Hz
13. 416 Hz   **15.**   549 Hz   **17.**   422 Hz   **19.**   466 Hz   **21.**   3 beats/s; 256 Hz

CHAPTER 21

Pages 419–421

1. 6.3×10^{18} electrons   **3.** -1.6×10^{-19} C   **5.** 2.2×10^{-2} N   **7.**   1.6 N out from triangle
9. (a) 4.3×10^3 N/C away from q_1 (b) -5.8×10^3 N/C toward q_2 (c) 1.7×10^{-4} N attraction
11. (a) 7.6×10^4 N/C toward the midpoint between negative charges
     (b) 2.3×10^{-3} N toward the midpoint between positive charges
13. (a) 4.1×10^4 V   (b) No   **15.**   (a) -5.7×10^3 V   (b) 1.1×10^{-3} J
17. (a) 4.5×10^3 V; -9.0×10^3 V; 1.4×10^4 V   (b) 1.0×10^4 V
19. (a) 1.1×10^{-11} F   (b) 9.0×10^{10} V
21. (a) 110 m^2 (b) 5.0×10^{-5} C   (c) 1.3×10^{-3} J
23. 9μF; 0.92μF; 5.2μF; 4.3μF; 3.1μF

CHAPTER 22

Page 427

1. 2.0 Ω   **3.** 1.72×10^{-4} Ω/m   **5.**   148°C   **7.**   144 Ω   **9.**   0.33 A

CHAPTER 23

Pages 445–449

1. (a) 48.0 Ω   (b) 0.500 A, 0.500 A, 0.500 A   (c) 5.00 V, 8.00 V, 11.0 V
3. (a) 13.5 V   (b) 18.00 Ω   (c) 3.75 V, 4.50 V, 5.25 V
5.

	V	I	R
Battery	12.0 V	0.50 A	24.0 Ω
R_1	4.0 V	0.50 A	8.0 Ω
R_2	6.0 V	0.50 A	12.0 Ω
R_3	2.0 V	0.50 A	4.0 Ω

7. (a) 4.00 Ω (b) 7.50 V (c) 0.500 A

9.

	V	I	R
Battery	27.0 V	3.00 A	9.00 Ω
R_1	27.0 V	1.00 A	27.0 Ω
R_2	27.0 V	1.50 A	18.0 Ω
R_3	27.0 V	0.50 A	54.0 Ω

11.

	V	I	R
Battery	24.0 V	6.00 A	4.00 Ω
R_1	8.00 V	6.00 A	1.33 Ω
R_2	16.0 V	4.00 A	4.00 Ω
R_3	16.0 V	2.00 A	8.00 Ω

13.

	V	I	R
Battery	30.0 V	5.00 A	6.00 Ω
R_1	6.00 V	3.00 A	2.00 Ω
R_2	6.00 V	2.00 A	3.00 Ω
R_3	15.0 V	5.00 A	3.00 Ω
R_4	8.00 V	1.00 A	8.00 Ω
R_5	8.00 V	4.00 A	2.00 Ω
R_6	1.00 V	5.00 A	0.200 Ω

15. (a) 0.500 A (b) 1.0 Ω **17.** 0.255 A, 0.273 A, 0.018 A
19. (a) 2.0 V (b) 4.0 V (c) 10.0 V
21. (a) 0.104 s (b) 0.043 s

CHAPTER 24

Pages 454–455

1. 1.49 V   **3.** (a) 0.850 A (b) 4.50 V (c) 0.1500 Ω
5. 0.405 A   **7.** 1.50 Ω

CHAPTER 25

Pages 475–477

1. 3.0×10^{-5} T   **3.** 7.96×10^{-5} Nm/T   **5.** 26 Nm/T   **7.** 0.25 A **9.** 3.1×10^{-4} Wb
11. (a) 0.582 V (b) 5.82×10^{-3} A   **13.** (a) 0.130 A (b) 18.1 ms (c) 3.75 V
15. (a) 8.3 W (b) 0 W (c) 0 J (d) 7.6×10^{-3} J
17. (a) 0 A (b) 0.667 A (c) 50.0 V (d) 2.00 A (e) 0.667 A
(f) 0 V (g) 2.00 A (h) 2.00 A (i) -50.0 V (j) 0 A
(k) 0 A (l) 0 V

CHAPTER 26

Pages 488–489

1. 240 V   **3.** 9.9 A   **5.** 440 V   **7.** 14 V   **9.** 340 V
11. 7.1 A   **13.** 7.2 V   **15.** 240 V   **17.** 5.00 turns   **19.** 0.50 A

CHAPTER 27

Pages 500–501

1. $2.51 \times 10^3\ \Omega$ **3.** $0.396\ \Omega$ **5.** $1.88\ \Omega$ **7.** $2.27 \times 10^{-2}\ \Omega$
9. $39.8\ \Omega$ **11.** $10\overline{0}0\ \Omega$; $2.9°$ **13.** $177\ \Omega$; $32°$ **15.** $254\ \Omega$; $38°$
17. $257\ \Omega$; $14°$ **19.** $151\ \Omega$; $6.7°$ **21.** $1470\ \Omega$; 3.40×10^{-3} A
23. $210\ \Omega$; 0.71 A   **25.** $4.42 \times 10^4\ \Omega$; 5.65×10^{-4} A   **27.** 1.92×10^6 Hz   **29.** 2.27×10^6 Hz

CHAPTER 28

Pages 509–510

1. 498 s or 8 min 18 s   **3.** 1.7×10^{-9} s   **5.** 0.11 s   **7.** (a) 1.99×10^{18} Hz (b) 1.32×10^{-15} J
9. 4260 lm   **11.** 368 lm   **13.** -2.50 **15.** 4.47 mm

CHAPTER 29

Pages 524–525

1. 0.25 m, 0.43 m, -0.72 **3.** 0.25 m, 1.5 m -5.0 **5.** -0.25 m, -0.15 m, $+0.43$
7. 0.20 m, 0.40 m, -1.0 **9.** 0.20 m, ∞, undetermined   **11.** -0.20 m, -0.086 m, $+0.57$
13. 6.00 m   **15.** (a) -0.60 m (b) $+5.0$ (c) Yes (d) No   **17.** 0.11 m   **19.** 0.10 m

CHAPTER 30

Pages 546–548

1. 2.25×10^8 m/s   **3.** 1.2   **5.** $25.4°$ **7.** $74.6°$ **9.** 3.01 m   **11.** $0.46°$
13. $19.15°$ **15.** $58.5°$ **17.** $+0.40$ m   **19.** -0.13 m   **21.** -0.15 m   **23.** 0.25 m
25. 0.40 m, 2.0   **27.** $+0.050$ m, -1.0 **29.** $+0.27$ m   **31.** Undetermined   **33.** $+1.2$ m
35. 0.063 m   **37.** (a) $S_i = -0.12$ m (b) No (c) Yes

CHAPTER 31

Page 559

1. 3.2×10^{-6} m   **3.** 0.36 m   **5.** $13.9°$, $24.8°$

CHAPTER 32

Page 570

1. 3.4×10^{-7} m, blue   3. 3.1×10^8 W/m^2 5. 6.5×10^{20} J
7. 1.8 eV   9. 12,400 V   11. 1.24×10^{-12} m

CHAPTER 33

Pages 577–578

1. (a) 2.46×10^{15} Hz, 2.92×10^{15} Hz, 3.08×10^{15} Hz, 3.15×10^{15} Hz,
   1.21×10^{-7} m, 1.02×10^{-7} m, 9.71×10^{-8} m, 9.49×10^{-8}m;   (b) 9.09×10^{-8} m
3. 4.05×10^{-6} m, 1.46×10^{-6} m   5. 8970 eV

INDEX